FIFTH EDITION

FUNDAMENTALS OF
PHYSICS

FIFTH EDITION

FUNDAMENTALS OF
PHYSICS

PART 4

DAVID HALLIDAY
University of Pittsburgh

ROBERT RESNICK
Rensselaer Polytechnic Institute

JEARL WALKER
Cleveland State University

JOHN WILEY & SONS, INC.

New York • Chichester • Brisbane • Toronto • Singapore

ACQUISITIONS EDITOR Stuart Johnson
DEVELOPMENTAL EDITOR Rachel Nelson
SENIOR PRODUCTION SUPERVISOR Cathy Ronda
PRODUCTION ASSISTANT Raymond Alvarez
MARKETING MANAGER Catherine Faduska
ASSISTANT MARKETING MANAGER Ethan Goodman
DESIGNER Dawn L. Stanley
MANUFACTURING MANAGER Mark Cirillo
PHOTO EDITOR Hilary Newman
COVER PHOTO William Warren/Westlight
ILLUSTRATION EDITOR Edward Starr
ILLUSTRATION Radiant/Precision Graphics

This book was set in Times Roman by Progressive Information Technologies, and printed and bound by Von Hofmann Press. The cover was printed by Phoenix Color Corp.

ISBN 0-471-14856-3

Printed in the United States of America

10 9 8 7 6 5 4 3 2 1

Hello There!

You are about to begin your first college level physics course. You may have heard from friends and fellow students that physics is a difficult course, especially if you don't plan to go on to a career in the hard sciences. But that doesn't mean it has to be difficult for you. The key to success in this course is to have a good understanding of each chapter before moving on to the next. When learned a little bit at a time, physics is straightforward and simple. Here are some ideas that can make this text and this class work for you:

- Read through the **Sample Problems** and solutions carefully. These problems are similar to many of the end-of-chapter exercises, so reading them will help you solve homework problems. In addition, they offer a look at how an experienced physicist would approach solving the problem.

- Try to answer the **Checkpoint** questions as you read through the chapter. Most of these can be answered by thinking through the problem, but it helps if you have some scratch paper and a pencil nearby to work out some of the harder questions. Hold the answer page at the back of the book with your thumb (or a tab or bookmark) so you can refer to it easily. The end-of-chapter **Questions** are very similar— you can use them to quiz yourself after you've read the whole chapter.

- Use the **Review & Summary** sections at the end of each chapter as the first place to look for formulas you might neeed to solve homework problems. These sections are also helpful in making study sheets for exams.

- The biggest tip I can give you is pretty obvious. Do your homework! Understanding the homework problems is the best way to master the material and do well in the course. Doing a lot of homework problems is also the best way to review for exams. And by all means, consult your classmates whenever you are stuck. Working in groups will make your studying more effective.

The study of basic physics is required for degrees in Engineering, Physics, Biology, Chemistry, Medicine, and many other sciences because the fundamentals of physics are the framework on which every other science is built. Therefore, a solid grasp of basic physical principles will help you understand upper-level science courses and make your study of these courses easier.

I took introductory physics because it was a prerequisite for my B.S. in Mechanical Engineering. Even though engineering and pure physics are worlds apart, I find myself using this book as a reference almost every day. I urge you to keep it after you have finished your course. The knowledge you will gain from this book and your introductory physics course is the foundation for all other sciences. This is the primary reason to take this class seriously and be successful in it.

Best of luck!

Josh Kane

Josh Kane

Preface

For four editions, *Fundamentals of Physics* has been successful in preparing physics students for careers in science and engineering. The first three editions were coauthored by the highly regarded team of David Halliday and Robert Resnick, who developed a groundbreaking text replete with conceptual structure and applications. In the fourth edition, the insights provided by new coauthor, Jearl Walker, took the text into the 1990s and met the challenge of guiding students through a time of tremendous advances and a ferment of activity in the science of physics. Now, in the fifth edition, we have expanded on the conventional strengths of the earlier editions and enhanced the applications that help students forge a bridge between concepts and reasoning. We not only *tell* students how physics works, we *show* them, and we give them the opportunity to show us what they have learned by testing their understanding of the concepts and applying them to real-world scenarios. Concept checkpoints, problem solving tactics, sample problems, electronic computations, exercises and problems—all of these skill-building signposts have been developed to help students establish a connection between conceptual theories and application. The students reading this text today are the scientists and engineers of tomorrow. It is our hope that the fifth edition of *Fundamentals of Physics* will help prepare these students for future endeavors by contributing to the enhancement of physics education.

CHANGES IN THE FIFTH EDITION

Although we have retained the basic framework of the fourth edition of *Fundamentals of Physics,* we have made extensive changes in portions of the book. Each chapter and element has been scrutinized to ensure clarity, currency, and accuracy, reflecting the needs of today's science and engineering students.

Content Changes

Mindful that textbooks have grown large and that they tend to increase in length from edition to edition, we have reduced the length of the fifth edition by combining several chapters and pruning their contents. In doing so, six chapters have been rewritten completely, while the remaining chapters have been carefully edited and revised, often ex-

tensively, to enhance their clarity, incorporating ideas and suggestions from dozens of reviewers.

- *Chapters 7 and 8 on energy* (and sections of later chapters dealing with energy) have been rewritten to provide a more careful treatment of energy, work, and the work–kinetic energy theorem. As the same time, the text material and problems at the end of each chapter still allow the instructor to present the more traditional treatment of these subjects.

- *Temperature, heat, and the first law of thermodynamics* have been condensed from two chapters to one chapter (*Chapter 19*).

- *Chapter 21 on entropy* now includes a statistical mechanical presentation of entropy that is tied to the traditional thermodynamical presentation.

- *Chapters on Faraday's law and inductance* have been combined into one new chapter (*Chapter 31*).

- *Treatment of Maxwell's equations* has been streamlined and moved up earlier into the chapter on magnetism and matter (*Chapter 32*).

- *Coverage of electromagnetic oscillations and alternating currents* has been combined into one chapter (*Chapter 33*).

- *Chapters 39, 40, and 41 on quantum mechanics* have been rewritten to modernize the subject. They now include experimental and theoretical results of the last few years. In addition, quantum physics and special relativity are introduced in some of the early chapters in short sections that can be covered quickly. These early sections lay some of the groundwork for the ''modern physics'' topics that appear later in the extended version of the text and add an element of suspense about the subject.

New Pedagogy

In the interest of addressing the needs of science and engineering students, we have added a number of new pedagogical features intended to help students forge a bridge between concepts and reasoning and to marry theory with practice. These new features are designed to help students test their understanding of the material. They were also developed to help students prepare to apply the information to exam questions and real-world scenarios.

- To provide opportunities for students to check their understanding of the physics concepts they have just read, we have placed **Checkpoint** questions within the chapter presentations. Nearly 300 Checkpoints have been added to help guide the student away from common errors and misconceptions. All of the Checkpoints require decision making and reasoning on the part of the student (rather than computations requiring calculators) and focus on the key points of the physics that students need to understand in order to tackle the exercises and problems at the end of each chapter. Answers to all of the Checkpoints are found in the back of the book, sometimes with extra guidance to the student.

- Continuing our focus on the key points of the physics, we have included additional **Checkpoint-type questions** in the Questions section at the end of each chapter. These new questions require decision making and reasoning on the part of the student; they ask the student to organize the physics concepts rather than just plug numbers into equations. Answers to the odd-numbered questions are now provided in the back of the book.

- To encourage the use of computer math packages and graphing calculators, we have added an **Electronic Computation problem section** to the Exercises and Problems sections of many of the chapters.

These new features are just a few of the pedagogical elements available to enhance the student's study of physics. A number of tried-and-true features of the previous edition have been retained and refined in the fifth edition, as described below.

CHAPTER FEATURES

The pedagogical elements that have been retained from previous editions have been carefully planned and crafted to motivate students and guide their reasoning process.

- *Puzzlers* Each chapter opens with an intriguing photograph and a "puzzler" that is designed to motivate the student to read the chapter. The answer to each puzzler is provided within the chapter, but it is not identified as such to ensure that the student reads the entire chapter.

- *Sample Problems* Throughout each chapter, sample problems provide a bridge from the concepts of the chapter to the exercises and problems at the end of the chapter. Many of the nearly 400 sample problems featured in the text have been replaced with new ones that more sharply focus on the common difficulties students experience in solving the exercises and problems. We have been especially mindful of the mathematical difficulties students face. The sample problems also provide

an opportunity for the student to see how a physicist thinks through a problem.

- *Problem Solving Tactics* To help further bridge concepts and applications and to add focus to the key physics concepts, we have refined and expanded the number of problem solving tactics that are placed within the chapters, particularly in the earlier chapters. These tactics provide guidance to the students about how to organize the physics concepts, how to tackle mathematical requirements in the exercises and problems, and how to prepare for exams.

- *Illustrations* Because the illustrations in a physics textbook are so important to an understanding of the concepts, we have altered nearly 30 percent of the illustrations to improve their clarity. We have also removed some of the less effective illustrations and added many new ones.

- *Review & Summary* A review and summary section is found at the end of each chapter, providing a quick review of the key definitions and physics concepts *without* being a replacement for reading the chapter.

- *Questions* Approximately 700 thought-provoking questions emphasizing the conceptual aspects of physics appear at the ends of the chapters. Many of these questions relate back to the checkpoints found throughout the chapters, requiring decision making and reasoning on the part of the student. Answers to the odd-numbered questions are provided in the back of the book.

- *Exercises & Problems* There are approximately 3400 end-of-chapter exercises and problems in the text, arranged in order of difficulty, starting with the exercises (labeled "E"), followed by the problems (labeled "P"). Particularly challenging problems are identified with an asterisk (*). Those exercises and problems that have been retained from previous editions have been edited for greater clarity; many have been replaced. Answers to the odd-numbered exercises and problems are provided in the back of the book.

VERSIONS OF THE TEXT

The fifth edition of *Fundamentals of Physics* is available in a number of different versions, to accommodate the individual needs of instructors and students alike. The Regular Edition consists of Chapters 1 through 38 (ISBN 0-471-10558-9). The Extended Edition contains seven additional chapters on quantum physics and cosmology (Chapters 1–45) (ISBN 0-471-10559-7). Both editions are available as single, hardcover books, or in the alternative versions listed on page ix:

- Volume 1—Chapters 1–21 (Mechanics/Thermodynamics), cloth, 0-471-15662-0
- Volume 2—Chapters 22–45 (E&M and Modern Physics), cloth, 0-471-15663-9
- Part 1—Chapters 1–12, paperback, 0-471-14561-0
- Part 2—Chapters 13–21, paperback, 0-471-14854-7
- Part 3—Chapters 22–33, paperback, 0-471-14855-5
- Part 4—Chapters 34–38, paperback, 0-471-14856-3
- Part 5—Chapters 39–45, paperback, 0-471-15719-8

The Extended edition of the text is also available on CD ROM.

SUPPLEMENTS

The fifth edition of *Fundamentals of Physics* is supplemented by a comprehensive ancillary package carefully developed to help teachers teach and students learn.

Instructor's Supplements

- *Instructor's Manual* by J. RICHARD CHRISTMAN, U.S. Coast Guard Academy. This manual contains lecture notes outlining the most important topics of each chapter, as well as demonstration experiments, and laboratory and computer exercises; film and video sources are also included. Separate sections contain articles that have appeared recently in the *American Journal of Physics* and *The Physics Teacher.*
- *Instructor's Solutions Manual* by JERRY J. SHI, Pasadena City College. This manual provides worked-out solutions for all the exercises and problems found at the end of each chapter within the text. *This supplement is available only to instructors.*
- *Solutions Disk.* An electronic version of the Instructor's Solutions Manual, for instructors only, available in TeX for Macintosh and Windows™.
- *Test Bank* by J. RICHARD CHRISTMAN, U.S. Coast Guard Academy. More than 2200 multiple-choice questions are included in the Test Bank for *Fundamentals of Physics.*

- *Computerized Test Bank.* IBM and Macintosh versions of the entire Test Bank are available with full editing features to help you customize tests.
- *Animated Illustrations.* Approximately 85 text illustrations are animated for enhanced lecture demonstrations.
- *Transparencies.* More than 200 four-color illustrations from the text are provided in a form suitable for projection in the classroom.

Student's Supplements

- *A Student's Companion* by J. RICHARD CHRISTMAN, U.S. Coast Guard Academy. Much more than a traditional study guide, this student manual is designed to be used in close conjunction with the text. The Student's Companion is divided into four parts, each of which corresponds to a major section of the text, beginning with an overview ''chapter.'' These overviews are designed to help students understand how the important topics are integrated and how the text is organized. For each chapter of the text, the corresponding Companion chapter offers: Basic Concepts, Problem Solving, Notes, Mathematical Skills, and Computer Projects and Notes.
- *Solutions Manual* by J. RICHARD CHRISTMAN, U.S. Coast Guard Academy and EDWARD DERRINGH, Wentworth Institute. This manual provides students with complete worked-out solutions to 30 percent of the exercises and problems found at the end of each chapter within the text.
- **Interactive Learningware** by JAMES TANNER, Georgia Institute of Technology, with the assistance of GARY LEWIS, Kennesaw State College. This software contains 200 problems from the end-of-chapter exercises and problems, presented in an interactive format, providing detailed feedback for the student. Problems from Chapter 1 to 21 are included in Part 1, from Chapters 22 to 38 in Part 2. The accompanying workbooks allow the student to keep a record of the worked-out problems. The Learningware is available in IBM 3.5″ and Macintosh formats.
- **CD Physics.** The entire Extended Version of the text (Chapters 1–45) is available on CD ROM, along with the student solutions manual, study guide, animated illustrations, and Interactive Learningware.

ROGER CLAPP
University of South Florida

W. R. CONKIE
Queen's University

PETER CROOKER
University of Hawaii at Manoa

WILLIAM P. CRUMMETT
Montana College of Mineral Science and Technology

EUGENE DUNNAM
University of Florida

ROBERT ENDORF
University of Cincinnati

F. PAUL ESPOSITO
University of Cincinnati

JERRY FINKELSTEIN
San Jose State University

ALEXANDER FIRESTONE
Iowa State University

ALEXANDER GARDNER
Howard University

ANDREW L. GARDNER
Brigham Young University

JOHN GIENIEC
Central Missouri State University

JOHN B. GRUBER
San Jose State University

ANN HANKS
American River College

SAMUEL HARRIS
Purdue University

EMILY HAUGHT
Georgia Institute of Technology

LAURENT HODGES
Iowa State University

JOHN HUBISZ
North Carolina State University

JOEY HUSTON
Michigan State University

DARRELL HUWE
Ohio University

CLAUDE KACSER
University of Maryland

LEONARD KLEINMAN
University of Texas at Austin

EARL KOLLER
Stevens Institute of Technology

ARTHUR Z. KOVACS
Rochester Institute of Technology

KENNETH KRANE
Oregon State University

SOL KRASNER
University of Illinois at Chicago

PETER LOLY
University of Manitoba

ROBERT R. MARCHINI
Memphis State University

DAVID MARKOWITZ
University of Connecticut

HOWARD C. MCALLISTER
University of Hawaii at Manoa

W. SCOTT MCCULLOUGH
Oklahoma State University

JAMES H. MCGUIRE
Tulane University

DAVID M. MCKINSTRY
Eastern Washington University

JOE P. MEYER
Georgia Institute of Technology

ROY MIDDLETON
University of Pennsylvania

IRVIN A. MILLER
Drexel University

EUGENE MOSCA
United States Naval Academy

MICHAEL O'SHEA
Kansas State University

PATRICK PAPIN
San Diego State University

GEORGE PARKER
North Carolina State University

ROBERT PELCOVITS
Brown University

OREN P. QUIST
South Dakota State University

JONATHAN REICHART
SUNY—Buffalo

MANUEL SCHWARTZ
University of Louisville

DARRELL SEELEY
Milwaukee School of Engineering

BRUCE ARNE SHERWOOD
Carnegie Mellon University

JOHN SPANGLER
St. Norbert College

ROSS L. SPENCER
Brigham Young University

HAROLD STOKES
Brigham Young University

JAY D. STRIEB
Villanova University

DAVID TOOT
Alfred University

J. S. TURNER
University of Texas at Austin

T. S. VENKATARAMAN
Drexel University

GIANFRANCO VIDALI
Syracuse University

FRED WANG
Prairie View A&M

ROBERT C. WEBB
Texas A&M University

GEORGE WILLIAMS
University of Utah

DAVID WOLFE
University of New Mexico

We hope that our words here reveal at least some of the wonder of physics, the fundamental clockwork of the universe. And, hopefully, those words might also reveal some of our awe of that clockwork.

DAVID HALLIDAY
6563 NE Windermere Road
Seattle, WA 98105

ROBERT RESNICK
Rensselaer Polytechnic Institute
Troy, NY 12181

JEARL WALKER
Cleveland State University
Cleveland, OH 44115

Chapter Opening Puzzlers

Each chapter opens with an intriguing example of physics in action. By presenting high-interest applications of each chapters concepts, the puzzlers are intended to peak your interest and motivate you to read the chapter.

In 1977, Kitty O'Neil set a dragster record by reaching 392.54 mi/h in a sizzling time of 3.72 s. In 1958, Eli Beeding Jr. rode a rocket sled from a standstill to a speed of 72.5 mi/h in an elapsed time of 0.04 s (less than an eye blink). How can we compare these two rides to see which was more exciting (or more frightening)—by final speeds, by elapsed times, or by some other quantity?

SAMPLE PROBLEM 2-6

(a) When Kitty O'Neil set the dragster records for the greatest speed and least elapsed time, she reached 392.54 mi/h in 3.72 s. What was her average acceleration?

SOLUTION: From Eq. 2-7, O'Neil's average acceleration was

$$\bar{a} = \frac{\Delta v}{\Delta t} = \frac{392.54 \text{ mi/h} - 0}{3.72 \text{ s} - 0}$$

$$= +106 \frac{\text{mi}}{\text{h} \cdot \text{s}},$$ (Answer)

where the motion is taken to be in the positive *x* direction. In

Answers to Puzzlers

All chapter-opening puzzlers are answered later in the chapter, either in text discussion or in a sample problem.

If the car
r, the bob
oninertial

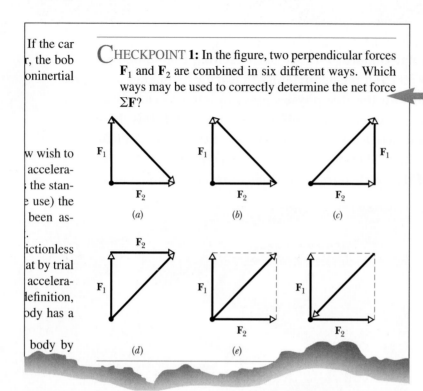

CHECKPOINT **1:** In the figure, two perpendicular forces F_1 and F_2 are combined in six different ways. Which ways may be used to correctly determine the net force ΣF?

(a)   (b)   (c)

(d)   (e)

w wish to
accelera-
the stan-
 use) the
been as-

ictionless
at by trial
accelera-
lefinition,
ody has a

body by

Checkpoints

Checkpoints appear throughout the text, focusing on the key points of physics you will need to tackle the exercises and problems found at the end of each chapter. These checkpoints help guide you away from common errors and misconceptions.

Checkpoint Questions

Checkpoint-type questions at the end of each chapter ask you to organize the physics concepts rather than plug numbers into equations. Answers to the odd-numbered questions are provided in the back of the book.

ey puck in

$v = -2t\mathbf{i}$

ponents of
ion vector
nd and t is
-2 and 3?

cal projec-
t identical
n the same
nal speeds

(c)

peed (a) a

$- 4.9\mathbf{j}$ (x is
). Has the

9. Figure 4-25 shows three paths for a kicked football. Ignoring the effects of air on the flight, rank the paths according to (a) time of flight, (b) initial vertical velocity component, (c) initial horizontal velocity component, and (d) initial speed. Place the greatest first in each part.

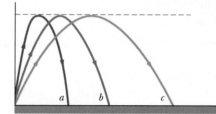

FIGURE 4-25   Question 9.

10. Figure 4-26 shows the velocity and acceleration of a particle at a particular instant in three situations. In which situation, and at that instant, is (a) the speed increasing, (b) the speed decreasing, (c) the speed not changing, (d) $\mathbf{v} \cdot \mathbf{a}$ positive, (e) $\mathbf{v} \cdot \mathbf{a}$ negative, and (f) $\mathbf{v} \cdot \mathbf{a} = 0$?

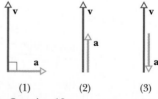

(1)   (2)   (3)

FIGURE 4-26   Question 10.

Sample Problems

The sample problems offer you the opportunity to work through the physics concepts just presented. Often built around real-world applications, they are closely coordinated with the end-of-chapter Questions, Exercises, and Problems.

SAMPLE PROBLEM 4-1

The position vector for a particle is initially

$$\mathbf{r}_1 = -3\mathbf{i} + 2\mathbf{j} + 5\mathbf{k}$$

and then later is

$$\mathbf{r}_2 = 9\mathbf{i} + 2\mathbf{j} + 8\mathbf{k}$$

(see Fig. 4-2). What is the displacement from $\mathbf{r}_1$ to $\mathbf{r}_2$?

SOLUTION: Recall from Chapter 3 that we add (or subtract) two vectors in unit-vector notation by combining the components, axis by axis. So Eq. 4-2 becomes

$$\Delta \mathbf{r} = (9\mathbf{i} + 2\mathbf{j} + 8\mathbf{k}) - (-3\mathbf{i} + 2\mathbf{j} + 5\mathbf{k})$$
$$= 12\mathbf{i} + 3\mathbf{k}. \qquad \text{(Answer)}$$

The displacement vector is parallel to the xz plane, because it lacks any y component, a fact that is easier to pick out in the numerical result than in Fig. 4-2.

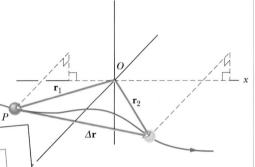

Sample Problem 4-1. The displacement $\Delta \mathbf{r} =$ ...d of $\mathbf{r}_1$ to the head of $\mathbf{r}_2$.

PROBLEM SOLVING TACTICS

TACTIC 1: *Reading Force Problems*
Read the problem statement several times until you have a good mental picture of what the situation is, what data are given, and what is requested. In Sample Problem 5-1, for example, you should tell yourself: "Someone is pushing a sled. Its speed changes, so acceleration is involved. The motion is along a straight line. A force is given in one part and asked for in the other, and so the situation looks like Newton's second law applied to one-dimensional motion."

If you know what the problem is about but don't know what to do next, put the problem aside and reread the text. If you are hazy about Newton's second law, reread that section. Study the sample problems. The one-dimensional-motion parts of Sample Problem 5-1 and the constant acceleration should send you back to Chapter 2 and especially to Table 2-1, which displays all the equations you are likely to need.

TACTIC 2: *Draw Two Types of Figures*
You may need two figures. One is a rough sketch of the actual real-world situation. When you draw the forces on it, place the tail of each force vector either on the boundary of or within the body feeling that force. The other figure is a free-body diagram in which the forces on a *single* body are drawn, with the body represented with a dot or a sketch. Place the tail of each force vector on the dot or sketch.

TACTIC 3: *What I_____em?*

Problem Solving Tactics

Careful attention has been paid to helping you develop your problem-solving skills. Problem-solving tactics are closely related to the sample problems and can be found throughout the text, though most fall within the first half. The tactics are designed to help you work through assigned homework problems and prepare for exams. Collectively, they represent the stock in trade of experienced problem solvers and practicing scientists and engineers.

REVIEW & SUMMARY

Conservative Forces

A force is a **conservative force** if the net work it does on a particle moving along a closed path from an initial point and then back to that point is zero. Or, equivalently, it is conservative if its work on a particle moving between two points does not depend on the path taken by the particle. The gravitational force (weight) and the spring force are conservative forces; the kinetic frictional force is a **nonconservative force**.

Potential Energy

A **potential energy** is energy that is associated with the configuration of a system in which a conservative force acts. When the conservative force does work W on a particle within the system, the change ΔU in the potential energy of the system is

$$\Delta U = -W. \tag{8-1}$$

If the particle moves from point x_i to point x_f, the change in the potential energy of the system is

$$\Delta U = -\int_{x_i}^{x_f} F(x)\, dx. \tag{8-6}$$

Gravitational Potential Energy

The potential energy associated with a system consisting of the Earth and a nearby particle is the **gravitational potential energy.** If the particle moves from height y_i to height y_f, the change in the gravitational potential energy of the particle–Earth system is

$$\Delta U = mg(y_f - y_i) = mg\, \Delta y. \tag{8-7}$$

If the **reference position** of the particle is set as $y_i = 0$ and the corresponding gravitational potential energy of the system is set as $U_i = 0$, then the gravitational potential energy U when the particle is at any position y is

$$U = mgy. \tag{8-9}$$

in which the subscripts refer to different instants during an transfer process. This conservation can also be written as

$$\Delta E = \Delta K + \Delta U = 0.$$

Potential Energy Curves

If we know the **potential energy function** $U(x)$ for a sy which a force F acts on a particle, we can find the force

$$F(x) = -\frac{dU(x)}{dx}.$$

If $U(x)$ is given on a graph, then at any value of x, the fo the negative of the slope of the curve there and the kinetic of the particle is given by

$$K(x) = E - U(x),$$

where E is the mechanical energy of the system. A **turnin** is a point x where the particle reverses its motion (there, The particle is in **equilibrium** at points where the slope $U(x)$ curve is zero (there, $F(x) = 0$).

Work by Nonconservative Forces

If a nonconservative applied force $\mathbf{F}$ does work on particl part of a system having a potential energy, then the wo done on the system by $\mathbf{F}$ is equal to the change ΔE in the m ical energy of the system:

$$W_{app} = \Delta K + \Delta U = \Delta E. \tag{8-24}$$

If a kinetic frictional force $\mathbf{f}_k$ does work on an obje change ΔE in the total mechanical energy of the object system containing it is given by

$$\Delta E = -f_k d,$$

in which d is the displacement of the object during the wo

Review and Summary

Review & Summary sections at the end of each chapter review the most important concepts and equations.

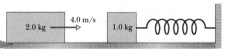

FIGURE 10-44   Problem 56.

57P. Two 22.7 kg ice sleds are placed a short distance apart, one directly behind the other, as shown in Fig. 10-45. A 3.63 kg cat, standing on one sled, jumps across to the other and immediately back to the first. Both jumps are made at a speed of 3.05 m/s relative to the ice. Find the final speeds of the two sleds.

FIGURE 10-45   Problem 57.

58P. The bumper of a 1200 kg car is designed so that it can just absorb all the energy when the car runs head-on into a solid wall at 5.00 km/h. The car is involved in a collision in which it runs at 70.0 km/h into the rear of a 900 kg car moving at 60.0 km/h in the same direction. The 900 kg car is accelerated to 70.0 km/h as a result of the collision. (a) What is the speed of the 1200 kg car immediately after impact? (b) What is the ratio of the kinetic energy absorbed in the collision to that which can be absorbed by the bumper of the 1200 kg car?

59P. A railroad freight car weighing 32 tons and traveling at 5.0

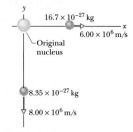

FIGURE 10-46   Exercise 62.

63E. In a game of pool, the cue ball strikes anothe at rest. After the collision, the cue ball moves at 3.5 line making an angle of 22.0° with its original dir tion, and the second ball has a speed of 2.00 m/s angle between the direction of motion of the secon original direction of motion of the cue ball and (b speed of the cue ball. (c) Is kinetic energy conserv

64E. Two vehicles A and B are traveling west and tively, toward the same intersection, where they co together. Before the collision, A (total weight 2700 with a speed of 40 mi/h and B (total weight 3600 lb of 60 mi/h. Find the magnitude and direction of the v (interlocked) vehicles immediately after the collisi

65E. In a game of billiards, the cue ball is given a V and strikes the pack of 15 stationary balls. All engage in nume... ...and ball–cushion col time lat... ...some accident) all

---

Exercises and Problems

A hallmark of this text, nearly 3400 end-of-chapter exercises and problems are arranged in order of difficulty, starting with the exercises (labeled "E"), followed by the problems (labeled "P"). Particularly difficult problems are identified with an asterisk (*). Answers to all the odd-numbered exercises and problems are provided in the back of the book. New electronic computation problems, which require the use of math packages and graphing calculators, have been added to many of the chapters.

Brief Contents

Contents

CHAPTER 38

SPECIAL THEORY OF RELATIVITY *958*

Why is special relativity so important in modern navigation?

APPENDICES *A1*

ANSWERS TO CHECKPOINTS AND ODD-NUMBERED QUESTIONS, EXERCISES AND PROBLEMS *AN1*

INDEX *I1*

Problem Solving Tactics

FIFTH EDITION

FUNDAMENTALS OF
PHYSICS

34
Electromagnetic Waves

As a comet swings around the Sun, ice on its surface vaporizes, releasing trapped dust and charged particles. The electrically charged "solar wind" forces the charged particles into a straight "tail" that points radially away from the Sun. But the dust is unaffected by the solar wind and seemingly should continue to travel along the comet's orbit. Why, instead, does much of the dust fashion the curved lower tail seen in the photograph?

34-1 MAXWELL'S RAINBOW

James Clerk Maxwell's crowning achievement was to show that a beam of light is a traveling wave of electric and magnetic fields—an **electromagnetic wave**—and thus that optics, the study of visible light, is a branch of electromagnetism. In this chapter we move from one to the other: we conclude our discussion of strictly electric and magnetic phenomena, and we build a foundation for optics.

In Maxwell's day (mid 1800s), the visible, infrared, and ultraviolet forms of light were the only electromagnetic waves known. Spurred on by Maxwell's work, however, Heinrich Hertz discovered what we now call radio waves and verified that they move through the laboratory at the same speed as visible light.

As Fig. 34-1 shows, we now know a wide *spectrum* (or range) of electromagnetic waves, referred to by one imaginative writer as "Maxwell's rainbow." Consider the extent to which we are bathed in electromagnetic waves from across this spectrum. The Sun, whose radiations define the environment in which we as a species have evolved and adapted, is the dominant source. We are also crisscrossed by radio and television signals. Microwaves from radar systems and from telephone relay systems may reach us. There are electromagnetic waves from lightbulbs, from the heated engine blocks of automobiles, from x-ray machines, from lightning flashes, and from buried radioactive materials. Beyond this, radiation reaches us from stars and other objects in our galaxy and from other galaxies.

Electromagnetic waves also travel in the other direction. Television signals, transmitted from Earth since about 1950, have now taken news about us (along with episodes of *I Love Lucy,* albeit *very* faintly) to whatever technically sophisticated inhabitants there may be on whatever planets may encircle the nearest 400 or so stars.

In the wavelength scale in Fig. 34-1 (and similarly the corresponding frequency scale), each scale marker represents a change in wavelength (and correspondingly in frequency) by a factor of 10. The scale is open-ended; the wavelengths of electromagnetic waves have no inherent upper or lower bounds.

Certain regions of the electromagnetic spectrum in Fig. 34-1 are identified by familiar labels, such as *x rays* and *radio waves*. These labels denote roughly defined wavelength ranges within which certain kinds of sources and detectors of electromagnetic waves are in common use. Other regions of Fig. 34-1, such as those labeled television and AM radio, represent specific wavelength bands assigned by law for certain commercial or other purposes. There are no gaps in the electromagnetic spectrum. And all electromagnetic waves, no matter where they lie in the spectrum, travel through *free space* (vacuum) with the same speed c.

The visible region of the spectrum is of course of particular interest to us. Figure 34-2 shows the relative sensitivity of the human eye to light of various wavelengths. The center of the visible region is about 555 nm, which produces the sensation that we call yellow-green.

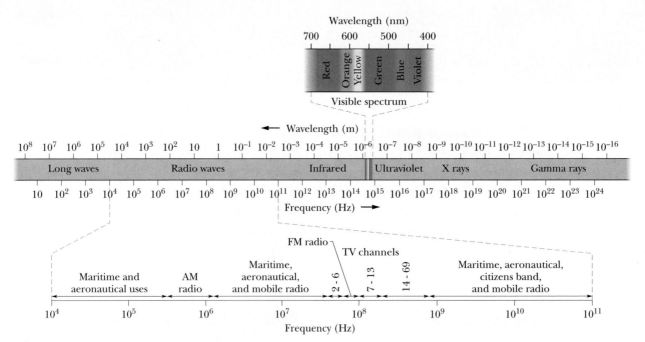

FIGURE 34-1 The electromagnetic spectrum.

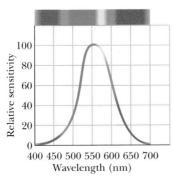

FIGURE 34-2 The relative sensitivity of the eye to electromagnetic waves at different wavelengths. This portion of the electromagnetic spectrum consists of *visible light*.

The limits of this visible spectrum are not well defined because the eye sensitivity curve approaches the zero-sensitivity line asymptotically at both long and short wavelengths. If we take the limits, arbitrarily, as the wavelengths at which eye sensitivity has dropped to 1% of its maximum value, these limits are about 430 and 690 nm; however, the eye can detect electromagnetic waves somewhat beyond these limits if they are intense enough.

34-2 THE TRAVELING ELECTROMAGNETIC WAVE, QUALITATIVELY

Some electromagnetic waves, including x rays, gamma rays, and visible light, are *radiated* (emitted) from sources that are of atomic or nuclear size, where quantum physics rules. Here we discuss how other electromagnetic waves are generated. To simplify matters, we restrict ourselves to that region of the spectrum (wavelength $\lambda \approx 1$ m) in which the source of the *radiation* (the emitted waves) is both macroscopic and of manageable dimensions.

Figure 34-3 shows, in broad outline, the generation of such waves. At its heart is an *LC oscillator,* which estab-

lishes an angular frequency ω $(= 1/\sqrt{LC})$. Charges and currents in this circuit vary sinusoidally at this frequency, as depicted in Fig. 33-1. An external source—possibly an ac generator—must be included to supply energy to compensate both for thermal losses in the circuit and for energy carried away by the radiated electromagnetic wave.

The *LC* oscillator of Fig. 34-3 is coupled by a transformer and a transmission line to an *antenna,* which consists essentially of two thin solid conducting rods. Through this coupling, the sinusoidally varying current in the oscillator causes charge to oscillate sinusoidally along the rods of the antenna at the angular frequency ω of the *LC* oscillator. The current in the rods associated with this movement of charge also varies sinusoidally, in magnitude and direction, at angular frequency ω. The antenna has the effect of an electric dipole whose electric dipole moment varies sinusoidally in magnitude and direction along the length of the antenna.

Because the dipole moment varies in magnitude and direction, the electric field produced by the dipole varies in magnitude and direction. And because the current varies, the magnetic field produced by the current varies in magnitude and direction. However, the changes in the electric and magnetic fields do not happen everywhere instantaneously; rather, the changes travel outward from the antenna at the speed of light c. Together the changing fields form an electromagnetic wave that travels away from the antenna at speed c. The angular frequency of this wave is ω, the same as that of the *LC* oscillator.

Figure 34-4 shows how the electric field **E** and the magnetic field **B** change with time as one wavelength of the wave sweeps past the distant point P of Fig. 34-3; in each part of Fig. 34-4, the wave is traveling directly out of the page. (We choose a distant point so that the curvature of the waves suggested in Fig. 34-3 is small enough to neglect. At such points, the wave is said to be a *plane wave,* and discussion of the wave is much simplified.) Note several key features in Fig. 34-4; they are present regardless of how the wave is created:

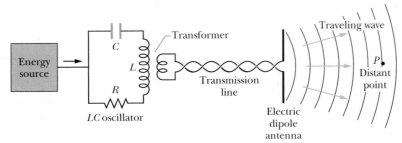

FIGURE 34-3 An arrangement for generating a traveling electromagnetic wave in the shortwave radio region of the spectrum: an *LC* oscillator produces a sinusoidal current in the antenna, which generates the wave. P is a distant point at which a detector can monitor the wave traveling past it.

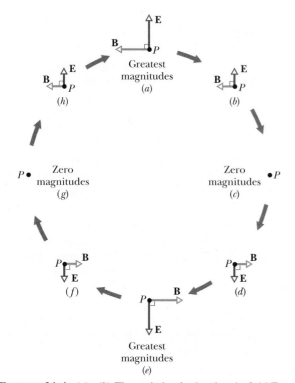

FIGURE 34-4 (a)–(h) The variation in the electric field **E** and the magnetic field **B** at the distant point P of Fig. 34-3 as one wavelength of the electromagnetic wave travels past it. In this perspective, the wave is traveling directly out of the page. The two fields vary sinusoidally in magnitude and direction. Note that they are always perpendicular to each other and to the direction of travel of the wave.

1. The electric and magnetic fields **E** and **B** are always perpendicular to the direction of travel of the wave. Thus the wave is a *transverse wave,* as discussed in Chapter 17.

2. The electric field is always perpendicular to the magnetic field.

3. The cross product **E** × **B** always gives the direction of travel of the wave.

4. The fields always vary sinusoidally, just like the transverse waves discussed in Chapter 17. Moreover, the fields vary with the same frequency and *in phase* (in step) with each other.

In keeping with these features, we can assume that the electromagnetic wave is traveling toward P in the positive direction of an x axis, that the electric field in Fig. 34-4 is oscillating parallel to the y axis, and that the magnetic field oscillates parallel to the z axis (using a right-handed coordinate system, of course). Then we can write the electric and magnetic fields as sinusoidal functions of position x

and time t:

$$E = E_m \sin(kx - \omega t), \tag{34-1}$$

$$B = B_m \sin(kx - \omega t), \tag{34-2}$$

in which E_m and B_m are the amplitudes of the fields and, as in Chapter 17, ω and k are the angular frequency and angular wave number of the wave, respectively. From these equations, note that not only do the two fields form the electromagnetic wave but each forms its own wave. Equation 34-1 gives the *electric wave component* of the electromagnetic wave, and Eq. 34-2 gives the *magnetic wave component.* As we shall discuss below, these two wave components cannot exist independently.

From Eq. 17-12, we know that the speed of the wave is ω/k. However, since this is an electromagnetic wave, its speed (in vacuum) is given the symbol c rather than v. In the next section you will see that c has the value

$$c = \frac{1}{\sqrt{\mu_0 \epsilon_0}} \quad \text{(wave speed),} \tag{34-3}$$

which is about 3.0×10^8 m/s. In other words:

All electromagnetic waves, including visible light, have the same speed c in vacuum.

You will also see that the wave speed c and the amplitudes of the electric and magnetic fields are related by

$$\frac{E_m}{B_m} = c \quad \text{(amplitude ratio).} \tag{34-4}$$

If we divide Eq. 34-1 by Eq. 34-2 and then substitute with Eq. 34-4, we find that the magnitudes of the fields at every instant are related by

$$\frac{E}{B} = c \quad \text{(magnitude ratio).} \tag{34-5}$$

We can represent the electromagnetic wave as in Fig. 34-5a, with a *ray* (a directed line showing the wave's direction of travel) and *wavefronts* (imaginary surfaces over which the wave has the same magnitude of electric field). The two wavefronts shown in Fig. 34-5a are separated by one wavelength $\lambda (= 2\pi/k)$ of the wave. (Waves traveling in approximately the same direction form a *beam,* such as a laser beam, which can be represented with a ray.)

We can also represent the wave as in Fig. 34-5b, which shows the electric and magnetic field vectors in a "snapshot" of the wave at a certain instant. The curves through the tips of the arrows represent the sinusoidal

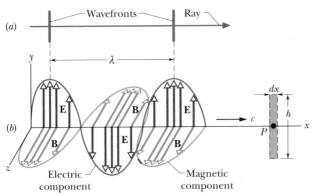

FIGURE 34-5 (*a*) An electromagnetic wave represented with a ray and two wavefronts; the wavefronts are separated by one wavelength λ. (*b*) The same wave represented in a "snapshot" of its electric field **E** and magnetic field **B** at points on the *x* axis, along which the wave travels at speed *c*. As it travels past point *P*, the fields vary as shown in Fig. 34-4. The electric component of the wave consists of only the electric fields; the magnetic component consists of only the magnetic fields. The dashed rectangle at *P* is used in Fig. 34-6.

oscillations given by Eqs. 34-1 and 34-2; the wave components **E** and **B** are in phase, perpendicular to each other, and perpendicular to the wave's direction of travel.

Interpretation of Fig. 34-5*b* requires some care. The similar drawings for a transverse wave on a taut string that we discussed in Chapter 17 represented the up and down displacement of sections of the string as the wave passed (*something actually moved*). Figure 34-5*b* is more abstract. At the instant shown, the electric and magnetic fields have certain magnitudes and directions (but are always perpendicular to the *x* axis) at each point along the *x* axis. We choose to represent these vector quantities with arrows, so we must draw arrows of different lengths, all pointing away from the *x* axis, like thorns on a rose stem. But the arrows represent only field values for points that are on the *x* axis. Neither the arrows nor the sinusoidal curves represent a sideways motion of anything, nor do the arrows connect points on the *x* axis with points off the axis.

Drawings like Fig. 34-5 help us visualize what is actually a very complicated situation. First consider the magnetic field. Because it varies sinusoidally, it induces (via Faraday's law of induction) a perpendicular electric field that also varies sinusoidally. But because that electric field is varying sinusoidally, it induces (via Maxwell's law of induction) a perpendicular magnetic field that also varies sinusoidally. And so on. The two fields continuously create each other via induction, and the resulting sinusoidal variations in the fields travel as a wave, the electromagnetic wave. Without this amazing result, we could not see; indeed, because we need electromagnetic waves from the

Sun to maintain Earth's temperature, we could not even exist without this result.

A Most Curious Wave

The waves we discussed in Chapters 17 and 18 require a *medium* (some material) through which or along which to travel. We had waves traveling along a string, through Earth, and through the air. But an electromagnetic wave (let's use the term *light wave* or *light*) is curiously different in that it requires no medium for its travel. It can, indeed, travel through a medium such as air or glass, but it can also travel through the vacuum of space between a star and us.

Once the special theory of relativity became accepted, long after Einstein published it in 1905, the speed of light waves was realized to be special. One reason was that light has the same speed regardless of the frame of reference from which it is measured. If you send a beam of light along an axis and ask several observers to measure its speed while they move at different speeds along that axis, either in the direction of the light or opposite it, they will all measure the *same speed* for the light. This result is an amazing one and quite different from what would have been found if those observers had measured the speed of any other type of wave; for other waves, the speed of the observers relative to the wave would have affected their measurements.

The meter has now been defined so that the speed of light (any electromagnetic wave) in vacuum has the exact value

$$c = 299{,}792{,}458 \text{ m/s},$$

which can be used as a standard. In fact, if you now measure the travel time of a pulse of light from one point to another, you are not really measuring the speed of the light but rather the distance between those two points.

34-3 THE TRAVELING ELECTROMAGNETIC WAVE, QUANTITATIVELY

We shall now derive Eqs. 34-3 and 34-4 and, even more important, explore the dual induction of electric and magnetic fields that gives us light.

Equation 34-4 and the Induced Electric Field

The dashed rectangle of dimensions *dx* and *h* in Fig. 34-6 is fixed at point *P* on the *x* axis and in the *xy* plane (it is shown at the right in Fig. 34-5*b*). As the electromagnetic wave moves rightward past the rectangle, the magnetic flux Φ_B through the rectangle changes and—according to

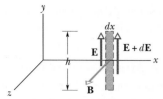

FIGURE 34-6 As the electromagnetic wave travels rightward past point P in Fig. 34-5, the sinusoidal variation of the magnetic field **B** through a rectangle centered at P induces electric fields along the rectangle. At the instant shown, **B** is decreasing in magnitude and the induced electric field is therefore greater in magnitude on the right side of the rectangle than on the left.

Faraday's law of induction—induced electric fields appear throughout the region of the rectangle. We take **E** and **E** + d**E** to be the induced fields along the two long sides of the rectangle. These induced electric fields are, in fact, the electric component of the electromagnetic wave.

Let us consider these fields at the instant when the magnetic wave component passing through the rectangle is the small section marked with red in Fig. 34-5b. Just then, the magnetic field through the rectangle points in the positive z direction and is decreasing in magnitude (the magnitude was greater just before the red section arrived). Because the magnetic field is decreasing, the magnetic flux Φ_B through the rectangle is also decreasing. According to Faraday's law, this change in flux is opposed by induced electric fields, which produce a magnetic field **B** in the positive z direction.

According to Lenz's law, this in turn means that if we imagine the boundary of the rectangle to be a conducting loop, a counterclockwise induced current would have to appear in it. There is, of course, no conducting loop; but this analysis shows that the induced electric field vectors **E** and **E** + d**E** are indeed oriented as shown in Fig. 34-6, with the magnitude of **E** + d**E** greater than that of **E**. Otherwise, the net induced electric field would not act counterclockwise around the rectangle.

Let us now apply Faraday's law of induction,

$$\oint \mathbf{E} \cdot d\mathbf{s} = -\frac{d\Phi_B}{dt}, \qquad (34\text{-}6)$$

counterclockwise around the rectangle of Fig. 34-6. There is no contribution to the integral from the top or bottom of the rectangle because **E** and d**s** are perpendicular there. The integral then has the value

$$\oint \mathbf{E} \cdot d\mathbf{s} = (E + dE)h - Eh = h\,dE. \qquad (34\text{-}7)$$

The flux Φ_B through this rectangle is

$$\Phi_B = (B)(h\,dx), \qquad (34\text{-}8)$$

where B is the magnitude of **B** within the rectangle and $h\,dx$ is the area of the rectangle. Differentiating Eq. 34-8 with respect to t gives

$$\frac{d\Phi_B}{dt} = h\,dx\,\frac{dB}{dt}. \qquad (34\text{-}9)$$

If we substitute Eqs. 34-7 and 34-9 into Eq. 34-6, we find

$$h\,dE = -h\,dx\,\frac{dB}{dt}$$

or

$$\frac{dE}{dx} = -\frac{dB}{dt}. \qquad (34\text{-}10)$$

Actually, both B and E are functions of *two* variables, x and t, as Eqs. 34-1 and 34-2 imply. However, in evaluating dE/dx, we must assume that t is constant because Fig. 34-6 is an "instantaneous snapshot." Also, in evaluating dB/dt we must assume that x is constant because we are dealing with the time rate of change of B at a particular place, the point P in Fig. 34-5b. The derivatives under these circumstances are *partial derivatives*, and Eq. 34-10 must be written

$$\frac{\partial E}{\partial x} = -\frac{\partial B}{\partial t}. \qquad (34\text{-}11)$$

The minus sign in this equation is appropriate and necessary because, although E is increasing with x at the site of the rectangle in Fig. 34-6, B is decreasing with t.

From Eq. 34-1, we have

$$\frac{\partial E}{\partial x} = kE_m \cos(kx - \omega t)$$

and from Eq. 34-2,

$$\frac{\partial B}{\partial t} = -\omega B_m \cos(kx - \omega t).$$

Then Eq. 34-11 reduces to

$$kE_m \cos(kx - \omega t) = \omega B_m \cos(kx - \omega t). \quad (34\text{-}12)$$

The ratio ω/k for a traveling wave is its speed, which we are calling c. Equation 34-12 then becomes

$$\frac{E_m}{B_m} = c \quad \text{(amplitude ratio)}, \qquad (34\text{-}13)$$

which is just Eq. 34-4.

Equation 34-3 and the Induced Magnetic Field

Figure 34-7 shows another dashed rectangle at point P of Fig. 34-5; this one is in the xz plane. As the electromagnetic wave moves rightward past this new rectangle, the

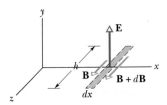

FIGURE 34-7 The sinusoidal variation of the electric field through this rectangle, located (but not shown) at point P in Fig. 34-5, induces magnetic fields along the rectangle. The instant shown is that of Fig. 34-6: **E** is decreasing in magnitude and the induced magnetic field is greater in magnitude on the right side of the rectangle than on the left.

electric flux Φ_E through the rectangle changes and—according to Maxwell's law of induction—induced magnetic fields appear throughout the region of the rectangle. These induced magnetic fields are, in fact, the magnetic component of the electromagnetic wave.

We see from Fig. 34-5 that at the instant chosen for the magnetic field in Fig. 34-6, the electric field through the rectangle of Fig. 34-7 is directed as shown. Recall that at the chosen instant, the magnetic field in Fig. 34-6 is decreasing. Because the two fields are in phase, the electric field in Fig. 34-7 must also be decreasing and so must the electric flux Φ_E through the rectangle. By applying the same reasoning we applied to Fig. 34-6, we see that the changing flux Φ_E will induce a magnetic field with vectors **B** and **B** + d**B** oriented as shown in Fig. 34-7, where **B** + d**B** is greater than **B**.

Let us apply Maxwell's law of induction,

$$\oint \mathbf{B} \cdot d\mathbf{s} = \mu_0 \epsilon_0 \frac{d\Phi_E}{dt}, \qquad (34\text{-}14)$$

by proceeding counterclockwise around the dashed rectangle of Fig. 34-7. Only the long sides of the rectangle contribute to the integral, whose value is

$$\oint \mathbf{B} \cdot d\mathbf{s} = -(B + dB)h + Bh = -h\, dB. \quad (34\text{-}15)$$

The flux Φ_E through the rectangle is

$$\Phi_E = (E)(h\, dx), \qquad (34\text{-}16)$$

where E is the average magnitude of **E** within the rectangle. Differentiating Eq. 34-16 with respect to t gives

$$\frac{d\Phi_E}{dt} = h\, dx\, \frac{dE}{dt}. \qquad (34\text{-}17)$$

If we substitute Eqs. 34-15 and 34-17 into Eq. 34-14, we find

$$-h\, dB = \mu_0 \epsilon_0 \left(h\, dx\, \frac{dE}{dt} \right)$$

or, changing to partial-derivative notation as we did before (Eq. 34-11),

$$-\frac{\partial B}{\partial x} = \mu_0 \epsilon_0 \frac{\partial E}{\partial t}. \qquad (34\text{-}18)$$

Again, the minus sign in this equation is necessary because, although B is increasing with x at point P in the rectangle in Fig. 34-7, E is decreasing with t.

Evaluating Eq. 34-18 by using Eqs. 34-1 and 34-2 leads to

$$-kB_m \cos(kx - \omega t) = -\mu_0 \epsilon_0 \omega E_m \cos(kx - \omega t),$$

which we can write as

$$\frac{E_m}{B_m} = \frac{1}{\mu_0 \epsilon_0 (\omega/k)} = \frac{1}{\mu_0 \epsilon_0 c}. \qquad (34\text{-}19)$$

Combining Eqs. 34-13 and 34-19 leads at once to

$$c = \frac{1}{\sqrt{\mu_0 \epsilon_0}} \qquad \text{(wave speed)}, \qquad (34\text{-}20)$$

which is exactly Eq. 34-3.

$\mathbf{C}$HECKPOINT 1: The magnetic field **B** through the rectangle of Fig. 34-6 is shown at a different instant in part 1 of the accompanying figure; **B** is directed in the xz plane, parallel to the z axis, and its magnitude is increasing. (a) Complete part 1 by drawing the induced electric fields, indicating both directions and relative magnitudes (as in Fig. 34-6). (b) For the same instant, complete part 2 of the figure by drawing the electric field of the electromagnetic wave. Also draw the induced magnetic fields, indicating both directions and relative magnitudes (as in Fig. 34-7).

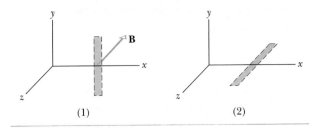

(1)   (2)

34-4 ENERGY TRANSPORT AND THE POYNTING VECTOR

All sunbathers know that an electromagnetic wave can transport energy and deliver it to a body on which it falls. The rate of energy transport per unit area in such a wave is described by a vector **S**, called the **Poynting vector** after John Henry Poynting (1852–1914), who first discussed its properties. **S** is defined as

$$\mathbf{S} = \frac{1}{\mu_0} \mathbf{E} \times \mathbf{B} \quad \text{(Poynting vector)} \quad (34\text{-}21)$$

and has the SI unit watt per square meter (W/m²).

> The direction of the Poynting vector **S** of an electromagnetic wave at any point gives the wave's direction of travel and the direction of energy transport at that point.

Because **E** and **B** are perpendicular to each other in an electromagnetic wave, the magnitude of **E** × **B** is EB. Then the magnitude of **S** is

$$S = \frac{1}{\mu_0} EB, \quad (34\text{-}22)$$

in which S, E, and B are instantaneous values. E and B are so closely coupled to each other that we need to deal with only one of them; we choose E, largely because most instruments for detecting electromagnetic waves deal with the electric component of the wave rather than the magnetic component. So, using $B = E/c$ from Eq. 34-5, we can rewrite Eq. 34-22 as

$$S = \frac{1}{c\mu_0} E^2 \quad \text{(instantaneous energy flow rate).} \quad (34\text{-}23)$$

By substituting $E = E_m \sin(kx - \omega t)$ into Eq. 34-23, we could obtain an equation for the energy transport rate as a function of time. More useful in practice, however, is the average energy transported over time; for that, we need to find the time-averaged value of S, written $\overline{S}$ and also called the **intensity** I of the wave. That is,

$$I = \overline{S} = \frac{1}{c\mu_0} \overline{E^2} = \frac{1}{c\mu_0} \overline{E_m^2 \sin^2(kx - \omega t)}, \quad (34\text{-}24)$$

where in each term the bar means "average value of." The average value of $\sin^2 \theta$, for any angular variable θ, is $\frac{1}{2}$ (see Fig. 33-14). In addition, we define a new quantity E_{rms}, the *root-mean-square* value of the electric field, as

$$E_{\text{rms}} = \frac{E_m}{\sqrt{2}}. \quad (34\text{-}25)$$

We can then rewrite Eq. 34-24 as

$$I = \frac{1}{c\mu_0} E_{\text{rms}}^2. \quad (34\text{-}26)$$

Because $E = cB$ and c is such a very large number, you might conclude that the energy associated with the electric field is much larger than that associated with the magnetic field. That conclusion is incorrect; the two energies are exactly equal. To show this, we start with Eq. 26-23, which gives the energy density u $(= \frac{1}{2}\epsilon_0 E^2)$ within an electric field, and substitute cB for E; then we can write

$$u_E = \frac{1}{2}\epsilon_0 E^2 = \frac{1}{2}\epsilon_0 (cB)^2.$$

If we now substitute for c with Eq. 34-3, we get

$$u_E = \frac{1}{2}\epsilon_0 \frac{1}{\mu_0 \epsilon_0} B^2 = \frac{B^2}{2\mu_0}.$$

But Eq. 31-56 tells us that $B^2/2\mu_0$ is the energy density u_B of the magnetic field, so we see that $u_E = u_B$.

Variation of Intensity with Distance

How intensity varies with distance from a real source of electromagnetic radiation is often complex—especially when the source (like a searchlight at a movie premier) beams the radiation in a particular direction. However, in some situations we can assume that the source is a *point source* that emits the light *isotropically,* that is, with equal intensity in all directions. The spherical wavefronts spreading from such an isotropic point source S at a particular instant are shown in cross section in Fig. 34-8.

Let us assume that the energy of the waves is conserved as they spread from this source. Let us also center an imaginary sphere of radius r on the source, as shown in Fig. 34-8. All the energy emitted by the source must pass through the sphere. Thus, the rate at which energy is transferred through the sphere by the radiation must equal the rate at which energy is emitted by the source, that is, the power P_s of the source. The intensity I at the sphere must

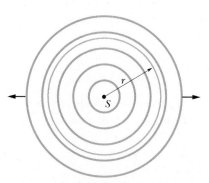

FIGURE 34-8 A point source S emits electromagnetic waves uniformly in all directions. The spherical wavefronts pass through an imaginary sphere of radius r that is centered on S.

then be

$$I = \frac{P_s}{4\pi r^2}, \qquad (34\text{-}27)$$

where $4\pi r^2$ is the area of the sphere. Equation 34-27 tells us that the intensity of the electromagnetic radiation from an isotropic point source decreases with the square of the distance r from the source.

Cʜᴇᴄᴋᴘᴏɪɴᴛ **2:** The figure gives the electric field of an electromagnetic wave at a certain point and a certain instant. The wave is transporting energy in the negative z direction. What is the direction of the magnetic field of the wave at that point and instant?

SAMPLE PROBLEM 34-1

An observer is 1.8 m from a point light source whose power P_s is 250 W. Calculate the rms values of the electric and magnetic fields due to the source at the position of the observer.

SOLUTION: Combining Eqs. 34-27 and Eq. 34-26 gives us

$$I = \frac{P_s}{4\pi r^2} = \frac{1}{c\mu_0} E_{\text{rms}}^2,$$

where $4\pi r^2$ is the area of a sphere of radius r centered on the source. The rms value of the electric field is then

$$E_{\text{rms}} = \sqrt{\frac{P_s c \mu_0}{4\pi r^2}}$$

$$= \sqrt{\frac{(250 \text{ W})(3.00 \times 10^8 \text{ m/s})(4\pi \times 10^{-7} \text{ H/m})}{(4\pi)(1.8 \text{ m})^2}}$$

$$= 48.1 \text{ V/m} \approx 48 \text{ V/m.} \qquad \text{(Answer)}$$

The rms value of the magnetic field follows from Eq. 34-5 and is

$$B_{\text{rms}} = \frac{E_{\text{rms}}}{c} = \frac{48.1 \text{ V/m}}{3.00 \times 10^8 \text{ m/s}}$$

$$= 1.6 \times 10^{-7} \text{ T.} \qquad \text{(Answer)}$$

Note that E_{rms} (= 48 V/m) is appreciable as judged by ordinary laboratory standards, but B_{rms} (= 1.6×10^{-7} T) is quite small. This discrepancy helps to explain why most instruments used for the detection and measurement of electromagnetic waves are designed to respond to the electric component of the wave. It is wrong, however, to say that the electric

component of an electromagnetic wave is "stronger" than the magnetic component. You cannot compare quantities that are measured in different units. As we have seen, the electric and magnetic components are on an absolutely equal basis as far as the propagation of the wave is concerned, because their average energies, which *can* be compared, are exactly equal.

34-5 RADIATION PRESSURE

Electromagnetic waves have linear momentum as well as energy. This means that we can exert a pressure—a **radiation pressure**—on an object by shining light on it. However, the pressure must be very small because, for example, you do not feel it when a camera flash is used to take your photograph.

To find an expression for the pressure, let us shine a beam of electromagnetic radiation—light, for example—on an object for a time interval Δt. Further, let us assume that the object is free to move and that the radiation is entirely **absorbed** (taken up) by the object. This means that during the interval Δt, the object gains an energy ΔU from the radiation. Maxwell showed that the object also gains linear momentum. The magnitude Δp of the momentum change of the object is related to the energy change ΔU by

$$\Delta p = \frac{\Delta U}{c} \quad \text{(total absorption),} \qquad (34\text{-}28)$$

where c is the speed of light. The direction of the momentum change of the object is the direction of the *incident* (incoming) beam that the object absorbs.

Instead of being absorbed, the radiation can be **reflected** by the object; that is, the radiation can be sent off in a new direction as if it bounced off the object. If the radiation is entirely reflected back along its original path, the magnitude of the momentum change of the object is twice that given above, or

$$\Delta p = \frac{2\,\Delta U}{c} \quad \begin{array}{l}\text{(total reflection} \\ \text{back along path).}\end{array} \qquad (34\text{-}29)$$

In the same way, an object undergoes twice as much momentum change when a perfectly elastic tennis ball is bounced from it as when it is struck by a perfectly inelastic ball (a lump of wet putty, say) of the same mass and velocity. If the incident radiation is partly absorbed and partly reflected, the momentum change of the object is between $\Delta U/c$ and $2\,\Delta U/c$.

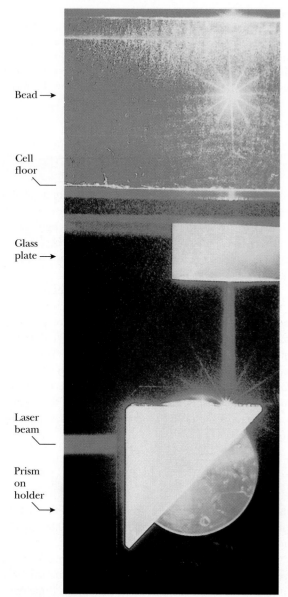

Bead →

Cell floor

Glass plate →

Laser beam

Prism on holder

An initially horizontal laser beam of green light is sent upward by a glass prism into an evacuated transparent cell and onto a glass sphere 20 μm in diameter. The sphere scatters the light, giving the starlike appearance in the upper part of the photograph. Before the laser was turned on, the glass sphere was on the floor of the cell. But the radiation pressure of the laser light has lifted the sphere by about 1 cm.

From Newton's second law, we know that a change in momentum is related to a force by

$$F = \frac{\Delta p}{\Delta t}. \tag{34-30}$$

To find expressions for the force exerted by radiation in terms of the intensity I of the radiation, suppose that a flat surface of area A, perpendicular to the path of the radiation, intercepts the radiation. In time interval Δt, the energy intercepted by area A is

$$\Delta U = IA\,\Delta t. \tag{34-31}$$

If the energy is completely absorbed, then Eq. 34-28 tells us that $\Delta p = IA\,\Delta t/c$ and, from Eq. 34-30, the magnitude of the force on the area A is

$$F = \frac{IA}{c} \quad \text{(total absorption)}. \tag{34-32}$$

Similarly, if the radiation is totally reflected back along its original path, Eq. 34-29 tells us that $\Delta p = 2IA\,\Delta t/c$ and, from Eq. 34-30,

$$F = \frac{2IA}{c} \quad \begin{array}{l}\text{(total reflection}\\ \text{back along path).}\end{array} \tag{34-33}$$

If the radiation is partly absorbed and partly reflected, the magnitude of the force on area A is between the values of IA/c and $2IA/c$.

The force per unit area on an object due to radiation is the radiation pressure p_r. We can find it for the situations of Eqs. 34-32 and 34-33 by dividing both sides of each equation by A. We obtain

$$p_r = \frac{I}{c} \quad \text{(total absorption)} \tag{34-34}$$

and

$$p_r = \frac{2I}{c} \quad \begin{array}{l}\text{(total reflection}\\ \text{back along path).}\end{array} \tag{34-35}$$

Be careful not to confuse the symbol p_r for radiation pressure with the symbol p for momentum.

The development of laser technology has permitted researchers to achieve radiation pressures much greater than, say, that due to a camera flashlamp. This comes about because a beam of laser light—unlike a beam of light from a small lamp filament—can be focused to a tiny spot only a few wavelengths in diameter. This permits the delivery of very large energy to small objects placed at that spot.

CHECKPOINT 3: Light of uniform intensity shines perpendicularly on a totally absorbing surface, fully illuminating the surface. If the area of the surface is decreased, do (a) the radiation pressure and (b) the radiation force on the surface increase, decrease, or stay the same?

SAMPLE PROBLEM 34-2

When dust is released by a comet, it does not continue along the comet's orbit because radiation pressure from sunlight pushes it radially outward from the Sun. Assume that a dust particle is spherical with radius R, has density $\rho = 3.5 \times 10^3$ kg/m³, and totally absorbs the sunlight it intercepts. For what value of R does the gravitational force F_g on the dust particle due to the Sun just balance the radiation force F_r on it from the sunlight?

SOLUTION: From Eq. 34-27, the intensity of sunlight on a dust particle (or anything else) at distance r from the Sun is

$$I = \frac{P_S}{4\pi r^2}, \qquad (34\text{-}36)$$

where $4\pi r^2$ is the surface area of a sphere of radius r centered on the sun, and P_S (= 3.9×10^{26} W) is the average power radiated by the Sun. From Eq. 34-32,

$$F_r = \frac{IA}{c} = \frac{I\pi R^2}{c}, \qquad (34\text{-}37)$$

where the area A of the particle that intercepts the sunlight is the particle's cross-sectional area πR^2 (and *not* half its surface area). Substituting Eq. 34-36 into Eq. 34-37 yields

$$F_r = \frac{P_S R^2}{4cr^2}. \qquad (34\text{-}38)$$

From Eq. 14-1 we can write the gravitational force F_g on the particle as

$$F_g = \frac{GM_S m}{r^2} = \frac{4GM_S \rho \pi R^3}{3r^2}, \qquad (34\text{-}39)$$

where M_S (= 1.99×10^{30} kg) is the Sun's mass and we have replaced the dust particle's mass m with $\rho(4/3)\pi R^3$. Setting $F_r = F_g$, and solving for R, we find

$$R = \frac{3P_S}{16\pi c\rho GM_S}.$$

The denominator is

$(16\pi)(3 \times 10^8 \text{ m/s})(3.5 \times 10^3 \text{ kg/m}^3)$
$\times (6.67 \times 10^{-11} \text{ N}\cdot\text{m}^2/\text{kg}^2)(1.99 \times 10^{30} \text{ kg})$
$= 7.0 \times 10^{33} \text{ N/s}.$

We then have

$$R = \frac{(3)(3.9 \times 10^{26} \text{ W})}{7.0 \times 10^{33} \text{ N/s}} = 1.7 \times 10^{-7} \text{ m.} \quad \text{(Answer)}$$

Note that this result is independent of the particle's distance r from the Sun.

Dust particles with $R \approx 1.7 \times 10^{-7}$ m follow an approximately straight path like path b in Fig. 34-9. For larger values of R, comparison of Eqs. 34-38 and 34-39 shows that, because F_g varies with R^3 and F_r varies with R^2, the gravitational force F_g dominates the radiation force F_r. Thus such particles follow a path that is curved toward the Sun like path c in Fig. 34-9. Similarly, for smaller values of R, the radiation force dominates, and the dust follows a path that is curved away from the Sun like path a. The composite of these dust particles is the dust tail of the comet.

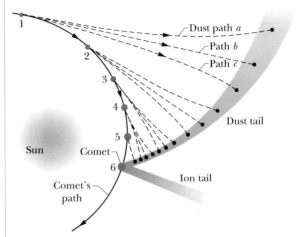

FIGURE 34-9 Sample Problem 34-2. A comet is now at position 6. Dust it has released at five previous positions has been pushed outward by radiation pressure from sunlight, has taken the dashed paths, and now forms the comet's curved dust tail.

34-6 POLARIZATION

VHF (very high frequency) television antennas in England are oriented vertically, but those in North America are horizontal. The difference is due to the direction of oscillation of the electromagnetic waves carrying the TV signal. In England, the transmitting equipment is designed to produce waves that are **polarized** vertically; that is, their electric field oscillates vertically. So, for the electric field of the incident television waves to drive a current along an antenna (and thus provide a signal to a television set), the antenna must be vertical. In North America, the waves are horizontally polarized.

Figure 34-10a shows an electromagnetic wave with its electric field oscillating parallel to the vertical y axis. The plane containing the **E** vectors is called the **plane of oscillation** of the wave (hence, the wave is said to be *plane-polarized* in the y direction). We can represent the wave's **polarization** (state of being polarized) by showing the extent of the electric field oscillations in a "head-on" view of the plane of oscillation, as in Fig. 34-10b.

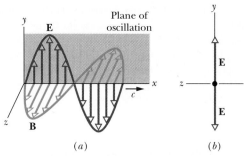

FIGURE 34-10 (*a*) The plane of oscillation of a polarized electromagnetic wave. (*b*) To represent the polarization, we view the plane of oscillation "head-on" and indicate the amplitude of the oscillating electric field.

Polarized Light

The electromagnetic waves emitted by a television station all have the same polarization, but the electromagnetic waves emitted by any common source of light (such as the Sun or a bulb) are **polarized randomly** or **unpolarized;** that is, the electric field at any given point is always perpendicular to the direction of travel of the waves but changes directions randomly. So if we try to represent a head-on view of the oscillations over some time period, we do not have a simple drawing like that of Fig. 34-10*b*; instead we have a mess like that in Fig. 34-11*a*.

In principle, we can simplify the mess by resolving each electric field of Fig. 34-11*a* into *y* and *z* components and then finding the net fields along the *y* axis and the *z* axis separately, as shown in Fig. 34-11*b*. In doing so, we mathematically change unpolarized light into the superposition of two polarized waves whose planes of oscillation are perpendicular to each other. The result is the double-

arrow representation of Fig. 34-11*b*, which simplifies drawings of unpolarized light. Similarly we can represent light that is **partially polarized** (its field oscillations are not completely random as in Fig. 34-11*a* nor are they parallel to a single axis as in Fig. 34-10*b*). For this situation, we can draw one of the arrows of the double-arrow representation longer than the other arrow.

We can actually transform unpolarized visible light into polarized light by sending it through a *polarizing sheet,* as is shown in Fig. 34-12. Such sheets, commercially known as Polaroids or Polaroid filters, were invented in 1932 by Edwin Land while he was an undergraduate student. A sheet consists of certain long molecules embedded in plastic. When the sheet is manufactured, it is stretched to align the molecules in parallel rows, like rows in a plowed field. When light is then sent through the sheet, electric field components along one direction pass through the sheet, while components perpendicular to that direction are absorbed by the molecules and disappear.

We shall not dwell on the molecules but, instead, shall assign to the sheet a *polarizing direction,* along which electric field components are passed:

> An electric field component parallel to the polarizing direction is passed (*transmitted*) by a polarizing sheet; a component perpendicular to it is absorbed.

Thus the light emerging from the sheet consists of only the components that are parallel to the polarizing direction of the sheet; hence the light must be polarized in that direction. In Fig. 34-12, the vertical components are transmitted by the sheet; the horizontal components are absorbed. The transmitted waves are then vertically polarized.

We next consider the intensity of the transmitted light. We start with unpolarized light, whose electric field oscillations we can resolve into *y* and *z* components as in Fig.

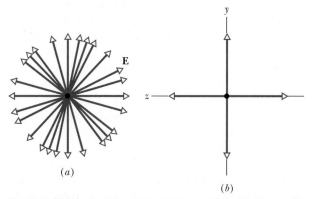

FIGURE 34-11 (*a*) Unpolarized light consists of waves with randomly directed electric fields. Here the waves are all traveling along the same axis, directly out of the page, and all have the same amplitude *E*. (*b*) A second way of representing unpolarized light: the light is the superposition of two polarized waves whose planes of oscillation are perpendicular to each other.

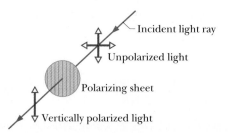

FIGURE 34-12 Unpolarized light becomes polarized when it is sent through a polarizing sheet. Its direction of polarization is then parallel to the polarizing direction of the sheet, which is represented here by the vertical lines drawn in the sheet.

34-11*b*. Further, we can arrange for the *y* axis to be parallel to the polarizing direction of the sheet. Then only the *y* components of the light are passed by the sheet; the *z* components are absorbed. As suggested by Fig. 34-11*b*, if the original waves are randomly oriented, the sum of the *y* components and the sum of the *z* components are equal. When the *z* components are absorbed, half the intensity I_0 of the original light is lost. The intensity I of the emerging polarized light is then

$$I = \tfrac{1}{2}I_0. \qquad (34\text{-}40)$$

Let us call this the *one-half rule;* we can use it *only* when the light reaching a polarizing sheet is unpolarized.

Suppose now that the light reaching a polarizing sheet is already polarized. Figure 34-13 shows a polarizing sheet in the plane of the page and the electric field **E** of such a polarized light wave traveling toward the sheet (and thus prior to any absorption). We can resolve **E** into two components relative to the polarizing direction of the sheet: parallel component E_y is transmitted by the sheet, and perpendicular component E_z is absorbed. Since θ is the angle between **E** and the polarizing direction of the sheet, the transmitted parallel component is

$$E_y = E \cos \theta. \qquad (34\text{-}41)$$

Recall that the intensity of an electromagnetic wave (such as our light wave) is proportional to the square of the electric field's magnitude (Eq. 34-26). In our present case then, the intensity I of the emerging wave is proportional to E_y^2 and the intensity I_0 of the original wave is proportional to E^2. Hence, from Eq. 34-41 we can write $I/I_0 = \cos^2 \theta$, or

$$I = I_0 \cos^2 \theta. \qquad (34\text{-}42)$$

Let us call this the *cosine-squared rule;* we can use it *only* when the light reaching a polarizing sheet is already polarized. The transmitted intensity I is a maximum and is equal to the original intensity I_0 when the original wave is polarized parallel to the polarizing direction of the sheet (when θ in Eq. 34-42 is $0°$ or $180°$). And I is zero when the original wave is polarized perpendicular to the polarizing direction of the sheet (when θ is $90°$).

Figure 34-14 shows an arrangement in which initially unpolarized light is sent through two polarizing sheets P_1 and P_2. (Often, the first sheet is called the *polarizer,* and the second the *analyzer.*) Because the polarizing direction of P_1 is vertical, the light transmitted by P_1 to P_2 is polarized vertically. If the polarizing direction of P_2 is also vertical, then all the light transmitted by P_1 is transmitted by P_2. If the polarizing direction of P_2 is horizontal, none of the light transmitted by P_1 is transmitted by P_2. We reach the same conclusions by considering only the *relative* orientations of the two sheets: if their polarizing directions are parallel, all the light passed by the first sheet is passed by the second sheet. If those directions are perpendicular (the sheets are said to be *crossed*), no light is passed by the second sheet. These two extremes are displayed with polarized sunglasses in Fig. 34-15.

Finally, if the two polarizing directions of Fig. 34-14 make an angle between $0°$ and $90°$, some of the light transmitted by P_1 will be transmitted by P_2. The intensity of that light is determined by Eq. 34-42.

Light can be polarized by means other than polarizing sheets, such as by reflection (discussed in Section 34-9) and by scattering from atoms or molecules. In *scattering,* light that is intercepted by an object, such as a molecule, is sent off in many, perhaps random, directions. An example is the scattering of sunlight by molecules in the atmosphere, which gives the sky its general glow.

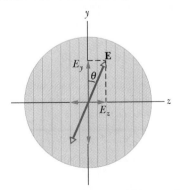

FIGURE 34-13 Polarized light approaching a polarizing sheet. The electric field **E** of the light can be resolved into components E_y (parallel to the polarizing direction of the sheet) and E_z (perpendicular to that direction). Component E_y will be transmitted by the sheet; component E_z will be absorbed.

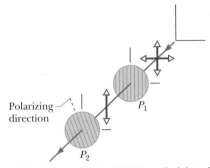

FIGURE 34-14 The light transmitted by polarizing sheet P_1 is vertically polarized, as represented by the vertical arrow. The amount of that light that is then transmitted by polarizing sheet P_2 depends on the angle between the polarization direction of that light and the polarizing direction of P_2 (indicated by the lines drawn in the sheet and by the dashed line).

(a)

(b)

FIGURE 34-15 Polarizing sunglasses consist of sheets whose polarizing directions are vertical when the sunglasses are worn. (a) Overlapping sunglasses transmit light fairly well when their polarizing directions have the same orientation, but (b) they block most of the light when they are crossed.

Although direct sunlight is unpolarized, light from much of the sky is at least partially polarized by such scattering. Bees use the polarization of sky light in navigating to and from their hives. Similarly, the Vikings used it to navigate across the North Sea when the daytime Sun was below the horizon (because of the high latitude of the North Sea). These early seafarers had discovered certain crystals (now called cordierite) that changed color when rotated in polarized light. By looking at the sky through such a crystal while rotating it about their line of sight, they could locate the hidden Sun and thus determine which way was south.

SAMPLE PROBLEM 34-3

Figure 34-16a shows a system of three polarizing sheets in the path of initially unpolarized light. The polarizing direction of the first sheet is parallel to the y axis, that of the second sheet is 60° counterclockwise from the y axis, and that of the third sheet is parallel to the x axis. What fraction of the initial intensity I_0 of the light emerges from the system, and how is that light polarized?

SOLUTION: We work the problem in steps, sheet by sheet. The original light wave is represented in Fig. 34-16b, using the head-on double-arrow representation of Fig. 34-11b. Because the light is initially unpolarized, the intensity I_1 of the light transmitted by the first sheet is given by the one-half rule (Eq. 34-40):

$$I_1 = \tfrac{1}{2}I_0.$$

The polarization of this transmitted light is (as always) parallel to the polarizing direction of the sheet transmitting it; here it is parallel to the y axis, as shown in the head-on view of Fig. 34-16c.

Because the light reaching the second sheet is polarized, the intensity I_2 of the light transmitted by that sheet is given by the cosine-squared rule (Eq. 34-42). The angle θ in the rule is the angle between the polarization direction of the entering light (parallel to the y axis) and the polarizing direction of the second sheet (60° counterclockwise from the y axis), and so θ is 60°. Then

$$I_2 = I_1 \cos^2 60°.$$

The polarization of this transmitted light is parallel to the polarizing direction of the sheet transmitting it, that is, 60° counterclockwise from the y axis, as shown in the head-on view of Fig. 34-16d.

Because this light is polarized, the intensity I_3 of the light transmitted by the third sheet is given by the cosine-squared rule. The angle θ is now the angle between the polarization direction of the entering light (Fig. 34-16d) and the polarizing direction of the third sheet (parallel to the x axis), and so $\theta = 30°$. Thus,

$$I_3 = I_2 \cos^2 30°.$$

This final transmitted light is polarized parallel to the x axis (Fig. 34-16e). We find its intensity by substituting first for I_2 and then for I_1 in the equation above:

$$I_3 = I_2 \cos^2 30° = (I_1 \cos^2 60°) \cos^2 30°$$
$$= (\tfrac{1}{2}I_0) \cos^2 60° \cos^2 30° = 0.094I_0.$$

Thus $\qquad\qquad \dfrac{I_3}{I_0} = 0.094.$ (Answer)

That is, 9.4% of the initial intensity emerges from the three-sheet system. (If we now remove the second sheet, what fraction of the initial intensity emerges from the system?)

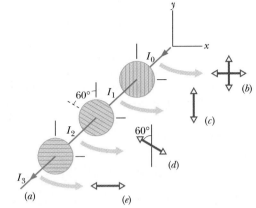

FIGURE 34-16 Sample Problem 34-3. (*a*) Initially unpolarized light of intensity I_0 is sent into a system of three polarizing sheets. The intensities I_1, I_2, and I_3 of the light transmitted by the sheets are indicated. Shown also are the polarizations, from head-on views, of (*b*) the initial light, as well as the light transmitted by (*c*) the first sheet, (*d*) the second sheet, and (*e*) the third sheet.

CHECKPOINT **4:** The figure shows four pairs of polarizing sheets, seen face-on. Each pair is mounted in the path of initially unpolarized light (like the three sheets in Fig. 34-16*a*). The polarizing direction of each sheet (indicated by the dashed line) is referenced to either a horizontal x axis or a vertical y axis. Rank the pairs according to the fraction of the initial intensity that they pass, greatest first.

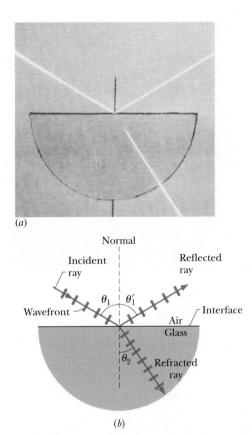

FIGURE 34-17 (*a*) A black-and-white photograph showing the reflection and refraction of an incident beam of light by a horizontal plane glass surface. (A portion of the refracted beam within the glass was not well photographed.) At the bottom surface, which is curved, the beam is perpendicular to the surface; so the refraction there does not bend the beam. (*b*) A representation of (*a*) using rays. The angles of incidence (θ_1), of reflection (θ_1'), and of refraction (θ_2) are marked.

34-7 REFLECTION AND REFRACTION

Although a light wave spreads as it moves away from its source, we can often approximate its travel as being in a straight line; we did so for the light wave in Fig. 34-5*a*. The study of the properties of light waves under that approximation is called *geometrical optics*. For the rest of this chapter and all of Chapter 35, we shall discuss the geometrical optics of visible light.

The black-and-white photograph in Fig. 34-17*a* shows an example of light waves traveling in approximately straight lines. A narrow beam of light, angled downward from the left and traveling through air, encounters a *plane* (flat) glass surface. Part of the light is **reflected** by the surface, forming a beam directed upward toward the right, traveling as if the original beam had bounced from the surface. The rest of the light travels through the surface and into the glass, forming a beam directed downward to the right. Because light can travel through the glass like this, the glass is said to be *transparent*; that is, we can see through it. (In this chapter we shall consider only transparent materials.)

The travel of light through a surface (or *interface*) that separates two media is called **refraction,** and the light is said to be *refracted*. Unless an incident beam of light is perpendicular to a surface, refraction by the surface changes the light's direction of travel. For this reason, the beam is said to be ''bent'' by the refraction. Note in Fig.

The stealth aircraft F-117A is virtually invisible to radar largely because of the flat panels that are angled so as to reflect incident radar signals up or down, rather than back to the radar station.

34-17a that the bending occurs only at the surface; within the glass, the light travels in a straight line.

In Figure 34-17b, the beams of light in the photograph are represented with an *incident ray*, a *reflected ray*, and a *refracted ray*. Each ray is oriented with respect to a line, called the *normal*, that is perpendicular to the surface at the point of reflection and refraction. In Fig. 34-17b, the **angle of incidence** is θ_1, the **angle of reflection** is θ_1', and the **angle of refraction** is θ_2, all measured *relative to the normal* as shown. The plane containing the incident ray and the normal is the *plane of incidence*, which is in the plane of the page in Fig. 34-17b.

Experiment shows that reflection and refraction are governed by two laws:

Law of reflection: A reflected ray lies in the plane of incidence and has an angle of reflection equal to the angle of incidence. In Fig. 34-17b, this means that

$$\theta_1' = \theta_1 \quad \text{(reflection)}. \qquad (34\text{-}43)$$

(We shall now drop the prime on the angle of reflection.)
Law of refraction: A refracted ray lies in the plane of incidence and has an angle of refraction that is related to the angle of incidence by

$$n_2 \sin \theta_2 = n_1 \sin \theta_1 \quad \text{(refraction)}. \qquad (34\text{-}44)$$

Here each of the symbols n_1 and n_2 is a dimensionless constant called the **index of refraction** that is associated with a medium involved in the refraction. We derive this equation, called Snell's law, in Chapter 36. As we shall discuss there, the index of refraction of a medium is equal to c/v, where v is the speed of light in that medium and c is its speed in vacuum.

Table 34-1 gives the indices of refraction of vacuum and some common substances. For vacuum, n is defined to be exactly 1; for air, n is very close to 1.0 (an approximation we shall often make). Nothing has an index of refraction below 1.

We can rearrange Eq. 34-44 as

$$\sin \theta_2 = \frac{n_1}{n_2} \sin \theta_1 \qquad (34\text{-}45)$$

to compare the angle of refraction θ_2 with the angle of incidence θ_1. We can then see that the relative value of θ_2 depends on the relative values of n_2 and n_1. In fact, we can have three basic results:

1. If n_2 is equal to n_1, then θ_2 is equal to θ_1. In this case, refraction does not bend the light beam, which continues in an *undeflected direction*, as in Fig. 34-18a.
2. If n_2 is greater than n_1, then θ_2 is less than θ_1. In this

TABLE 34-1 **SOME INDICES OF REFRACTION**[a]

MEDIUM	INDEX	MEDIUM	INDEX
Vacuum	exactly 1	Typical crown glass	1.52
Air (STP)[b]	1.00029	Sodium chloride	1.54
Water (20°C)	1.33	Polystyrene	1.55
Acetone	1.36	Carbon disulfide	1.63
Ethyl alcohol	1.36	Heavy flint glass	1.65
Sugar solution (30%)	1.38	Sapphire	1.77
Fused quartz	1.46	Heaviest flint glass	1.89
Sugar solution (80%)	1.49	Diamond	2.42

[a]For a wavelength of 589 nm (yellow sodium light).

[b]STP means "standard temperature (0°C) and pressure (1 atm)."

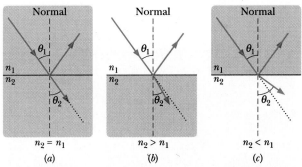

FIGURE 34-18 Light refracting from a medium with an index of refraction n_1 and into a medium with an index of refraction n_2. (*a*) The beam does not bend when $n_2 = n_1$; the refracted light then travels in the *undeflected direction* (the dotted line), which is the same as the direction of the incident beam. The beam bends (*b*) toward the normal when $n_2 > n_1$ and (*c*) away from the normal when $n_2 < n_1$.

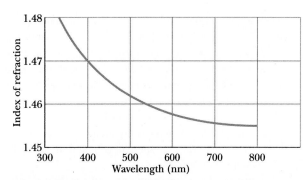

FIGURE 34-19 The index of refraction as a function of wavelength for fused quartz. The graph indicates that a beam of short-wavelength light, for which the index of refraction is higher, is bent more upon entering or leaving quartz than a beam of long-wavelength light.

case, refraction bends the light beam away from the undeflected direction and toward the normal, as in Fig. 34-18*b*.

3. If n_2 is less than n_1, then θ_2 is greater than θ_1. In this case, refraction bends the light beam away from the undeflected direction and away from the normal, as in Fig. 34-18*c*.

Refraction cannot bend a beam so much that the refracted ray is on the same side of the normal as the incident ray.

Chromatic Dispersion

The index of refraction n encountered by light in any medium except vacuum depends on the wavelength of the light. The dependence of n on wavelength implies that when a light beam consists of rays of different wavelengths, the rays will be refracted at different angles by a surface. That is, the light will be spread out by the refraction. This spreading of light is called **chromatic dispersion,** in which ''chromatic'' refers to the colors associated with the individual wavelengths and ''dispersion'' refers to the spreading of the light according to its wavelengths or colors. The refractions of Figs. 34-17 and 34-18 do not show chromatic dispersion because the beams are *monochromatic* (of a single wavelength or color).

Generally, the index of refraction in a given medium is *greater* for a shorter wavelength (corresponding to, say, blue light) than for a longer wavelength (say, red light). As an example, Fig. 34-19 shows how the index of refraction for fused quartz depends on the wavelength of light. Such dependence means that when a beam with waves of both blue and red light is refracted through a surface, such as from air into quartz or vice versa, the blue *component* (the ray corresponding to the wave of blue light) bends more than the red component.

A beam of *white light* consists of components of all (or nearly all) the colors in the visible spectrum with approximately uniform intensities. When you see such a beam, you perceive white rather than the individual colors. In Fig. 34-20*a*, a beam of white light in air is incident on a glass surface. (Because the pages of this book are white, a beam of white light is represented with a gray ray here. Also, a

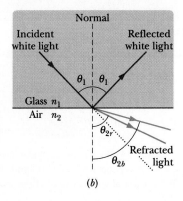

FIGURE 34-20 Chromatic dispersion of white light. The blue component is bent more than the red component. (*a*) Passing from air to glass, the blue component ends up with the smaller angle of refraction. (*b*) Passing from glass to air, the blue component ends up with the larger angle of refraction.

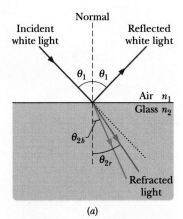

beam of monochromatic light is generally represented with a red ray.) Of the refracted light in Fig. 34-20a, only the red and blue components are shown. Because the blue component is bent more than the red component, the angle of refraction θ_{2b} for the blue component is *smaller* than the angle of refraction θ_{2r} for the red component. (Remember, angles are measured relative to the normal.) In Fig. 34-20b, a ray of white light in glass is incident on a glass–air interface. Again, the blue component is bent more than the red component, but now $\theta_{2b} > \theta_{2r}$.

To increase the color separation, we can use a solid glass prism with a triangular cross section, as in Fig. 34-21a. The dispersion at the first surface (on the left in Fig. 34-21a,b) is then enhanced by that at the second surface.

The most charming example of chromatic dispersion is a rainbow. When white sunlight is intercepted by a falling raindrop, some of the light refracts into the drop, reflects from the drop's inner surface, and then refracts out of the drop (Fig. 34-22). As with a prism, the first refraction separates the sunlight into its component colors, and the second refraction increases the separation.

The rainbow you see is formed by light refracted by many such drops: the red comes from drops angled slightly higher in the sky, the blue from drops angled slightly lower, and the intermediate colors from drops at interme-

(a)

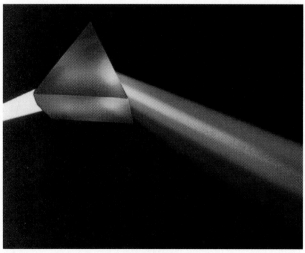

(a)

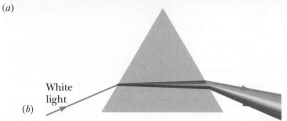

White light

(b)

FIGURE 34-21   (a) A triangular prism separating white light into its component colors. (b) Chromatic dispersion occurs at the first surface and is increased at the second surface.

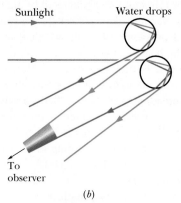

(b)

FIGURE 34-22   (a) A rainbow is always a circular arc that is centered on the direction you would look if you looked directly away from the Sun. Under normal conditions, you are lucky if you see a long arc, but if you are looking downward from an elevated position, you might actually see a full circle. (b) The separation of colors when sunlight refracts into and out of falling raindrops leads to a rainbow. The figure shows the situation for the Sun on the horizon (the rays of sunlight are then horizontal). The paths of red and blue rays from two drops are indicated. Many other drops also contribute red and blue rays, as well as the intermediate colors of the visible spectrum.

diate angles. All the drops sending separated colors to you are angled at about 42° from a point that is directly opposite the Sun in your view. If the rainfall is extensive and brightly lit, you see a circular arc of color, with red on top and blue on bottom. Your rainbow is a personal one, because another observer intercepts light from other drops.

CHECKPOINT **5:** Which of the three drawings (if any) show physically possible refraction?

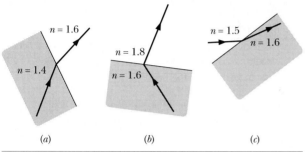

(a)   (b)   (c)

SAMPLE PROBLEM 34-4

A submerged swimmer is looking directly upward through the air–water interface in a pool.

(a) Over what range of angles do rays reach the swimmer's eyes from light sources external to the water? Assume that the light is monochromatic and that the index of refraction of water is 1.33.

SOLUTION: Light reaches the swimmer's eyes from external sources by refracting through the air–water interface, in accordance with Eq. 34-44. Let us associate subscript 1 in that equation with air and subscript 2 with water. Then, with $n_1 =$ 1.00 and $n_2 = 1.33$, we have $n_2 > n_1$. So, the refraction of light rays at the air–water interface bends the rays *toward* the normal, as in Fig. 34-18b. The bending of an arbitrary light ray, with angle of incidence θ_1 and angle of refraction θ_2, is shown in Fig. 34-23a; the swimmer's eyes are located at point E. Note that the refracted ray also makes an angle θ_2 with the vertical at E.

From Eq. 34-44 we know that the angle of refraction θ_2 depends on the angle of incidence θ_1 and is given by

$$\sin \theta_2 = \frac{n_1}{n_2} \sin \theta_1. \qquad (34\text{-}46)$$

So, to find the angles at which rays reach point E from external sources, we must consider the range of values of θ_1. That will give us the range of values of θ_2.

The least value of θ_1 is 0°, which is the value for an incident ray that is perpendicular to the air–water interface. For that value Eq. 34-46 gives us

$$\sin \theta_2 = \frac{1.00}{1.33} \sin 0° = 0,$$

or $\qquad \theta_2 = 0°.$

Incident ray A in Fig. 34-23b shows this refraction situation: the incident ray is not bent; it reaches E along the vertical through E.

The maximum value of θ_1 is approximately 90°, which is the value for an incident ray that is almost parallel to the air–water interface. Equation 34-46 now gives us

$$\sin \theta_2 = \frac{1.00}{1.33} \sin 90° = 0.752,$$

or $\qquad \theta_2 = 48.8°.$

In Fig. 34-23b, the two incident rays B show this refraction situation: these two rays are incident on the interface at the greatest possible angle (90°) but are 48.8° to the vertical when they reach the swimmer.

Figure 34-23b shows only one plane of the swimmer's field of view. But if we rotate that plane about the vertical, we

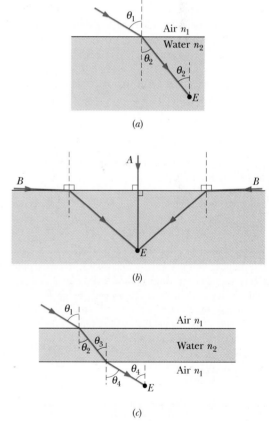

(a)

(b)

(c)

FIGURE 34-23 Sample Problem 34-4. (*a*) A ray of light, refracted into water, reaches a swimmer's eyes located at point E. The angle of the ray at E is measured relative to the vertical through E. (*b*) Ray A, perpendicular to the air–water interface, and rays B, almost parallel to the interface, reach point E. (*c*) The swimmer's eyes at point E are now in air that is trapped by a mask of transparent plastic. A ray reaching E refracts at two air–water interfaces.

see that it then gives the swimmer's entire field of view. Thus, all the refracted rays reaching the swimmer from external sources are contained within a vertical cone that has its apex located at E and intersects the air–water interface in a circle directly above E. Moreover, the apex angle is

$$2\theta_2 = 97.6° \approx 100°. \qquad \text{(Answer)}$$

The swimmer's entire view of the external world comes through the overhead circle, which acts as a personal window for the swimmer.

(b) The swimmer now wears a swimming mask over his eyes; the thin flat layer of transparent plastic through which the swimmer looks is horizontal; and air fills the interior of the mask. Over what range of angles do rays now reach the swimmer's eyes from light sources external to the water? (Neglect refraction by the plastic faceplate of the mask; its inclusion would not alter the result.)

SOLUTION: To answer, we examine the refraction of an arbitrary light ray as shown in Figure 34-23c. As before, the ray refracts from air into water, but now, to reach point E, it must refract a second time, from the water into the air held by the mask. Let the angle of incidence at the second refraction be θ_3 and the angle of refraction be θ_4. We seek the angle at which the ray reaches point E, which is equal to θ_4.

Because the two air–water interfaces in Fig. 34-23c are parallel, $\theta_3 = \theta_2$. From Eq. 34-44, the angle of refraction θ_4 is given by

$$\sin \theta_4 = \frac{n_2}{n_1} \sin \theta_3.$$

Substituting θ_2 for θ_3 and then substituting for $\sin \theta_2$ from Eq. 34-46, we find

$$\sin \theta_4 = \frac{n_2}{n_1} \sin \theta_2 = \frac{n_2 n_1}{n_1 n_2} \sin \theta_1 = \sin \theta_1,$$

which (in this situation) gives us

$$\theta_4 = \theta_1.$$

In words, the arbitrary ray in Fig. 34-23c reaches point E traveling parallel to its original direction. So do all other rays reaching E from external sources. Thus the range of those rays is from about 90° on one side of the vertical to about 90° on the other side. This means that with the mask, the swimmer sees the external world as if the water were not present, and not compressed into a 100° cone.

34-8 TOTAL INTERNAL REFLECTION

Figure 34-24 shows rays of monochromatic light from a point source S in glass incident on the interface between the glass and air. For ray a, which is perpendicular to the interface, part of the light reflects at the interface and the rest travels through it with no change in direction.

For rays b through e, which have progressively larger angles of incidence at the interface, there are also both

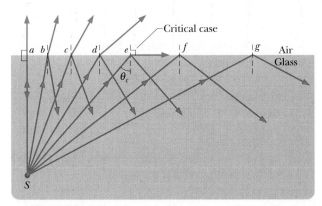

FIGURE 34-24 Total internal reflection of light from a point source S in glass occurs for all angles of incidence greater than the critical angle θ_c. At the critical angle, the refracted ray points along the air–glass interface.

reflection and refraction at the interface. As the angle of incidence increases, the angle of refraction increases; for ray e it is 90°, which means that the refracted ray points directly along the interface. The angle of incidence giving this situation is called the **critical angle** θ_c. For angles of incidence larger than θ_c, such as for rays f and g, there is no refracted ray and *all* the light is reflected; this effect is called **total internal reflection.**

To find θ_c, we use Eq. 34-44: we arbitrarily associate subscript 1 with the glass and subscript 2 with the air, and then we substitute θ_c for θ_1 and 90° for θ_2, finding

$$n_1 \sin \theta_c = n_2 \sin 90°,$$

which gives us

$$\theta_c = \sin^{-1} \frac{n_2}{n_1} \qquad \text{(critical angle)}. \qquad (34\text{-}47)$$

Because the sine of an angle cannot exceed unity, n_2 cannot exceed n_1 in this equation. This restriction tells us that total internal reflection cannot occur when the incident light is in the medium of lower index of refraction. If source S were in the air in Fig. 34-24, all its rays that are incident on the air–glass interface (including f and g) would be both reflected *and* refracted at the interface.

Total internal reflection has found many applications in medical technology. For example, a physician can search for an ulcer in the stomach of a patient by running two thin bundles of *optical fibers* (Fig. 34-25) down the patient's throat. Light introduced at the outer end of one bundle undergoes repeated total internal reflection within the fibers so that, even though the bundle provides a curved path, the light ends up illuminating the interior of the stomach. Some of the light reflected from the interior then comes back up the second bundle in a similar way, to be detected and converted to an image on a monitor's screen for the physician to view.

FIGURE 34-25 Light sent into one end of an optical fiber is transmitted to the opposite end with little loss of the light through the sides of the fiber, because most of the light undergoes repeated total internal reflection along those sides.

SAMPLE PROBLEM 34-5

Figure 34-26 shows a triangular prism of glass in air; a ray incident perpendicularly to one face is totally reflected at the glass–air interface indicated. If θ_1 is 45°, what can you say about the index of refraction n of the glass?

SOLUTION: Using Eq. 34-47, approximating the index of refraction n_2 of air as unity, and substituting the index of refraction n of the glass for n_1, we find for the critical angle θ_c:

$$\theta_c = \sin^{-1}\frac{n_2}{n_1} = \sin^{-1}\frac{1}{n}.$$

Since total internal reflection occurs, θ_c must be less than θ_1, which is 45°. Then

$$\sin^{-1}\frac{1}{n} < 45°,$$

which gives us

$$\frac{1}{n} < \sin 45°,$$

or $\qquad\qquad n > \dfrac{1}{\sin 45°} = 1.4.$ \qquad (Answer)

The index of refraction of the glass must be greater than 1.4; otherwise total internal reflection would not occur for the incident ray shown.

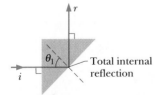

FIGURE 34-26 Sample Problem 34-5. The incident ray i is totally internally reflected at the glass–air interface, becoming the reflected ray r.

CHECKPOINT **6:** Suppose the prism in Sample Problem 34-5 has the index of refraction $n = 1.4$. Does the light still totally internally reflect if we keep the incident ray horizontal but rotate the prism (a) 10° clockwise and (b) 10° counterclockwise in Fig. 34-26?

34-9 POLARIZATION BY REFLECTION

You can increase and decrease the glare you see in sunlight that has been reflected from, say, water by looking through a polarizing sheet (such as a polarizing sunglass lens) and then rotating the sheet's polarizing axis around your line of sight. You can do so because reflected light is fully or partially polarized by the reflection from a surface.

Figure 34-27 shows a ray of unpolarized light incident on a glass surface. Let us resolve the electric field vectors of the light into two components. The *perpendicular components* are perpendicular to the plane of incidence and thus also to the page in Fig. 34-27; these components are represented with dots (as if we see the tips of the vectors). The *parallel components* are parallel to the plane of incidence and the page; they are represented with double-headed arrows. Because the light is unpolarized, these two components are of equal magnitude.

In general, the reflected light also has both components but with unequal magnitudes. This means that the reflected light is partially polarized—the electric fields

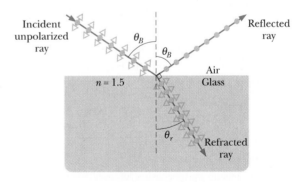

● Component perpendicular to page
◁—▷ Component parallel to page

FIGURE 34-27 A ray of unpolarized light in air is incident on a glass surface at the Brewster angle θ_B. The electric field vectors are resolved into components perpendicular to the page (the plane of incidence) and components parallel to the page. The reflected light consists only of components perpendicular to the page and is thus polarized in that direction. The refracted light consists of the original components parallel to the page and weaker components perpendicular to the page; this light is partially polarized.

oscillating along one direction have greater amplitudes than those oscillating along other directions. However, when the light is incident at a particular incident angle, called the *Brewster angle* θ_B, the reflected light has only perpendicular components, as shown in Fig. 34-27. The reflected light is then fully polarized perpendicular to the plane of incidence. The parallel components of the incident light do not disappear; they and perpendicular components form the light that is refracted through the glass surface.

Glass, water, and the other dielectric materials discussed in Section 26-7 can partially and fully polarize light by reflection. When you intercept sunlight reflected from such a surface, you see a bright spot (the glare) on the surface where the reflection takes place. If the surface is horizontal as in Fig. 34-27, the reflected light is partially or fully polarized horizontally. To eliminate such glare from horizontal surfaces, the lenses in polarizing sunglasses are mounted with their polarizing direction vertical.

Brewster's Law

For light incident at the Brewster angle θ_B, we find experimentally that the reflected and refracted rays are perpendicular to each other. Because the reflected ray is reflected at the angle θ_B in Fig. 34-27 and the refracted ray is at angle θ_r, we have

$$\theta_B + \theta_r = 90°.$$

These two angles can also be related with Eq. 34-44. Arbitrarily assigning subscript 1 in Eq. 34-44 to the material through which the incident and reflected rays travel, we have, from that equation,

$$n_1 \sin \theta_B = n_2 \sin \theta_r.$$

Combining these equations leads to

$$n_1 \sin \theta_B = n_2 \sin(90° - \theta_B) = n_2 \cos \theta_B,$$

which gives us

$$\theta_B = \tan^{-1} \frac{n_2}{n_1} \quad \text{(Brewster angle).} \quad (34\text{-}48)$$

(Note carefully that the subscripts in Eq. 34-48 are *not* arbitrary because of our decision as to their meanings.) If the incident and reflected rays travel *in air,* we can approximate n_1 as unity and let n represent n_2 in order to write Eq. 34-48 as

$$\theta_B = \tan^{-1} n \quad \text{(Brewster's law).} \quad (34\text{-}49)$$

This simplified version of Eq. 34-48 is known as **Brewster's law.** Like θ_B, it is named after Sir David Brewster, who found both experimentally in 1812.

SAMPLE PROBLEM 34-6

We wish to use a glass plate with index of refraction $n = 1.57$ to polarize light in air.

(a) At what angle of incidence is the light reflected by the glass fully polarized?

SOLUTION: Because the glass is in air, we can use Eq. 34-49 to find the Brewster angle:

$$\theta_B = \tan^{-1} n = \tan^{-1} 1.57 = 57.5°. \quad \text{(Answer)}$$

(b) What is the corresponding angle of refraction?

SOLUTION: Since $\theta_B + \theta_r = 90°$, we have

$$\theta_r = 90° - \theta_B = 90° - 57.5° = 32.5°. \quad \text{(Answer)}$$

REVIEW & SUMMARY

Electromagnetic Waves

An electromagnetic wave consists of oscillating electric and magnetic fields. The various possible frequencies of electromagnetic waves form a *spectrum*, a small part of which is visible light. An electromagnetic wave traveling along an x axis has an electric field $\mathbf{E}$ and a magnetic field $\mathbf{B}$ with magnitudes that depend on x and t:

$$E = E_m \sin(kx - \omega t)$$

and

$$B = B_m \sin(kx - \omega t), \quad (34\text{-}1, 34\text{-}2)$$

where E_m and B_m are the amplitudes of $\mathbf{E}$ and $\mathbf{B}$. The electric field induces the magnetic field and vice versa. The speed of any elec-

tromagnetic wave in vacuum is c, which can be written as

$$c = \frac{E}{B} = \frac{1}{\sqrt{\mu_0 \epsilon_0}}, \quad (34\text{-}5, 34\text{-}3)$$

where E and B are the simultaneous magnitudes of the fields.

Energy Flow

The rate per unit area at which energy is transported via an electromagnetic wave is given by the Poynting vector $\mathbf{S}$:

$$\mathbf{S} = \frac{1}{\mu_0} \mathbf{E} \times \mathbf{B}. \quad (34\text{-}21)$$

The direction of **S** (and thus of the wave's travel and the energy transport) is perpendicular to the directions of both **E** and **B**. The time-averaged rate per unit area at which energy is transported is $\bar{S}$, which is called the *intensity I* of the wave:

$$I = \frac{1}{c\mu_0} E_{\text{rms}}^2, \qquad (34\text{-}26)$$

in which $E_{\text{rms}} = E_m/\sqrt{2}$. A *point source* of electromagnetic waves emits the waves *isotropically,* that is, with equal intensity in all directions. The intensity of the waves at distance r from a point source of power P_s is

$$I = \frac{P_s}{4\pi r^2}. \qquad (34\text{-}27)$$

Radiation Pressure

When a surface intercepts electromagnetic radiation, a force and a pressure are exerted on the surface. If the radiation is totally absorbed by the surface, the force is

$$F = \frac{IA}{c} \quad \text{(total absorption),} \qquad (34\text{-}32)$$

in which I is the intensity of the radiation and A is the area of the surface perpendicular to the path of the radiation. If the radiation is totally reflected back along its original path, the force is

$$F = \frac{2IA}{c} \quad \text{(total reflection back along path).} \qquad (34\text{-}33)$$

The radiation pressure p_r is the force per unit area:

$$p_r = \frac{I}{c} \quad \text{(total absorption)} \qquad (34\text{-}34)$$

and

$$p_r = \frac{2I}{c} \quad \begin{array}{l}\text{(total reflection)} \\ \text{back along path).}\end{array} \qquad (34\text{-}35)$$

Polarization

Electromagnetic waves are **polarized** if their electric field vectors are all in a single plane, called the *plane of oscillation*. Light waves from common sources are not polarized, that is, they are **unpolarized** or **randomly polarized.**

Polarizing Sheets

When a polarizing sheet is placed in the path of light, only electric field components of the light parallel to the sheet's **polarizing direction** are *transmitted* by the sheet; components perpendicular to the polarizing direction are absorbed. The light that emerges from a polarizing sheet is polarized parallel to the polarizing direction of the sheet.

If the original light is initially unpolarized, the transmitted intensity I is half the original intensity I_0:

$$I = \tfrac{1}{2}I_0. \qquad (34\text{-}40)$$

If the original light is initially polarized, the transmitted intensity depends on the angle θ between the polarization direction of the original light and the polarizing direction of the sheet:

$$I = I_0 \cos^2 \theta. \qquad (34\text{-}42)$$

Geometrical Optics

Geometrical optics is an approximate treatment in which light waves are represented as straight-line rays.

Reflection and Refraction

When a light ray encounters a boundary between two transparent media, a **reflected** ray and a **refracted** ray generally appear. Both rays remain in the plane of incidence. The **angle of reflection** is equal to the angle of incidence, and the **angle of refraction** is related to the angle of incidence by

$$n_1 \sin \theta_1 = n_2 \sin \theta_2 \quad \text{(refraction),} \qquad (34\text{-}44)$$

where n_1 and n_2 are the indices of refraction of the media in which the incident and refracted rays travel.

Total Internal Reflection

A wave encountering a boundary across which the index of refraction decreases will experience **total internal reflection** if the angle of incidence exceeds a **critical angle** θ_c, where

$$\theta_c = \sin^{-1} \frac{n_2}{n_1} \quad \text{(critical angle).} \qquad (34\text{-}47)$$

Polarization by Reflection

A reflected wave will be fully **polarized,** with its **E** vectors perpendicular to the plane of incidence, if it strikes a boundary at the **Brewster angle** θ_B, where

$$\theta_B = \tan^{-1} \frac{n_2}{n_1} \quad \text{(Brewster angle).} \qquad (34\text{-}48)$$

QUESTIONS

1. If the magnetic field of a light wave oscillates parallel to a y axis and is given by $B_y = B_m \sin(kz - \omega t)$, (a) in what direction does the wave travel and (b) parallel to which axis does the associated electric field oscillate?

2. Figure 34-28 shows the electric and magnetic fields of an electromagnetic wave at a certain instant. Is the wave traveling into the page or out of it?

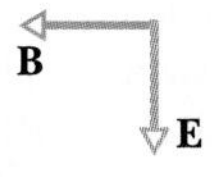

FIGURE 34-28 Question 2.

3. (a) Fig. 34-29 shows light reaching a polarizing sheet whose polarizing direction is parallel to a y axis. We shall rotate the

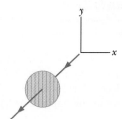

FIGURE 34-29 Question 3.

sheet 40° clockwise about the light's indicated line of travel. During this rotation, does the fraction of the initial light intensity passed by the sheet increase, decrease, or remain the same if the light is (a) initially unpolarized, (b) initially polarized parallel to the x axis, and (c) initially polarized parallel to the y axis?

4. Light is sent through the two-sheet polarizing system of Fig. 34-30. If the ratio of the emerging intensity to the initial intensity is 0.7, is the light initially polarized or unpolarized?

5. Initially unpolarized light is sent through the two-sheet polarizing system of Fig. 34-30. The emerging light is polarized 20° clockwise from the y axis and has half of the initial intensity. What are the polarizing directions of the sheets?

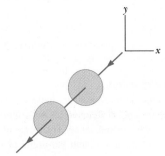

FIGURE 34-30 Questions 4, 5, and 7.

6. In Fig. 34-16a, start with light that is initially polarized parallel to the x axis, and write the ratio of its final intensity I_3 to its initial intensity I_0 as $I_3/I_0 = A \cos^n \theta$. What are A, n, and θ if we rotate the polarizing direction of the first sheet (a) 60° counterclockwise and (b) 90° clockwise from what is shown?

7. Three polarizing sheets are positioned like the two sheets in Fig. 34-30, and initially unpolarized light is sent into the system. How many different final intensities can be produced if the polarizing directions of the sheets are the following: one is parallel to the y axis, one is rotated 20° clockwise from the y axis about the light's line of travel, and one is rotated 20° in the opposite direction about that line of travel.

8. Suppose we rotate the second sheet in Fig. 34-16a, starting with its polarization direction aligned with the y axis ($\theta = 0$) and ending with its polarization direction aligned with the x axis ($\theta = 90°$). Which of the curves in Fig. 34-31 best shows the intensity of the light through the three-sheet system during this 90° rotation?

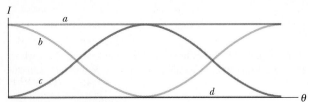

FIGURE 34-31 Question 8.

9. Figure 34-32 shows the multiple reflections of a light ray along a glass corridor where the walls are either parallel or perpendicular to one another. If the angle of incidence at point a is 30°, what are the angles of reflection at points b, c, d, e, and f?

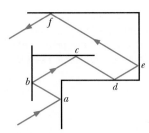

FIGURE 34-32 Question 9.

10. Figure 34-33 shows rays of monochromatic light passing through three materials a, b, and c. Rank the materials according to their indices of refraction, greatest first.

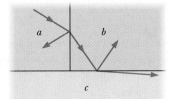

FIGURE 34-33 Question 10.

11. In Fig. 34-34, light travels from material a, through three layers of other materials with surfaces parallel to one another, and then back into another layer of material a. The refractions (but not the associated reflections) at the surfaces are shown. Rank the materials according to their indices of refraction, greatest first.

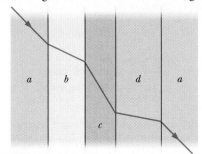

FIGURE 34-34 Question 11.

12. Which of the three parts of Fig. 34-35 show physically possible refraction?

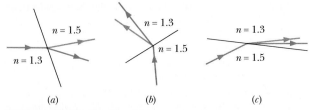

FIGURE 34-35 Question 12.

13. (a) Figure 34-36a shows a ray of sunlight that just barely passes a vertical stick in a pool of water. Does that ray end in the general region of point a or point b? (b) Does the red or the blue component of the ray end up closer to the stick? (c) Figure 34-36b shows a flat object (such as a double-edged razor blade) floating in shallow water and illuminated vertically. The object's weight

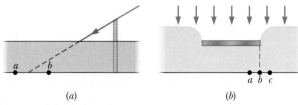

FIGURE 34-36 Question 13.

causes the water surface to curve as shown. In which general region (*a*, *b*, or *c*) is the edge of the object's shadow? (To the right of the edge of the shadow, many rays of sunlight are concentrated and produce an especially bright region, said to be a *caustic*.)

14. Figure 34-22 shows some of the rays of sunlight responsible for the *primary rainbow* (which involves one reflection inside each water drop). A fainter, less frequent *secondary rainbow* (involving two reflections inside each water drop) can appear above a primary rainbow, formed by rays entering and exiting water drops as shown in Fig. 34-37 (without color indicated). Which ray, *a* or *b*, corresponds to red light?

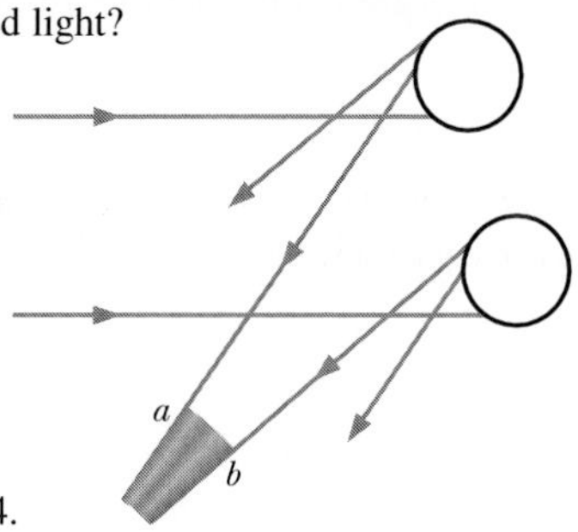

FIGURE 34-37 Question 14.

15. Figure 34-38 shows four long horizontal layers of different materials, with air above and below them. The index of refraction of each material is given. Rays of light are sent into the left ends of each layer as shown. In which layer (give the index of refrac-

tion) is there the possibility of totally trapping the light in that layer so that, after many reflections, all the light reaches the right end of the layer?

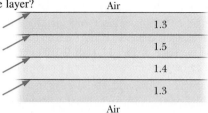

FIGURE 34-38 Question 15.

16. Figure 34-39*a* shows an overhead view of a rectangular room with fully reflecting walls. The room length *L* and width *W* are both integer numbers of units. You fire a laser from corner *a* at 45° to the walls. Small clay armadillos are located in the other corners. You are concerned as to whether your shot will hit one of those targets or yourself. Figure 34-39*b* shows a way to find out by drawing repeated reflections of the room and extending a straight light path through them. Which corner is hit depends on the ratio *L/W*, once that ratio is reduced to lowest terms. (For example, 4/2 reduces to 2/1.) Figure 34-39*b* is for *L/W* = 2/1. We see that the target in corner *d* is hit after one reflection. Determine which corner is hit for any (reduced) *L/W* in the form of (a) even number/odd number, (b) odd number/even number, and (c) odd number/odd number.

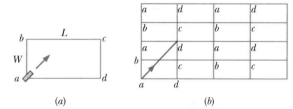

FIGURE 34-39 Question 16.

EXERCISES & PROBLEMS

SECTION 34-1 Maxwell's Rainbow

1E. Project Seafarer was an ambitious program to construct an enormous antenna, buried underground on a site about 4000 square miles in area. Its purpose was to transmit signals to submarines while they were deeply submerged. If the effective wavelength was 1.0×10^4 Earth radii, what would be (a) the frequency and (b) the period of the radiations emitted? Ordinarily electromagnetic radiations do not penetrate very far into conductors such as seawater.

2E. (a) How long does it take a radio signal to travel 150 km from a transmitter to a receiving antenna? (b) We see a full Moon by reflected sunlight. How much earlier did the light that enters our eye leave the Sun? The Earth–Moon and Earth–Sun distances are 3.8×10^5 km and 1.5×10^8 km. (c) What is the round-trip travel time for light between Earth and a spaceship

orbiting Saturn, 1.3×10^9 km distant? (d) The Crab nebula, which is about 6500 light-years (ly) distant, is thought to be the result of a supernova explosion recorded by Chinese astronomers in A.D. 1054. In approximately what year did the explosion actually occur?

3E. (a) The wavelength of the most energetic x rays produced when electrons are accelerated to a kinetic energy of 18 GeV in the Stanford Linear Accelerator and then slam into a solid target is 0.067 fm. What is the frequency of these x rays? (b) A VLF (very low frequency) radio wave has a frequency of only 30 Hz. What is its wavelength?

4E. (a) At what wavelengths does the eye of a standard observer have half its maximum sensitivity? (b) What are the wavelength, the frequency, and the period of the light for which the eye is the most sensitive?

5E. In Fig. 34-1, verify that the uniform spaces between successive powers of 10 must be the same on the wavelength scale and on the frequency scale.

6E. A certain helium–neon laser emits red light in a narrow band of wavelengths centered at 632.8 nm and with a "width" (such as on the scale of Fig. 34-1) of 0.0100 nm. What is the corresponding range of frequencies for the emission?

7P. One method for measuring the speed of light, based on observations by Roemer in 1676, consisted in observing the apparent times of revolution of one of the moons of Jupiter. The true period of revolution is 42.5 h. (a) Taking into account the finite speed of light, how would you expect the apparent time for one revolution to change as Earth moves in its orbit from point x to point y in Fig. 34-40? (b) What observations would be needed to compute the speed of light? Neglect the motion of Jupiter in its orbit. Figure 34-40 is not drawn to scale.

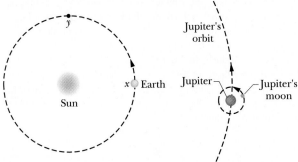

FIGURE 34-40 Problem 7.

SECTION 34-2 The Traveling Electromagnetic Wave, Qualitatively

8E. What is the wavelength of the electromagnetic wave emitted by the oscillator–antenna system of Fig. 34-3 if $L = 0.253$ μH and $C = 25.0$ pF?

9E. What inductance must be connected to a 17 pF capacitor in an oscillator capable of generating 550 nm (i.e., visible) electromagnetic waves? Comment on your answer.

10P. Figure 34-41 shows an LC oscillator connected by a transmission line to an antenna of a so-called *magnetic* dipole type. Compare it with Fig. 34-3, which shows a similar arrangement but with an *electric* dipole type of antenna. (a) What is the basis for the names of these two antenna types? (b) Draw a figure like Fig. 34-4 to describe the electromagnetic wave that sweeps past an observer at point P in Fig. 34-41.

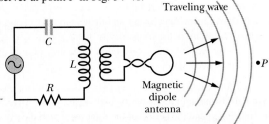

FIGURE 34-41 Problem 10.

SECTION 34-3 The Traveling Electromagnetic Wave, Quantitatively

11E. A plane electromagnetic wave has a maximum electric field of 3.20×10^{-4} V/m. Find the maximum magnetic field.

12E. The electric field of a certain plane electromagnetic wave is given by $E_x = 0$; $E_y = 0$; $E_z = 2.0 \cos[\pi \times 10^{15}(t - x/c)]$, with $c = 3.0 \times 10^8$ m/s and all quantities in SI units. The wave is propagating in the positive x direction. Write expressions for the components of the magnetic field of the wave.

13P. Start from Eqs. 34-11 and 34-18 and show that $E(x, t)$ and $B(x, t)$, the electric and magnetic field components of a plane traveling electromagnetic wave, must satisfy the "wave equations"

$$\frac{\partial^2 E}{\partial t^2} = c^2 \frac{\partial^2 E}{\partial x^2} \quad \text{and} \quad \frac{\partial^2 B}{\partial t^2} = c^2 \frac{\partial^2 B}{\partial x^2}.$$

14P. (a) Show that Eqs. 34-1 and 34-2 satisfy the wave equations displayed in Problem 13. (b) Show that any expressions of the form

$$E = E_m f(kx \pm \omega t) \quad \text{and} \quad B = B_m f(kx \pm \omega t),$$

where $f(kx \pm \omega t)$ denotes an arbitrary function, also satisfy these wave equations.

SECTION 34-4 Energy Transport and the Poynting Vector

15E. Show, by finding the direction of the Poynting vector **S**, that the directions of the electric and magnetic fields at all points in Figs. 34-4 to 34-7 are consistent at all times with the assumed directions of propagation.

16E. Currently operating neodymium–glass lasers can provide 100 TW of power in 1.0 ns pulses at a wavelength of 0.26 μm. How much energy is contained in a single pulse?

17E. Our closest stellar neighbor, Proxima Centauri, is 4.3 ly away. It has been suggested that TV programs from our planet have reached this star and may have been viewed by the hypothetical inhabitants of a hypothetical planet orbiting it. Suppose a television station on Earth has a power of 1.0 MW. What is the intensity of its signal at Proxima Centauri?

18E. An electromagnetic wave is traveling in the negative y direction. At a particular position and time, the electric field is along the positive z axis and has a magnitude of 100 V/m. What are the direction and magnitude of the magnetic field at that position and at that time?

19E. Earth's mean radius is 6.37×10^6 m and the mean Earth–Sun distance is 1.50×10^8 km. What fraction of the radiation emitted by the Sun is intercepted by the disk of Earth?

20E. The radiation emitted by a laser spreads out in the form of a narrow cone with circular cross section. The angle θ of the cone (see Fig. 34-42) is called the *full-angle beam divergence*. An argon laser, radiating at 514.5 nm, is aimed at the Moon in a ranging experiment. If the beam has a full-angle beam divergence of 0.880 μrad, what area on the Moon's surface is illuminated by the laser?

FIGURE 34-42
Exercise 20.

21E. The intensity of direct solar radiation that is not absorbed by the atmosphere on a particular summer day is 100 W/m². How close would you have to stand to a 1.0 kW electric heater to feel the same intensity? Assume that the heater radiates uniformly in all directions.

22E. Show that in a plane traveling electromagnetic wave the intensity, that is, the average rate of energy transport per unit area, is given by

$$\bar{S} = \frac{E_m^2}{2\mu_0 c} = \frac{cB_m^2}{2\mu_0}.$$

23E. What is the intensity of a plane traveling electromagnetic wave if B_m is 1.0×10^{-4} T?

24E. In a plane radio wave the maximum value of the electric field component is 5.00 V/m. Calculate (a) the maximum value of the magnetic field component and (b) the wave intensity.

25P. You walk 150 m directly toward a street lamp and find that the intensity increases to 1.5 times the intensity at your original position. How far from the lamp were you first standing? (Assume that the lamp is an isotropic point source of light.)

26P. Prove that the intensity of an electromagnetic wave is the product of the wave's energy density and its speed.

27P. Sunlight just outside Earth's atmosphere has an intensity of 1.40 kW/m². Calculate E_m and B_m for sunlight, assuming it to be a plane wave.

28P. The maximum electric field at a distance of 10 m from a point light source is 2.0 V/m. What are (a) the maximum value of the magnetic field and (b) the average intensity of the light there? (c) What is the power of the source?

29P. Frank D. Drake, an active investigator in the SETI (Search for Extra-Terrestrial Intelligence) program, has said that the large radio telescope in Arecibo, Puerto Rico, "can detect a signal which lays down on the entire surface of the earth a power of only

one picowatt." See Fig. 34-43. (a) What is the power actually received by the Arecibo antenna for such a signal? The antenna diameter is 1000 ft. (b) What would be the power of a source at the center of our galaxy that could provide such a signal? The galactic center is 2.2×10^4 ly away. Take the source as radiating uniformly in all directions.

30P. A helium–neon laser, radiating at 632.8 nm, has a power output of 3.0 mW and a full-angle beam divergence (see Exercise 20) of 0.17 mrad. (a) What is the intensity of the beam 40 m from the laser? (b) What is the power of a point source that provides this same intensity at the same distance?

31P. An airplane flying at a distance of 10 km from a radio transmitter receives a signal of power 10 μW/m². Calculate (a) the amplitude of the electric field at the airplane due to this signal; (b) the amplitude of the magnetic field at the airplane; and (c) the total power of the transmitter, assuming the transmitter to radiate uniformly in all directions.

32P. During a test, a NATO surveillance radar system, operating at 12 GHz at 180 kW of power, attempts to detect an incoming stealth aircraft at 90 km. Assume that the radar beam is emitted uniformly over a hemisphere. (a) What is the intensity of the beam at the aircraft's location? The aircraft reflects radar waves as though it has a cross-sectional area of only 0.22 m². (b) What is the power of the aircraft's reflection? Assume that the beam is reflected uniformly over a hemisphere. Back at the radar site, what are (c) the intensity, (d) the maximum value of the electric field vector, and (e) the rms value of the magnetic field of the reflected radar beam?

SECTION 34-5   Radiation Pressure

33E. A black, totally absorbing piece of cardboard of area $A = 2.0$ cm² intercepts light with an intensity of 10 W/m² from a camera strobe light. What radiation pressure is produced on the cardboard by the light?

34E. High-power lasers are used to compress a plasma (a gas of charged particles) by radiation pressure. A laser generating pulses of radiation of peak power 1.5×10^3 MW is focused onto 1.0 mm² of high-electron-density plasma. Find the pressure exerted on the plasma if the plasma reflects perfectly.

35E. The average intensity of the solar radiation that falls normally on a surface just outside Earth's atmosphere is 1.4 kW/m². (a) What radiation pressure is exerted on this surface, assuming complete absorption? (b) How does this pressure compare with Earth's sea-level atmospheric pressure, which is 1.0×10^5 Pa?

36E. Radiation from the Sun reaching Earth (just outside the atmosphere) has an intensity of 1.4 kW/m². (a) Assuming that Earth (and its atmosphere) behaves like a flat disk perpendicular to the Sun's rays and that all the incident energy is absorbed, calculate the force on Earth due to radiation pressure. (b) Compare it with the force due to the Sun's gravitational attraction.

37E. What is the radiation pressure 1.5 m away from a 500 W lightbulb? Assume that the surface on which the pressure is exerted faces the bulb and is perfectly absorbing and that the bulb radiates uniformly in all directions.

FIGURE 34-43   Problem 29.

38P. A plane electromagnetic wave, with wavelength 3.0 m, travels in vacuum in the positive x direction with its electric vector $\mathbf{E}$, of amplitude 300 V/m, directed along the y axis. (a) What is the frequency f of the wave? (b) What are the direction and amplitude of the magnetic field associated with the wave? (c) If $E = E_m \sin(kx - \omega t)$, what are the values of k and ω? (d) What is the time-averaged rate of energy flow in watts per square meter associated with this wave? (e) If the wave falls on a perfectly absorbing sheet of area 2.0 m^2, at what rate would momentum be delivered to the sheet and what is the radiation pressure exerted on the sheet?

39P. A helium–neon laser of the type often found in physics laboratories has a beam power of 5.00 mW at a wavelength of 633 nm. The beam is focused by a lens to a circular spot whose effective diameter may be taken to be equal to 2.00 wavelengths. Calculate (a) the intensity of the focused beam, (b) the radiation pressure exerted on a tiny perfectly absorbing sphere whose diameter is that of the focal spot, (c) the force exerted on this sphere, and (d) the acceleration imparted to it. Assume a sphere density of 5.00×10^3 kg/m^3.

40P. In Fig. 34-44, a laser beam of power 4.60 W and diameter 2.60 mm is directed upward at one circular face (of diameter $d < 2.60$ mm) of a perfectly reflecting cylinder, which is made to "hover" by the beam's radiation pressure. The cylinder's density is 1.20 g/cm^3. What is the cylinder's height H?

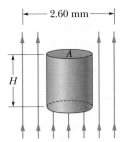

FIGURE 34-44 Problem 40.

41P. Radiation of intensity I is normally incident on an object that absorbs a fraction *frac* of it and reflects the rest back along the original path. What is the radiation pressure on the object?

42P. Prove, for a plane wave that is normally incident on a plane surface, that the radiation pressure on the surface is equal to the energy density in the beam outside the surface. (This relation holds no matter what fraction of the incident energy is reflected.)

43P. A laser beam of intensity I reflects from a flat, totally reflecting surface of area A whose normal makes an angle θ with the direction of the beam. Write an expression for the radiation pressure $p_r(\theta)$ exerted on the surface, in terms of the pressure $p_{r\perp}$ that would be exerted if the beam were perpendicular to the surface.

44P. Prove that the average pressure of a stream of bullets striking a plane surface perpendicularly is twice the kinetic energy density in the stream above the surface. Assume that the bullets are completely absorbed by the surface. Contrast this with the behavior of light in Problem 42.

45P. A small spaceship whose mass is 1.5×10^3 kg (including an astronaut) is drifting in outer space with negligible gravitational forces acting on it. If the astronaut turns on a 10 kW laser beam, what speed will the ship attain in 1.0 day because of the momentum carried away by the beam?

46P. It has been proposed that a spaceship might be propelled in the solar system by radiation pressure, using a large sail made of foil. How large must the sail be if the radiation force is to be equal in magnitude in the Sun's gravitational attraction? Assume that the mass of the ship + sail is 1500 kg, that the sail is perfectly reflecting, and that the sail is oriented perpendicularly to the Sun's rays. See Appendix C for needed data. (With a larger sail, the ship is continually driven away from the Sun.)

47P. A particle in the solar system is under the combined influence of the Sun's gravitational attraction and the radiation force due to the Sun's rays. Assume that the particle is a sphere of density 1.0×10^3 kg/m^3 and that all the incident light is absorbed. (a) Show that, if its radius is less than some critical radius R, the particle will be blown out of the solar system. (b) Calculate the critical radius.

SECTION 34-6 Polarization

48E. The magnetic field equations for an electromagnetic wave in vacuum are $B_x = B \sin(ky + \omega t)$, $B_y = B_z = 0$. (a) What is the direction of propagation? (b) Write the electric field equations. (c) Is the wave polarized? If so, in what direction?

49E. A beam of unpolarized light of intensity 10 mW/m^2 is sent through a polarizing sheet as in Fig. 34-12. (a) Find the maximum value of the electric field of the transmitted beam. (b) What radiation pressure is exerted on the polarizing sheet?

50E. A beam of unpolarized light is sent through two polarizing sheets placed one on top of the other. What must be the angle between the polarizing directions of the sheets if the intensity of the transmitted light is to be one-third the incident intensity?

51E. Three polarizing plates are stacked. The first and third are crossed; the one between has its polarizing direction at 45° to the polarizing directions of the other two. What fraction of the intensity of an originally unpolarized beam is transmitted by the stack?

52E. In Fig. 34-45, initially unpolarized light is sent through three polarizing sheets whose polarizing directions make angles of $\theta_1 = \theta_2 = \theta_3 = 50°$ with the direction of the y axis. What percentage of the initial intensity is transmitted by the system of the three sheets?

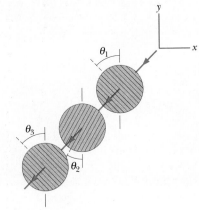

FIGURE 34-45
Exercise 52 and
Problem 53.

53P. In Fig. 34-45, initially unpolarized light is sent through three polarizing sheets whose polarizing directions make angles of $\theta_1 = 40°$, $\theta_2 = 20°$, and $\theta_3 = 40°$ with the direction of the y axis. What percentage of the light's initial intensity is transmitted by the system?

54P. An unpolarized beam of light is sent through a stack of four polarizing sheets, oriented so that the angle between the polarizing directions of adjacent sheets is 30°. What fraction of the incident intensity is transmitted by the system?

55P. A beam of polarized light is sent through a system of two polarizing sheets. Relative to the polarization direction of that incident light, the polarizing directions of the sheets are at angles θ for the first sheet and 90° for the second sheet. If 0.10 of the incident intensity is transmitted by the two sheets, what is θ?

56P. A horizontal beam of vertically polarized light of intensity 43 W/m² is sent through two polarizing sheets. The polarizing direction of the first is at 70° to the vertical, and that of the second is horizontal. What is the intensity of the light transmitted by the pair of sheets?

57P. Suppose that in Problem 56 the initial beam is unpolarized. What then is the intensity of the transmitted light?

58P. A beam of partially polarized light can be considered to be a mixture of polarized and unpolarized light. Suppose we send such a beam through a polarizing filter and then rotate the filter through 360° while keeping it perpendicular to the beam. If the transmitted intensity varies by a factor of 5.0 during the rotation, what fraction of the intensity of the original beam is associated with the beam's polarized light?

59P. We want to rotate the direction of polarization of a beam of polarized light through 90° by sending the beam through one or more polarizing sheets. (a) What is the minimum number of sheets required? (b) What is the minimum number of sheets required if the transmitted intensity is to be more than 60% of the original intensity?

60P. At a beach the light is generally partially polarized owing to reflections off sand and water. At a particular beach on a particular day near sundown, the horizontal component of the electric field vector is 2.3 times the vertical component. A standing sunbather puts on polarizing sunglasses; the glasses eliminate the horizontal field component. (a) What fraction of the light intensity received before the glasses were put on now reaches the sunbather's eyes? (b) The sunbather, still wearing the glasses, lies on his side. What fraction of the light intensity received before the glasses were put on now reaches his eyes?

SECTION 34-7 Reflection and Refraction

61E. Figure 34-46 shows light reflecting from two perpendicular reflecting surfaces A and B. Find the angle between the incoming ray i and the outgoing ray r'.

62E. Light in vacuum is incident on the surface of a glass slab. In the vacuum the beam makes an angle of 32.0° with the normal to the surface, while in the glass it makes an angle of 21.0° with the normal. What is the index of refraction of the glass?

63E. When the rectangular metal tank in Fig. 34-47 is filled to

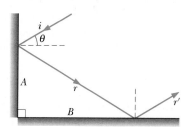

FIGURE 34-46
Exercise 61.

the top with an unknown liquid, an observer with eyes level with the top of the tank can just see the corner E; a ray that refracts toward the observer at the top surface of the liquid is shown. Find the index of refraction of the liquid.

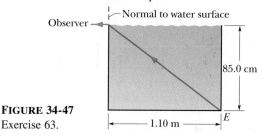

FIGURE 34-47
Exercise 63.

64E. In about A.D. 150, Claudius Ptolemy gave the following measured values for the angle of incidence θ_1 and the angle of refraction θ_2 for a light beam passing from air to water:

θ_1	θ_2	θ_1	θ_2
10°	8°	50°	35°
20°	15°30′	60°	40°30′
30°	22°30′	70°	45°30′
40°	29°	80°	50°

(a) Are these data consistent with the law of refraction? (b) If so, what index of refraction results? These data are interesting as perhaps the oldest recorded physical measurements.

65P. In Fig. 34-48, a 2.00 m long vertical pole extends from the bottom of a swimming pool to a point 50.0 cm above the water. Sunlight is incident at 55.0° above the horizon. What is the length of the shadow of the pole on the level bottom of the pool?

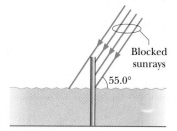

FIGURE 34-48
Problem 65.

66P. A catfish is 2.00 m below the surface of a smooth lake. (a) What is the diameter of the circle on the surface through which the fish can see the world outside the water? (b) If the fish descends, does the diameter of the circle increase, decrease, or remain the same?

67P. Prove that a ray of light incident on the surface of a sheet of plate glass of thickness t emerges from the opposite face parallel to its initial direction but displaced sideways, as in Fig. 34-49. Show that, for small angles of incidence θ, this displacement is given by

$$x = t\theta \frac{n-1}{n},$$

where n is the index of refraction of the glass and θ is measured in radians.

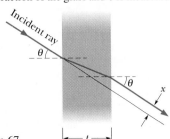

FIGURE 34-49 Problem 67.

68P. A ray of white light makes an angle of incidence of 35° on one face of a prism of fused quartz; the prism's cross section is an equilateral triangle. Sketch the light as it passes through the prism, showing the paths traveled by rays representing (a) blue light, (b) yellow-green light, and (c) red light.

69P. In Figure 34-50, two light rays pass from air through five transparent layers of plastic whose boundaries are parallel, whose indexes of refraction are as given, and whose thickness are unknown. The rays emerge back into air at the right. With respect to a normal taken on the last boundary, what is the angle of (a) emerging ray a and (b) emerging ray b? (c) What are your answers if there is glass, with $n = 1.5$, instead of air on the left and right sides of the plastic layers? (*Hint:* Save yourself much time by first solving the problems algebraically.)

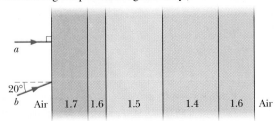

FIGURE 34-50 Problem 69.

70P. In Fig. 34-51, two perpendicular mirrors form the sides of a vessel filled with water. (a) A light ray is incident from above, normal to the water surface. Show that the emerging ray is parallel to the incident ray. Assume that there are two reflections at the mirror surfaces. (b) Repeat the analysis for the case of oblique incidence, with the ray lying in the plane of the figure.

FIGURE 34-51 Problem 70.

71P. In Fig. 34-52, a ray is incident on one face of a triangular glass prism in air. The angle of incidence θ is chosen so that the emerging ray also makes the same angle θ with the normal to the

other face. Show that the index of refraction n of the glass prism is given by

$$n = \frac{\sin \frac{1}{2}(\psi + \phi)}{\sin \frac{1}{2}\phi},$$

where ϕ is the vertex angle of the prism and ψ is the *deviation angle*, the total angle through which the beam is turned in passing through the prism. (Under these conditions the deviation angle ψ has the smallest possible value, which is called the *angle of minimum deviation*.)

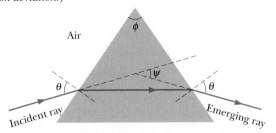

FIGURE 34-52 Problems 71 and 81.

72P. When the atmosphere is cold, moisture can form ice crystals of various shapes. If the atmosphere in the direction of the Sun happens to contain a sufficient number of ice crystals in the shape of flat hexagonal plates, a bright (perhaps colorful) region, called a *sun dog*, appears to the left or right of the Sun. A sun dog is formed by rays of sunlight that pass through the ice plates. These rays are parallel with one another when they arrive at Earth. The rays that pass through an ice plate are redirected by refraction, and those that pass through at the angle of minimum deviation ψ (shown from overhead in Fig. 34-53; see Problem 71) can form a sun dog. The sun dog can then be seen at an angle ψ away from the Sun. If the index of refraction of ice is 1.31, what is ψ?

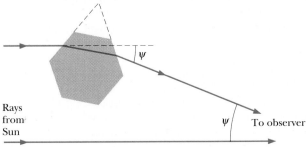

FIGURE 34-53 Problem 72.

73P. A ray of light goes through an equilateral triangular prism that is in the orientation for minimum deviation (see Problem 71). The total deviation is $\psi = 30.0°$. What is the index of refraction of the prism?

SECTION 34-8 Total Internal Reflection

74E. The refractive index of benzene is 1.8. What is the critical angle for a light ray traveling in benzene toward a plane layer of air above the benzene?

75E. In Fig. 34-54, a light ray enters a glass slab at point A and then undergoes total internal reflection at point B. What minimum value for the index of refraction of the glass can be inferred from this information?

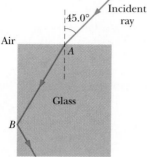

FIGURE 34-54 Exercise 75.

76E. In Fig. 34-55, a ray of light is perpendicular to the face ab of a glass prism ($n = 1.52$). Find the largest value for the angle ϕ so that the ray is totally reflected at face ac if the prism is immersed (a) in air and (b) in water.

FIGURE 34-55
Exercise 76.

77E. A point source of light is 80.0 cm below the surface of a body of water. Find the diameter of the circle at the surface through which light emerges from the water.

78P. A solid glass cube, of edge length 10 mm and index of refraction 1.5, has a small spot at its center. (a) What parts of the cube face must be covered to prevent the spot from being seen, no matter what the direction of viewing? (Neglect the subsequent behavior of internally reflected rays.) (b) What fraction of the cube surface must be so covered?

79P. A ray of white light traveling in fused quartz strikes a plane surface of the quartz with an angle of incidence θ. Is it possible for the internally reflected beam to appear (a) bluish or (b) reddish? (c) If so, what value of θ is required? (*Hint:* White light will appear bluish if wavelengths corresponding to red are removed from the spectrum, and vice versa.)

80P. In Fig. 34-56, light enters a 90° triangular prism at point P with incident angle θ and then some of it refracts at point Q with an angle of refraction of 90°. (a) What is the index of refraction of the prism in terms of θ? (b) What, numerically, is the maximum value that the index of refraction can have? Explain what happens to the light at Q if the incident angle at Q is (c) increased slightly and (d) decreased slightly.

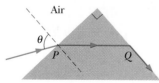

FIGURE 34-56
Problem 80.

81P. Suppose the prism of Fig. 34-52 has apex angle $\phi = 60.0°$ and index of refraction $n = 1.60$. (a) What is the smallest angle of incidence θ for which a ray can enter the left face of the prism and

exit the right face? (b) What angle of incidence θ is required for the ray to exit the prism with an identical angle θ for its refraction, as it does in Fig. 34-52? (See Problem 71.)

82P. A point source of light is placed a distance h below the surface of a large deep lake. (a) Neglecting reflection at the surface except where it is total, show that the fraction $frac$ of the light energy that escapes directly from the water surface is independent of h and is given by

$$frac = \tfrac{1}{2}(1 - \sqrt{1 - 1/n^2}),$$

where n is the index of refraction of the water. (b) Evaluate this fraction for $n = 1.33$.

83P. An optical fiber consists of a glass core (index of refraction n_1) surrounded by a coating (index of refraction $n_2 < n_1$). Suppose a beam of light enters the fiber from air at an angle θ with the fiber axis as shown in Fig. 34-57. (a) Show that the greatest possible value of θ for which a ray can travel down the fiber is given by $\theta = \sin^{-1} \sqrt{n_1^2 - n_2^2}$. (b) If the indices of refraction of the glass and coating are 1.58 and 1.53, respectively, what is this value of the incident angle θ?

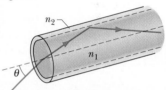

FIGURE 34-57
Problems 83 and 84.

84P. In an optical fiber (see Problem 83), different rays travel different paths along the fiber, leading to different travel times. This causes a light pulse to spread out as it travels along the fiber, resulting in information loss. The delay time should be minimized by the design of the fiber. Consider a ray that travels a distance L directly along a fiber axis and another that is repeatedly reflected, at the critical angle, as it travels to the same point as the first ray. (a) Show that the difference Δt in the times of arrival is given by

$$\Delta t = \frac{L}{c} \frac{n_1}{n_2} (n_1 - n_2),$$

where n_1 is the index of refraction of the glass core and n_2 is the index of refraction of the fiber coating. (b) Evaluate Δt for the fiber of Problem 83, with $L = 300$ m.

SECTION 34-9 Polarization by Reflection

85E. (a) At what angle of incidence will the light reflected from water be completely polarized? (b) Does this angle depend on the wavelength of the light?

86E. Light traveling in water of refractive index 1.33 is incident on a plate of glass of refractive index 1.53. At what angle of incidence will the reflected light be fully polarized?

87E. Calculate the upper and lower limits of the Brewster angles for white light incident on fused quartz. Assume that the wavelength limits of the light are 400 and 700 nm.

88P. When red light in vacuum is incident at the Brewster angle on a certain glass slab, the angle of refraction is 32.0°. What are (a) the index of refraction of the glass and (b) the Brewster angle?

Edouard Manet's <u>A Bar at the Folies-Bergère</u> has enchanted viewers ever since it was painted in 1882. Part of its appeal lies in the contrast between an audience ready for entertainment and a bartender whose eyes betray her fatigue. But its appeal also depends on a subtle distortion of reality that Manet hid in the painting—a distortion that gives an eerie feel to the scene even before you recognize what is "wrong." Can you find it?

35-1 TWO TYPES OF IMAGE

For you to see, say, a penguin, your eye must intercept some of the light rays spreading from the penguin and then redirect them onto the retina at the rear of the eye. Your visual system, starting with the retina and ending with the visual cortex at the rear of your brain, automatically and subconsciously processes the information provided by the light. That system identifies edges, orientations, textures, shapes, and colors and then rapidly brings to your consciousness an **image** (a reproduction derived from light) of the penguin: you perceive and recognize the penguin as being in the direction from which the light rays came and at the proper distance.

Your visual system goes through this processing and recognition even if the light rays do not come directly from the penguin, but instead reflect toward you from a mirror or refract through the lenses in a pair of binoculars. However, you now see the penguin in the direction from which the light rays came after they reflected or refracted, and the distance you perceive may be quite different from the penguin's true distance.

For example, if the light rays have been reflected toward you from a standard flat mirror, the penguin appears to be behind the mirror because the rays you intercept come from that direction. Of course, the penguin is not back there. This type of image, which is called a **virtual image,** truly exists only within the brain but nevertheless is *said* to exist at the perceived location.

A **real image** differs in that it can be formed on a surface, such as a card or a movie screen. You can see a real image (otherwise movie theaters would be empty), but the existence of the image does not depend on your seeing it and it is present even if you are not.

In this chapter we explore several ways in which virtual and real images are formed by reflection (as with mirrors) and refraction (as with lenses). We also distinguish between the two types of image more clearly, but here first is an example of a natural virtual image.

A Common Mirage

A common example of a virtual image is a pool of water that appears to lie on the road some distance ahead of you on a sunny day but which you can never reach. The pool is a *mirage* (a type of illusion), formed by light rays coming from the low section of the sky in front of you (Fig. 35-1a). As the rays approach the road, they travel through progressively warmer air that has been heated by the road, which is usually relatively warm. With an increase in air temperature, the speed of light in air increases slightly and, correspondingly, the index of refraction of the air decreases slightly. So, as the rays descend, encountering progres-

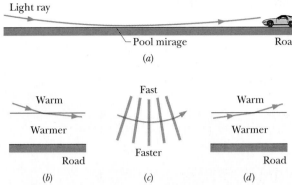

FIGURE 35-1 (*a*) A ray from a low section of the sky refracts through air that is heated by a road (without reaching the road). An observer who intercepts the light perceives it to be from a pool of water on the road. (*b*) Bending (exaggerated) of a light ray descending across an imaginary boundary from warm air to warmer air. (*c*) Bending of a ray when the ray is horizontal, which occurs because the lower ends of wavefronts move faster in warmer air. (*d*) Bending of a ray ascending across an imaginary boundary to warm air from warmer air.

sively smaller indices of refraction, they continuously bend toward the horizontal (Fig. 35-1b).

Once a ray is horizontal, somewhat above the road's surface, it still bends because the lower portion of each associated wavefront is in slightly warmer air and is moving slightly faster than the upper portion of the wavefront (Fig. 35-1c). This nonuniform motion of the wavefronts bends the ray upward. As the ray then ascends, it continues to bend upward through progressively larger indices of refraction (Fig. 35-1d).

If you intercept some of this light, your visual system automatically infers that it originated along a backward extension of the rays you have intercepted and, to make sense of the light, assumes that it came from the road surface. If the light happens to be bluish from blue sky, the mirage appears bluish, like water. Because the air is probably turbulent due to the heating, the mirage shimmies, as if water waves were present. The bluish coloring and the shimmy enhance the illusion of a pool of water, but you are actually seeing a virtual image of a low section of the sky.

35-2 PLANE MIRRORS

A **mirror** is a surface that can reflect a beam of light in one direction instead of either scattering it widely into many directions or absorbing it. A shiny metal surface acts as a mirror; a concrete wall does not. In this section we examine the images that a **plane mirror** (a flat reflecting surface) can produce.

Figure 35-2 shows a point source of light *O*, which we shall call the *object*, at a perpendicular distance *p* in front of a plane mirror. The light that is incident on the mirror is

What clues tell you whether this photograph is upside down? There are several.

A section of what you see in a kaleidoscope is a direct view of what lies at the far end of the kaleidoscope tube; the rest consists of images of the direct view that are produced by mirrors extending along the tube. How many mirrors are in this kaleidoscope, and how are they arranged? (The answer is given with the Checkpoint answers.)

represented with rays spreading from O. The reflection of that light is represented with reflected rays spreading from the mirror. If we extend the reflected rays backward (behind the mirror), we find that the extensions intersect at a point that is a perpendicular distance i behind the mirror.

If you look into the mirror of Fig. 35-2, your eyes intercept some of the reflected light. To make sense of what you see, you perceive a point source of light located at the point of intersection of the extensions. This point source is the image I of object O. It is called a *point image* because it is a point, and it is a virtual image because the rays do not actually pass through it. (As you will see, rays *do* pass through a point of intersection for a real image.)

Figure 35-3 shows two rays selected from the many rays in Fig. 35-2. One reaches the mirror at point b, perpendicularly. The other reaches it at an arbitrary point a, with an angle of incidence θ. The extensions of the two reflected rays are also shown. The right triangles $aOba$ and

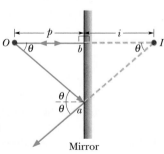

FIGURE 35-2 A point source of light O, called the *object*, is a perpendicular distance p in front of a plane mirror. Light rays reaching the mirror from O reflect from the mirror. If your eye intercepts some of the reflected rays, you perceive a point source of light I to be behind the mirror, at a perpendicular distance i. The perceived source I is a virtual image of object O.

FIGURE 35-3 Two rays from Fig. 35-2. Ray Oa makes an arbitrary angle θ with the normal to the mirror surface. Ray Ob is perpendicular to the mirror.

alba have a common side and three equal angles and are thus congruent. So their horizontal sides are congruent. That is,

$$Ib = Ob, \qquad (35\text{-}1)$$

where Ib and Ob are the distances from the mirror to the image and the object, respectively. Equation 35-1 tells us that the image is as far behind the mirror as the object is in front of it. By convention (that is, to get our equations to work out), *object distances* p are taken to be positive quantities, and *image distances* i for virtual images (as here) are taken to be negative quantities. Thus Eq. 35-1 can be written as $|i| = p$, or as

$$i = -p \quad \text{(plane mirror)}. \qquad (35\text{-}2)$$

Only rays that are fairly close together can enter the eye after reflection at a mirror. For the eye position shown in Fig. 35-4, only a small portion of the mirror near point a (a portion smaller than the pupil of the eye) is useful in forming the image. To find this portion, close one eye and look at the mirror image of a small object such as the tip of a pencil. Then move your fingertip over the mirror surface until you cannot see the image. Only that small portion of the mirror under your fingertip produced the image.

Extended Objects

In Fig. 35-5, an extended object O, represented by an upright arrow, is at perpendicular distance p in front of a plane mirror. Each small portion of the object that faces the mirror acts like the point source O of Figs. 35-2 and 35-3. If you intercept the light reflected by the mirror, you perceive a virtual image I that is a composite of the virtual point images of all those portions of the object and seems to be at distance i behind the mirror. Distances i and p are related by Eq. 35-2.

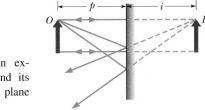

FIGURE 35-5 An extended object O and its virtual image I in a plane mirror.

We can also locate the image of an extended object as we did for a point object in Fig. 35-2: we draw some of the rays that reach the mirror from the top of the object, draw the corresponding reflected rays, and then extend those reflected rays behind the mirror until they intersect to form an image of the top of the object. We then do the same for rays from the bottom of the object. As shown in Fig. 35-5, we find that virtual image I has the same orientation and *height* (measured parallel to the mirror) as object O.

Manet's "Folies-Bergère"

In *A Bar at the Folies-Bergère* you see the barroom via reflection by a large mirror on the wall behind the woman tending bar, but the reflection is subtly wrong in three ways. First note the bottles at the left. Manet painted their reflections in the mirror but misplaced them, painting them farther toward the front of the bar than they really were.

Now note the reflection of the woman. Since your view is from directly in front of the woman, her reflection should be behind her, with only a little of it (if any) visible to you; yet Manet painted her reflection well off to the right. Finally, note the reflection of the man facing her. He must be you, because the reflection shows that he is directly in front of the woman, and thus he must be the viewer of the painting. You are looking into Manet's work and seeing your reflection well off to your right. The effect is eerie because it is not what we expect from a painting or from a mirror.

CHECKPOINT **1:** In the figure you look into a system of two vertical parallel mirrors A and B separated by distance d. A grinning gargoyle is perched at point O, a distance $0.2d$ from mirror A. Each mirror produces a *first* (least deep) image of the gargoyle. Then each mirror produces a *second* image with the object being the first image in the opposite mirror. Then each mirror produces a *third* image with the object being

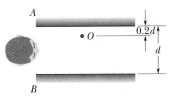

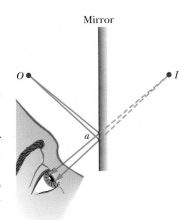

FIGURE 35-4 A "pencil" of rays from O enters the eye after reflection at the mirror. Only a small portion of the mirror near a is involved in this reflection. The light appears to originate at point I behind the mirror.

the second image in the opposite mirror. And so on—you might see hundreds of grinning gargoyle images. How deep behind mirror *A* are the first, second, and third images in mirror *A*?

SAMPLE PROBLEM 35-1

Charles Barkley is 198 cm tall. How tall must a vertical mirror be if he is to be able to see his entire length in it?

SOLUTION: In Fig. 35-6, the heights of the top of Barkley's head (*h*), his eyes (*e*), and the bottoms of his feet (*f*) are marked by dots. (Dot *h* has been drawn slightly too high for clarity.) The figure shows the paths followed by rays that leave his head and his feet and enter his eyes, reflecting from the mirror at points *a* and *c*, respectively. The mirror need occupy only the vertical distance *H* between those points.

From the geometry and Eq. 34-43,

$$ab = \tfrac{1}{2}he \quad \text{and} \quad bc = \tfrac{1}{2}ef.$$

So the required height is

$$H = ab + bc = \tfrac{1}{2}(he + ef)$$
$$= (\tfrac{1}{2})(198 \text{ cm}) = 99 \text{ cm.} \qquad \text{(Answer)}$$

Thus the mirror need be no taller than half the athlete's height. And this result is independent of his distance from the mirror. (If you have a full-length mirror available, you might experiment by taping newspaper over the portions of the mirror that do not contribute to your image. You will find that what you have left is just half your height. Mirrors that extend below point *c* just allow you to look at an image of the floor.)

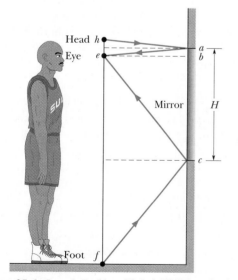

FIGURE 35-6   Sample Problem 35-1. A "full-length mirror" need be only half a person's height.

35-3 SPHERICAL MIRRORS

We turn now from images produced by plane mirrors to images produced by mirrors with curved surfaces. In particular, we shall consider a spherical mirror, which is simply a mirror in the shape of a small section of the surface of a sphere. A plane mirror is in fact a spherical mirror with an infinitely large *radius of curvature.*

Making a Spherical Mirror

We start with the plane mirror of Fig. 35-7*a*, which faces leftward toward an object *O* that is shown and an observer that is not shown. We make a **concave mirror** by curving the mirror's surface so it is *concave* ("caved in") as in Fig. 35-7*b*. Curving the surface in this way changes several characteristics of the mirror and the image it produces of the object:

1. The *center of curvature C* (the center of the sphere of which the mirror's surface is part) was infinitely far from the plane mirror; it is now closer but still in front of the concave mirror.

2. The *field of view*—the extent of the scene that is reflected to the observer—was wide; it is now smaller.

3. The image of the object was as far behind the plane mirror as the object was in front; the image is farther behind the concave mirror; that is, |*i*| is greater.

4. The height of the image was equal to the height of the object; the height of the image is now greater. This feature is why many makeup mirrors and shaving mirrors are concave—they produce a larger image of a face.

We can make a **convex mirror** by curving a plane mirror so its surface is *convex* ("flexed out") as in Fig. 35-7*c*. Curving the surface in this way moves the center of curvature *C* to behind the mirror and increases the field of view. It also moves the image of the object closer to the mirror and shrinks it. Store surveillance mirrors are usually convex to take advantage of the increase in the field of view—more of the store can then be monitored.

Focal Points of Spherical Mirrors

For a plane mirror, the magnitude of the image distance *i* is always equal to the object distance *p*. Before we can determine how these two distances are related for a spherical mirror, we must consider the reflection of light from an object *O* located an effectively infinite distance in front of a spherical mirror, on the mirror's *central axis*. That axis extends through the center of curvature *C* and the center *c*

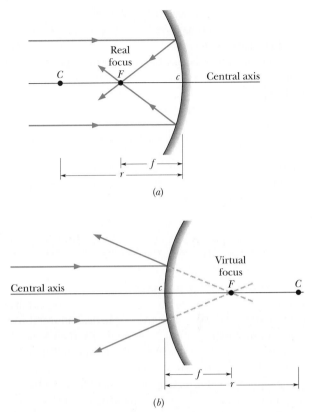

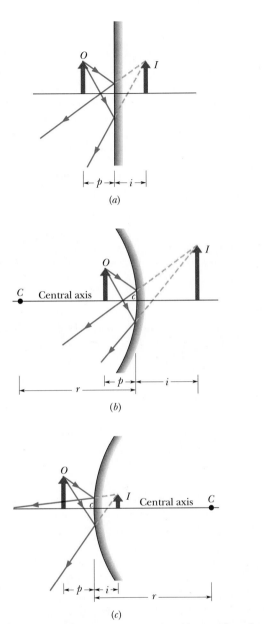

FIGURE 35-8 (*a*) In a concave mirror, incident parallel light rays are brought to a real focus at *F*, on the same side of the mirror as the light rays. (*b*) In a convex mirror, incident parallel light rays seem to diverge from a virtual focus at *F*, on the side of the mirror opposite the light rays.

FIGURE 35-7 (*a*) An object *O* forms a virtual image *I* in a plane mirror. (*b*) If the mirror is bent so that it becomes *concave,* the image moves farther away and becomes larger. (*c*) If the plane mirror is bent so that it becomes *convex,* the image moves closer and becomes smaller.

flected through a common point *F*; two of these reflected rays are shown in the figure. If we placed a card at *F*, a point image of the infinitely distant object *O* would appear on the card. (This would occur for any infinitely distant object.) Point *F* is called the **focal point** (or **focus**) of the mirror, and its distance from the center of the mirror is the **focal length** *f* of the mirror.

If we now substitute a convex mirror for the concave mirror, we find that the parallel rays are no longer reflected through a common point. Instead, they diverge as shown in Fig. 35-8*b*. However, if your eye intercepts some of the reflected light, you perceive the light as originating from a point source behind the mirror. This perceived source is located where extensions of the reflected rays pass through a common point (*F* in Fig. 35-8*b*). That point is the focal point (or focus) *F* of the convex mirror, and its distance from the mirror surface is the focal length *f* of the mirror. If we placed a card at this focal point, an image of object *O* would *not* appear on the card. So, this focal point is not like that of a concave mirror.

To distinguish the actual focal point of a concave

of the mirror. Because of the great distance between the object and the mirror, the light waves spreading from the object are plane waves when they reach the mirror along the central axis. This means that the rays representing the light waves are all parallel to the central axis when they reach the mirror.

When these parallel rays reach a concave mirror like that of Fig. 35-8*a*, those close to the central axis are re-mirror from the perceived focal point of a convex mirror, the former is said to be a *real focal point* and the latter is said to be a *virtual focal point*. Moreover, the focal length f of a concave mirror is taken to be a positive quantity, and that of a convex mirror a negative quantity. For mirrors of both types, the focal length f is related to the radius of curvature r of the mirror by

$$f = \tfrac{1}{2}r \quad \text{(spherical mirror)}, \quad (35\text{-}3)$$

where, consistent with the signs for the focal length, r is a positive quantity for a concave mirror and a negative quantity for a convex mirror.

35-4 IMAGES FROM SPHERICAL MIRRORS

With the focal point of a spherical mirror defined, we can find the relation between image distance i and object distance p for concave and convex spherical mirrors. We begin by placing the object O *inside the focal point* of the concave mirror, that is, between the mirror and its focal point F (Fig. 35-9a). An observer can then see a virtual image of O in the mirror: The image appears to be behind the mirror, and it has the same orientation as the object.

If we now move the object away from the mirror until it is at the focal point, the image moves farther back from the mirror until it is at infinity (Fig. 35-9b). The image is then ambiguous and imperceptible because neither the rays reflected by the mirror nor the ray extensions behind the mirror cross to form an image of O.

If we next move the object *outside the focal point*, that is, farther away from the mirror than the focal point, the rays reflected by the mirror converge to form an *inverted* image of object O (Fig. 35-9c) in front of the mirror. That image moves in from infinity as we move the object farther outside F. If you were to hold a card at the position of the image, the image would show up on the card—the image is said to be *focused* on the card by the mirror. (The verb ''focus,'' which in this context means to produce an image, differs from the noun ''focus,'' which is another name for the focal point.) Because this image can actually appear on a surface, it is a real image—the rays actually intersect to create the image, regardless of whether an observer is present. The image distance i of a real image is a positive quantity, in contrast to that for a virtual image. We also see that:

> Real images form on the same side of a mirror as where the object is, and virtual images form on the opposite side.

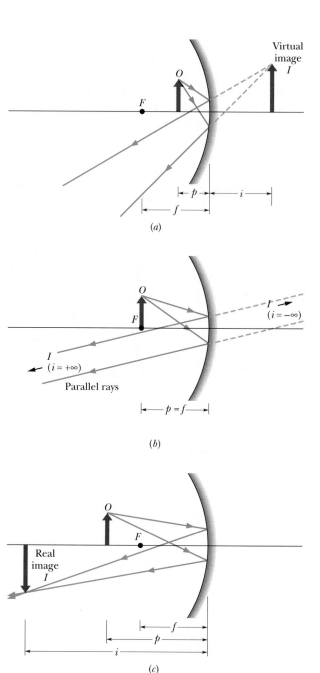

FIGURE 35-9 (a) An object O inside the focal point of a concave mirror, and its virtual image I. (b) The object at the focal point F. (c) The object outside the focal point, and its real image I.

As we shall prove in Section 35-8, when light rays from an object make only small angles with the central axis of a spherical mirror, a simple equation relates the object distance p, the image distance i, and the focal length f:

$$\frac{1}{p} + \frac{1}{i} = \frac{1}{f} \quad \text{(spherical mirror).} \quad (35\text{-}4)$$

(For clarity in figures such as Fig. 35-9, the rays are drawn with exaggerated angles.)

The size of an object or image, as measured *perpendicular* to the mirror's central axis, is called the object or image *height*. Let h represent the height of the object, and h' the height of the image. Then the ratio h'/h is called the **lateral magnification** m produced by the mirror. However, by convention, the lateral magnification always includes a plus sign when the image orientation is that of the object and a minus sign when the image orientation is opposite that of the object. For this reason, we write the formula for m as

$$|m| = \frac{h'}{h} \quad \text{(lateral magnification).} \quad (35\text{-}5)$$

We shall soon prove that the lateral magnification can also be written as

$$m = -\frac{i}{p} \quad \text{(lateral magnification).} \quad (35\text{-}6)$$

For a plane mirror, for which $i = -p$, we have $m = +1$. The magnification of 1 means that the image is the same size as the object. The plus sign means that the image and the object have the same orientation. For the concave mirror of Fig. 35-9c, $m \approx -1.5$.

Equations 35-3 through 35-6 hold for all plane mirrors, concave spherical mirrors, and convex spherical mirrors. In addition to those equations, you have been asked to absorb a lot of information about these mirrors, and you should organize it for yourself by filling in Table 35-1. Under Image Location, decide if the image is on the *same* side of the mirror as the object or on the *opposite* side. Under Image Type, decide if the image is *real* or *virtual*. Under Image Orientation, decide if the image has the *same* orientation as the object or is *inverted*. Under Sign, give the sign of the quantity or fill in $\pm$ if the sign is ambiguous. You will need this organization to tackle homework or a test.

Locating Images by Drawing Rays

Figures 35-10a and b show an object O in front of a concave mirror. We can graphically locate the image of any off-axis point of the object by drawing a *ray diagram* with any two of four special rays through the point:

1. A ray that is initially parallel to the central axis reflects through the focal point F (ray 1 in Fig. 35-10a).

2. A ray that reflects from the mirror after passing through the focal point emerges parallel to the central axis (ray 2 in Fig. 35-10a).

3. A ray that reflects from the mirror after passing through the center of curvature C returns along itself (ray 3 in Fig. 35-10b).

4. A ray that reflects from the mirror at its intersection c with the central axis is reflected symmetrically about that axis (ray 4 in Fig. 35-10b).

The image of the point is at the intersection of the two special rays you choose. The image of the object can then be found by locating the images of two or more of its off-axis points. You need to modify the descriptions of the rays slightly to apply them to convex mirrors, as in Figs. 35-10c and d.

Proof of Equation 35-6

We are now in a position to derive Eq. 35-6 ($m = -i/p$), the expression for the lateral magnification of an object reflected in a mirror. Consider ray 4 in Fig. 35-10b. It is reflected at point c so that the incident and reflected rays make equal angles with the axis of the mirror at that point.

The two right triangles abc and edc in the figure are similar, so we can write

$$\frac{de}{ab} = \frac{cd}{ca}.$$

TABLE 35-1 **YOUR ORGANIZING TABLE FOR MIRRORS**

MIRROR TYPE	OBJECT LOCATION	IMAGE			SIGN			
		LOCATION	TYPE	ORIENTATION	OF f	OF r	OF i	OF m
Plane	Anywhere							
Concave	Inside F							
	Outside F							
Convex	Anywhere							

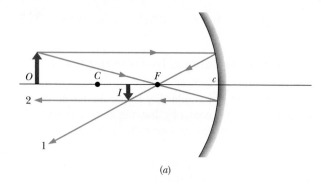

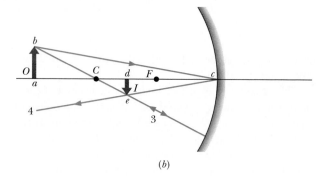

(a)   (b)

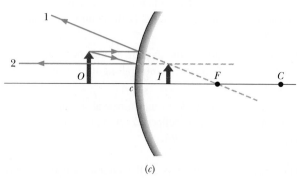

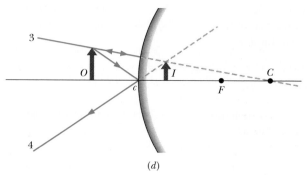

(c)   (d)

FIGURE 35-10 (a, b) Four rays that may be drawn to find the image of an object in a concave mirror. For the object position shown, the image is real, inverted, and smaller than the object. (c, d) Four similar rays for the case of a convex mirror. For a convex mirror, the image is always virtual, oriented like the object, and smaller than the object. [In (c), ray 2 is initially directed toward focal point F.]

The quantity on the left (apart from the question of sign) is the lateral magnification m produced by the mirror. Since we indicate an inverted image as a *negative* magnification, we symbolize this as $-m$. But $cd = i$ and $ca = p$, so we have at once

$$m = -\frac{i}{p} \quad \text{(magnification)}, \quad (35\text{-}7)$$

which is the relation we set out to prove.

SAMPLE PROBLEM 35-2

A tarantula of height h sits cautiously before a spherical mirror whose focal length has absolute value $|f| = 40$ cm. The image of the tarantula produced by the mirror has the same orientation as the tarantula and has height $h' = 0.20h$.

(a) Is the image real or virtual, and is it on the same side of the mirror as the tarantula or the opposite side?

SOLUTION: Because the image has the same orientation as the tarantula (the object), it must be virtual and on the opposite side of the mirror. (You can easily see this result if you have filled out Table 35-1.)

(b) Is the mirror concave or convex, and what is its focal length f, sign included?

SOLUTION: Can we tell the type of mirror by the type of image it produces? No, because both types of mirror can produce a virtual image. Can we tell the type of mirror by finding the sign of f with our only two equations that involve f (Eqs. 35-3 and 35-4)? No, we don't have enough information. The only approach left is to consider the magnification information. We know that the ratio of image height h' to object height h is 0.20. So, from Eq. 35-5 we have

$$|m| = \frac{h'}{h} = 0.20.$$

Because the object and image have the same orientation, we know that m must be positive: $m = +0.20$. Substituting this into Eq. 35-6 and solving for, say, i gives us

$$i = -0.20p,$$

which does not appear to be of help in finding f. But it is helpful if we substitute it into Eq. 35-4. That equation gives us

$$\frac{1}{f} = \frac{1}{i} + \frac{1}{p} = \frac{1}{-0.20p} + \frac{1}{p} = \frac{1}{p}(-5 + 1),$$

from which we find

$$f = -p/4.$$

Now we have it: because p is positive, f must be negative, which means that the mirror is convex with

$$f = -40 \text{ cm}. \quad \text{(Answer)}$$

CHECKPOINT 2: A Central American vampire bat, dozing on the central axis of a spherical mirror, is magnified by $m = -4$. Is its image (a) real or virtual, (b) inverted or of the same orientation as the bat, and (c) on the same side of the mirror as the bat or on the opposite side?

35-5 SPHERICAL REFRACTING SURFACES

We now turn from images formed by reflections to images formed by refraction through surfaces of transparent materials, such as glass. We shall consider only spherical surfaces, with radius of curvature r and center of curvature C. The light will be emitted by a point object O in a medium with index of refraction n_1; it will refract through a spherical surface into a medium of index of refraction n_2.

Our concern is whether the light rays, after refracting through the surface, form a real image (no observer necessary) or a virtual image (assuming that an observer intercepts the rays). The answer depends on the relative values of n_1 and n_2 and on the geometry of the situation.

Six possible results are shown in Fig. 35-11. In each part of the figure, the medium with the greater index of refraction is shaded, and object O is always in the medium with index of refraction n_1, to the left of the refracting surface. And in each part, a representative ray is shown refracting through the surface. (That ray and a ray along the central axis suffice to determine the position of the image in each case.)

At the point of refraction of the representative ray, the normal to the refracting surface is a radial line through the center of curvature C. Because of the refraction, the ray bends toward the normal if it is entering a medium of larger index of refraction, and away from the normal if it is entering a medium of smaller index of refraction. If the refracted ray is then directed toward the central axis, it and other (undrawn) rays will form a real image on that axis. If it is directed away from the central axis, it cannot form a real image; however, backward extensions of it and other refracted rays can form a virtual image, provided (as with mirrors) some of those rays are intercepted by an observer.

Real images I are formed (at image distance i) in parts a and b of Fig. 35-11, where the refraction directs the ray *toward* the central axis. Virtual images are formed in parts c and d, where the refraction directs the ray *away* from the central axis. Note, in these four parts, that real images are formed when the object is relatively far from the refracting surface, and virtual images are formed when the object is nearer the refracting surface. In the final situations (Fig.

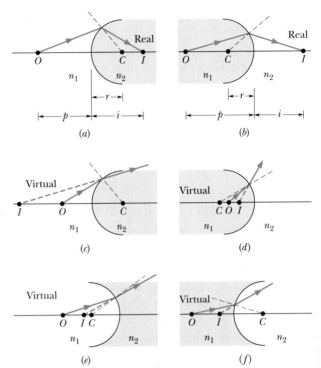

FIGURE 35-11 Six possible ways in which an image can be formed by refraction through a spherical surface of radius r and center of curvature C. The surface separates a medium with index of refraction n_1 from a medium with index of refraction n_2. The point object O is always in the medium with n_1, to the left of the surface. The material with the smaller index of refraction is unshaded (think of it as being air, and the other material as being glass). Real images are formed in (a) and (b); virtual images are formed in the other four situations.

35-11e and f), refraction always directs the ray away from the central axis and virtual images are always formed, regardless of the object distance.

Note the following major difference from reflected images:

Real images form on the side of a refracting surface that is opposite the object, and virtual images form on the same side as the object.

In Section 35-8, we shall show that (for light rays making only small angles with the central axis)

$$\frac{n_1}{p} + \frac{n_2}{i} = \frac{n_2 - n_1}{r}. \tag{35-8}$$

Just as with mirrors, the object distance p is positive, and the image distance i is positive for a real image and negative for a virtual image. However, to keep all the signs correct in Eq. 35-8, we must use the following rule for the sign of the radius of curvature r:

When the object faces a convex refracting surface, the radius of curvature r is positive. When it faces a concave surface, r is negative.

Be careful: this is just the reverse of the sign convention we have for mirrors.

SAMPLE PROBLEM 35-3

A Jurassic mosquito is discovered embedded in a chunk of amber, which has index of refraction 1.6. One surface of the amber is spherically convex with radius of curvature 3.0 mm (Fig. 35-12). The mosquito head happens to be along the central axis of that surface and, when viewed along the axis, appears to be buried 5.0 mm into the amber. How deep is it really?

SOLUTION: First we must realize what is meant by "appears": the observer (in air) sees an image of the mosquito head in the amber and 5.0 mm from the spherical surface of the amber. Because the object (the head) and its image are on the same side of the refracting surface, the image must be virtual and so $i = -5.0$ mm. Also, because the object is always taken to be in the medium of index of refraction n_1, we must have $n_1 = 1.6$ and $n_2 = 1.0$. Finally, because the object faces a concave refracting surface, the radius of curvature r is negative and so $r = -3.0$ mm. What we seek is the object distance p. Substituting the data into Eq. 35-8,

$$\frac{n_1}{p} + \frac{n_2}{i} = \frac{n_2 - n_1}{r},$$

so

$$\frac{1.6}{p} + \frac{1.0}{-5.0 \text{ mm}} = \frac{1.0 - 1.6}{-3.0 \text{ mm}}$$

and

$$p = 4.0 \text{ mm.} \qquad \text{(Answer)}$$

FIGURE 35-12 Sample Problem 35-3. A piece of amber with a mosquito from the Jurassic period, with the head buried at point O. The spherical refracting surface at the right end, with center of curvature C, provides an image I to an observer intercepting rays from the object at O.

This insect has been entombed in amber for about 25 million years. Because we view the insect through a curved refracting surface, the image we see does not coincide with the insect.

CHECKPOINT 3: A bee is hovering in front of the concave spherical refracting surface of a glass sculpture. (a) Which of the general situations of Fig. 35-11 is like this situation? (b) Is the image produced by the surface real or virtual, and is it on the same side as the bee or the opposite side?

35-6 THIN LENSES

A **lens** is a transparent object with two refracting surfaces whose central axes coincide. The common central axis is the central axis of the lens. When a lens is surrounded by air, light refracts from the air into the lens, crosses through the lens, and then refracts back into the air. Each refraction can change the direction of travel of the light.

A lens that causes light rays initially parallel to the central axis to converge is (reasonably) called a **converging lens.** If, instead, it causes such rays to diverge, the lens is a **diverging lens.** When an object is placed in front of a lens of either type, refraction by the lens of light rays from the object can produce an image of the object.

We shall consider only the special case of a **thin lens,** that is, a lens in which the thickest part is thin compared to the object distance p, the image distance i, and the radii of curvature r_1 and r_2 of the two surfaces of the lens. We shall also consider only light rays that make small angles with the central axis (they are exaggerated in the figures here). In Section 35-8 we shall prove that in such light, a thin lens has a focal length f. Moreover, i and p are related to each other by

$$\frac{1}{f} = \frac{1}{p} + \frac{1}{i} \quad \text{(thin lens)}, \quad (35\text{-}9)$$

which is the same form of equation we had for mirrors. We shall also prove that when a thin lens with index of refraction n is surrounded by air, this focal length f is given by

$$\frac{1}{f} = (n - 1)\left(\frac{1}{r_1} - \frac{1}{r_2}\right) \quad \text{(thin lens in air)}, \quad (35\text{-}10)$$

which is often called the *lens maker's equation.* Here r_1 is the radius of curvature of the lens surface nearer the object, and r_2 is that of the other surface. The signs of these radii are found with the rules in Section 35-5 for the radii of spherical refracting surfaces. If the lens is surrounded by some medium other than air (say, corn oil) with index of refraction n_{medium}, we replace n in Eq. 35-10 with n/n_{medium}. Keep in mind the basis of Eqs. 35-9 and 35-10:

> A lens can produce an image of an object only because it can bend light rays; but it can bend light rays only if its index of refraction differs from that of the surrounding medium.

Figure 35-13a shows a thin lens with convex refracting surfaces, or *sides.* When rays that are parallel to the central axis of the lens are sent through the lens, they refract twice, as is shown enlarged in Fig. 35-13b. This double refraction causes the rays to converge and pass through a common point F_2 at a distance f from the center of the lens. Hence this lens is a converging lens; further, a *real* focal point (or focus) exists at F_2 (because the rays really do pass through it), and the associated focal length is f. When rays parallel to the central axis are sent in the opposite direction through the lens, we find another real focal point at F_1 on the other side of the lens. For a thin lens, these two focal points are equidistant from the lens.

Because the focal points of a converging lens are real, we take the associated focal lengths f to be positive, just as we do with a real focus of a concave mirror. But signs in optics can be tricky, so we had better check this in Eq. 35-10. The left side of that equation is positive if f is positive; how about the right side? We examine it term by term. Because the index of refraction n of glass or any other material is greater than 1, the term $(n - 1)$ must be positive. Because the source of the light (which is the object) is at the left and faces the convex left side of the lens, the radius of curvature r_1 of that side must be positive according to the sign rule for refracting surfaces. Similarly, because the object faces a concave right side of the lens, the radius of curvature r_2 of that side must be negative. Thus, the term $(1/r_1 - 1/r_2)$ is positive, the whole right side of Eq. 35-10 is positive, and all the signs are consistent.

Figure 35-13c shows a thin lens with concave sides. When rays that are parallel to the central axis of the lens are sent through this lens, they refract twice, as is shown enlarged in Fig. 35-13d; these rays *diverge,* never passing

FIGURE 35-13 (a) Rays initially parallel to the central axis of a converging lens are made to converge to a real focal point F_2 by the lens. The lens is thinner than drawn, with a width like the vertical line through it, where we consider the bending of rays to occur. (b) An enlargement of the top part of the lens of (a); normals to the surfaces are shown dashed. Note that both refractions of the ray at the surfaces bend the ray downward, toward the central axis. (c) The same initially parallel rays are made to diverge by a diverging lens. Extensions of the diverging rays pass through a virtual focal point F_2. (d) An enlargement of the top part of the lens of (c). Note that both refractions of the ray at the surfaces bend the ray upward, away from the central axis.

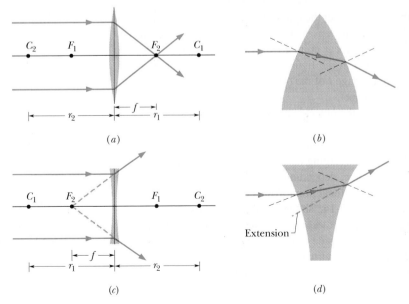

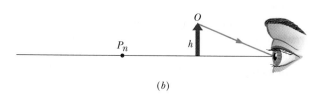

(a)

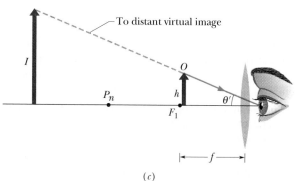

(b)

(c)

FIGURE 35-17 (a) An object O of height h, placed at the near point of a human eye, occupies angle θ in the eye's view. (b) The object is moved closer to increase the angle, but now the observer cannot bring the object into focus. (c) A converging lens is placed between the object and the eye, with the object just inside the focal point F_1 of the lens. The image produced by the lens is then far enough away to be focused by the eye, and the image occupies a larger angle θ' than object O does in (a).

increase the angle, hence the possibility of distinguishing details of the object. However, because the object is then closer than the near point, it is no longer *in focus*; that is, the image is no longer clear.

You can restore the clarity by looking at O through a converging lens, placed so that O is just inside the focal point F_1 of the lens, which is at focal length f (Fig. 35-17c). What you then see is the virtual image of O produced by the lens. That image is farther away than the near point; thus the eye can see it clearly.

Moreover, the angle θ' occupied by the virtual image is larger than the largest angle θ that the object alone can occupy and still be seen clearly. The *angular magnification* m_θ (not to be confused with lateral magnification m) of what is seen is

$$m_\theta = \theta'/\theta.$$

In words, the angular magnification of a simple magnifying lens is a comparison of the angle occupied by the image the lens produces with the angle occupied by the

object when the object is moved to the near point of the viewer.

From Fig. 35-17, assuming that O is at the focal point of the lens, and approximating tan θ as θ and tan θ' as θ', for small angles, we have

$$\theta \approx h/25 \text{ cm} \quad \text{and} \quad \theta' \approx h/f.$$

We then find that

$$m_\theta \approx \frac{25 \text{ cm}}{f} \quad \text{(simple magnifier).} \quad (35\text{-}12)$$

Compound Microscope

Figure 35-18 shows a thin-lens version of a compound microscope. The instrument consists of an *objective* (the front lens) of focal length f_{ob} and an *eyepiece* (the lens near the eye) of focal length f_{ey}. It is used for viewing small objects that are very close to the objective.

The object O to be viewed is placed just outside the first focal point F_1 of the objective, close enough to F_1 that we can approximate its distance p from the lens as being f_{ob}. The separation between the lenses is then adjusted so that the enlarged, inverted, real image I produced by the objective is located just inside the first focal point F_1' of the eyepiece. The *tube length s* shown in Fig. 35-18 is actually large relative to f_{ob}, and we can approximate the distance i between the objective and the image I as being length s.

From Eq. 35-6, and using our approximations for p and i, we can write the lateral magnification produced by the objective as

$$m = -\frac{i}{p} = -\frac{s}{f_{ob}}. \quad (35\text{-}13)$$

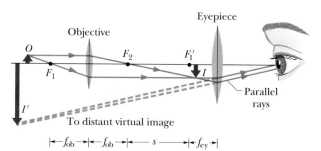

FIGURE 35-18 A thin-lens representation of a compound microscope (not to scale). The objective produces a real image I of object O just inside the focal point F_1' of the eyepiece. Image I then acts as an object for the eyepiece, which produces a virtual final image I' that is seen by the observer. The objective has focal length f_{ob}; the eyepiece has focal length f_{ey}; and s is the tube length.

Since the image I is located just inside the focal point F_1' of the eyepiece, the eyepiece acts as a simple magnifying lens, and an observer sees a final (virtual, inverted) image I' through it. The overall magnification of the instrument is the product of the lateral magnification m produced by the objective, given by Eq. 35-13, and the angular magnification m_θ produced by the eyepiece, given by Eq. 35-12. That is,

$$M = mm_\theta = -\frac{s}{f_{ob}}\frac{25\text{ cm}}{f_{ey}} \qquad \text{(microscope).} \quad (35\text{-}14)$$

Refracting Telescope

Telescopes come in a variety of forms. The form we describe here is the simple refracting telescope that consists of an objective and an eyepiece, both represented in Fig. 35-19 with simple lenses, although in practice, as is also true for most microscopes, each lens is actually a compound lens system.

The lens arrangements for telescopes and for microscopes are similar, but telescopes are designed to view

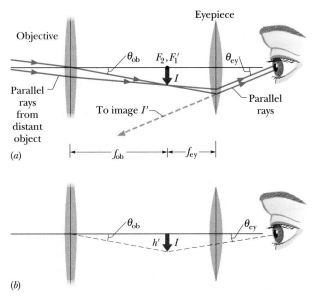

(a)

(b)

FIGURE 35-19 (a) A thin-lens representation of a refracting telescope. The objective produces a real image I of a distant source of light (the object), with approximately parallel light rays at the objective. (One end of the object is assumed to lie on the central axis.) Image I, formed at the common focal points F_2 and F_1', acts as an object for the eyepiece, which produces a virtual final image I' at a great distance from the observer. The objective has focal length f_{ey}; the eyepiece has focal length f_{ey}. (b) Image I has height h' and takes up angle θ_{ob} measured from the objective and angle θ_{ey} measured from the eyepiece. [Angles θ_{ey} in (a) and (b) are equal for rays closer to the central axis than what is drawn here.]

large objects, such as galaxies, stars, and planets, at large distances, whereas microscopes are designed for just the opposite purpose. This difference requires that in the telescope of Fig. 35-19 the second focal point of the objective F_2 coincide with the first focal point of the eyepiece F_1', whereas in the microscope of Fig. 35-18 these points are separated by the tube length s.

In Fig. 35-19a parallel rays from a distant object strike the objective, making an angle θ_{ob} with the telescope axis and forming a real, inverted image at the common focal point F_2, F_1'. This image acts as an object I for the eyepiece, and an observer sees a distant (still inverted) virtual image I' through it. The rays defining the image make an angle θ_{ey} with the telescope axis.

The angular magnification m_θ of the telescope is θ_{ey}/θ_{ob}. From Fig. 35-19b, for rays close to the central axis, we can write $\theta_{ob} = h'/f_{ob}$ and $\theta_{ey} = h'/f_{ey}$, which gives us

$$m_\theta = -\frac{f_{ob}}{f_{ey}} \qquad \text{(telescope),} \quad (35\text{-}15)$$

where the minus sign indicates that I' is inverted. In words, the angular magnification of a telescope is a comparison of the angle occupied by the image the telescope produces with the angle occupied by the distant object as seen without the telescope.

Magnification is only one of the design factors for an astronomical telescope and is indeed easily achieved. A good telescope needs *light-gathering power*, which determines how bright the image is. This is important for viewing faint objects such as distant galaxies and is accomplished by making the objective diameter as large as possible. A telescope also needs *resolving power*, which is the ability to distinguish between two distant objects (stars, say) whose angular separation is small. Field of view is another important parameter. A telescope designed to look at galaxies (which occupy a tiny field of view) is much different from one designed to track meteors (which move over a wide field of view).

The telescope designer must also take into account the difference between real lenses and the ideal thin lenses we have discussed. A real lens with spherical surfaces does not form sharp images, a flaw called *spherical aberration*. And because refraction by the two surfaces of a real lens depends on wavelength, a real lens does not focus light of different wavelengths to the same point, a flaw called *chromatic aberration*.

This brief discussion by no means exhausts the design parameters of astronomical telescopes—many others are involved. And we could make a similar listing for any other high-performance optical instrument.

35-8 THREE PROOFS (OPTIONAL)

The Spherical Mirror Formula (Eq. 35-4)

Figure 35-20 shows a point object O placed on the central axis of a concave spherical mirror, outside its center of curvature C. A ray from O that makes an angle α with the axis intersects the axis at I after reflection from the mirror at a. A ray that leaves O along the axis is reflected back along itself at c and also passes through I. Thus I is the image of O; it is a *real* image because light actually passes through it. Let us find the image distance i.

A theorem that is useful here tells us that an exterior angle of a triangle is equal to the sum of the two opposite interior angles. Applying this to triangles OaC and OaI in Fig. 35-20 yields

$$\beta = \alpha + \theta \quad \text{and} \quad \gamma = \alpha + 2\theta.$$

If we eliminate θ between these two equations, we find

$$\alpha + \gamma = 2\beta. \tag{35-16}$$

We can write angles α, β, and γ, in radian measure, as

$$\alpha \approx \frac{\widehat{ac}}{cO} = \frac{\widehat{ac}}{p}, \qquad \beta = \frac{\widehat{ac}}{cC} = \frac{\widehat{ac}}{r},$$

and

$$\gamma \approx \frac{\widehat{ac}}{cI} = \frac{\widehat{ac}}{i}. \tag{35-17}$$

Only the equation for β is exact, because the center of curvature of arc ac is at C. However, the equations for α and γ are approximately correct if these angles are small enough (that is, for rays close to the central axis). Substituting Eqs. 35-17 into Eq. 35-16, using Eq. 35-3 to replace r with $2f$, and canceling $\widehat{ac}$ lead exactly to Eq. 35-4, the relation that we set out to prove.

The Refracting Surface Formula (Eq. 35-8)

The incident ray from point object O in Fig. 35-21 that falls on point a is refracted there according to Eq. 34-44,

$$n_1 \sin \theta_1 = n_2 \sin \theta_2.$$

If α is small, θ_1 and θ_2 will also be small and we can replace the sines of these angles with the angles themselves. Thus the equation above becomes

$$n_1 \theta_1 \approx n_2 \theta_2. \tag{35-18}$$

We again use the fact that an exterior angle of a triangle is equal to the sum of the two opposite interior angles. Applying this to triangles COa and ICa yields

$$\theta_1 = \alpha + \beta \quad \text{and} \quad \beta = \theta_2 + \gamma. \tag{35-19}$$

If we use Eqs. 35-19 to eliminate θ_1 and θ_2 from Eq. 35-18, we find

$$n_1 \alpha + n_2 \gamma = (n_2 - n_1)\beta. \tag{35-20}$$

In radian measure the angles α, β, and γ are

$$\alpha \approx \frac{\widehat{ac}}{p}; \quad \beta = \frac{\widehat{ac}}{r}; \quad \gamma \approx \frac{\widehat{ac}}{i}. \tag{35-21}$$

Only the second of these equations is exact. The other two are approximate because I and O are not the centers of circles of which $\widehat{ac}$ is a part. However, for α small enough (for rays close to the axis), the inaccuracies in Eqs. 35-21 are small. Substituting Eqs. 35-21 into Eq. 35-20 leads directly to Eq. 35-8, the relation we set out to prove.

The Thin-Lens Formulas (Eqs. 35-9 and 35-10)

Our plan is to consider each lens surface as a separate refracting surface, and to use the image formed by the first surface as the object for the second.

We start with the thick glass "lens" of length L in Fig. 35-22a whose left and right refracting surfaces are ground to radii r' and r''. A point object O' is placed near the left surface as shown. A ray leaving O' along the central axis is not deflected on entering or leaving the lens.

A second ray leaving O' at an angle α with the central axis intersects the left surface at point a', is refracted, and intersects the second (right) surface at point a''. The ray is again refracted and crosses the axis at I'', which, being the intersection of two rays from O', is the image of point O', formed after refraction at two surfaces.

Figure 35-22b shows that the first (left) surface also forms a virtual image of O' at I'. To locate I', we use Eq. 35-8,

$$\frac{n_1}{p} + \frac{n_2}{i} = \frac{n_2 - n_1}{r}.$$

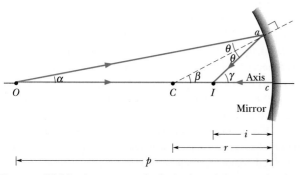

FIGURE 35-20 A concave spherical mirror forms a real point image I by reflecting light rays from a point object O.

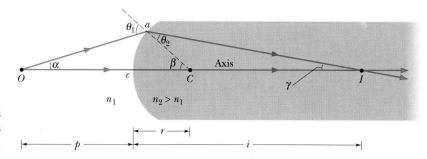

FIGURE 35-21 A real point image I of a point object O is formed by refraction at a spherical convex surface between two media.

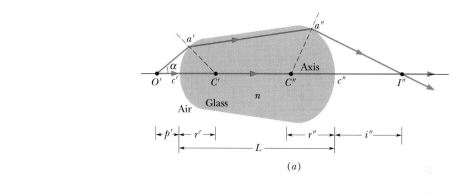

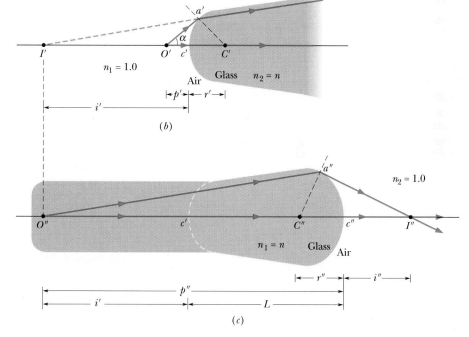

FIGURE 35-22 (*a*) Two rays from point object O' form a real image I'' after refracting through two spherical surfaces of a "lens." The object faces a convex surface at the left side of the lens and a concave surface at the right side. The ray traveling through points a' and a'' is actually close to the central axis through the lens. (*b*) The left side and (*c*) the right side, shown separately.

Putting $n_1 = 1$ and $n_2 = n$ and bearing in mind that the image distance is negative (that is, $i = -i'$ in Fig. 35-22*b*), we obtain

$$\frac{1}{p'} - \frac{n}{i'} = \frac{n-1}{r'}. \tag{35-22}$$

In this equation i' will be a positive number because we have already introduced the minus sign appropriate to a virtual image.

Figure 35-22*c* shows the second surface again. Unless an observer at point a'' were aware of the existence of the first surface, the observer would think that the light striking that point originated at point I' in Fig. 35-22*b* and that the region to the left of the surface was filled with glass as indicated. Thus the (virtual) image I' formed by the first surface serves as a real object O'' for the second surface. The distance of this object from the second surface is

$$p'' = i' + L. \qquad (35\text{-}23)$$

To apply Eq. 35-8 to the second surface, we must insert $n_1 = n$ and $n_2 = 1$ because the object now is effectively imbedded in glass. If we substitute with Eq. 35-23, then Eq. 35-8 becomes

$$\frac{n}{i' + L} + \frac{1}{i''} = \frac{1 - n}{r''}. \qquad (35\text{-}24)$$

Let us now assume that the thickness L of the "lens" in Fig. 35-22a is so small that we can neglect it in comparison with our other linear quantities (such as p', i', p'', i'', r', and r''). In all that follows we make this *thin-lens approximation*. Putting $L = 0$ in Eq. 35-24 and rearranging the right side lead to

$$\frac{n}{i'} + \frac{1}{i''} = -\frac{n - 1}{r''}. \qquad (35\text{-}25)$$

Adding Eqs. 35-22 and 35-25 leads to

$$\frac{1}{p'} + \frac{1}{i''} = (n - 1)\left(\frac{1}{r'} - \frac{1}{r''}\right).$$

Finally, calling the original object distance simply p and the final image distance simply i leads to

$$\frac{1}{p} + \frac{1}{i} = (n - 1)\left(\frac{1}{r'} - \frac{1}{r''}\right), \qquad (35\text{-}26)$$

which, with a small change in notation, is Eqs. 35-9 and 35-10, the relations we set out to prove.

REVIEW & SUMMARY

Real and Virtual Images

An *image* is a reproduction of an object via light. If the image can form on a surface, it is a *real image* and can exist even if no observer is present. If the image requires the visual system of an observer, it is a *virtual image*.

Image Formation

Spherical mirrors, spherical refracting surfaces, and *thin lenses* can form images of a source of light—the object—by redirecting rays emerging from the source. The image occurs where the redirected rays cross (forming a real image) or where backward extensions of those rays cross (forming a virtual image). If the rays are sufficiently close to the *central axis* through the spherical mirror, refracting surface, or thin lens, we have the following relations between the *object distance p* (which is positive) and the *image distance i* (which is positive for real images and negative for virtual images):

1. Spherical Mirror:

$$\frac{1}{p} + \frac{1}{i} = \frac{1}{f} = \frac{2}{r}, \qquad (35\text{-}4, 35\text{-}3)$$

where f is the mirror's focal length and r is the mirror's radius of curvature. A *plane mirror* is a special case for which $r \to \infty$, so that $p = -i$. Real images form on the side of a mirror where the object is located, and virtual images form on the opposite side.

2. Spherical Refracting Surface:

$$\frac{n_1}{p} + \frac{n_2}{i} = \frac{n_2 - n_1}{r} \quad \text{(single surface)}, \qquad (35\text{-}8)$$

where n_1 is the index of refraction of the material where the object is located, n_2 is the index of refraction of the material on the other side of the refracting surface, and r is the radius of curvature of the surface. When the object faces a convex refracting surface, the radius r is positive. When it faces a concave surface, r is negative. Real images form on the side of a refracting surface that is opposite the object, and virtual images form on the same side as the object.

3. Thin Lens:

$$\frac{1}{p} + \frac{1}{i} = \frac{1}{f} = (n - 1)\left(\frac{1}{r_1} - \frac{1}{r_2}\right), \qquad (35\text{-}9, 35\text{-}10)$$

where f is the lens's focal length, n is the index of refraction of the lens material, and r_1 and r_2 are the radii of curvature of the two sides of the lens, which are spherical surfaces. A convex lens surface that faces the object has a positive radius of curvature; a concave lens surface that faces the object has a negative radius of curvature. Real images form on the side of a lens that is opposite the object, and virtual images form on the same side as the object.

Lateral Magnification

The *lateral magnification m* produced by a spherical mirror or a thin lens is

$$m = -\frac{i}{p}. \qquad (35\text{-}6)$$

The magnitude of m is given by

$$|m| = \frac{h'}{h}, \qquad (35\text{-}5)$$

where h and h' are the heights (measured perpendicular to the central axis) of the object and image, respectively.

Optical Instruments

Three optical instruments that extend human vision are:

1. The *simple magnifying lens,* which produces an *angular magnification* m_θ given by

$$m_\theta = \frac{25 \text{ cm}}{f}, \qquad (35\text{-}12)$$

where f is the focal length of the magnifying lens.

2. The *compound microscope,* which produces an *overall magnification M* given by

$$M = mm_\theta = -\frac{s}{f_{ob}} \frac{25 \text{ cm}}{f_{ey}}, \qquad (35\text{-}14)$$

where m is the lateral magnification produced by the objective, m_θ is the angular magnification produced by the eyepiece, s is the tube length, and f_{ob} and f_{ey} are the focal lengths of the objective and eyepiece, respectively.

3. The *refracting telescope,* which produces an *angular magnification* m_θ given by

$$m_\theta = -\frac{f_{ob}}{f_{ey}}. \qquad (35\text{-}15)$$

QUESTIONS

1. Lake monsters, mermen, and mermaids have long been ''sighted'' by observers located either on a shore or on a low deck of a ship. From such a low point, an observer can intercept rays of light that leave a floating object (say, a log or a porpoise) and bend slightly back downward toward the observer (one is shown in Fig. 35-23a). The observer then perceives the object as being elongated upward from the water (and probably oscillating because of air turbulence) in a mirage that might easily resemble one of the fabled creatures. Figure 35-23b gives several plots of height from the water surface versus air temperature. Which one best illustrates the conditions giving rise to this mirage?

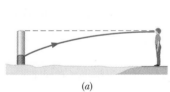

FIGURE 35-23 Questions 1 and 2.

2. When Erik the Red was exiled from Iceland by the other Vikings, he headed directly toward the nearest part of Greenland, apparently knowing where that undiscovered land was because of an occasional mirage that brought a virtual image of Greenland around the curve of Earth (Fig. 35-24). Figure 35-23b gives plots of height from Earth's surface versus air temperature. Which one best shows the conditions giving rise to that mirage?

FIGURE 35-24 Question 2.

3. Figure 35-25 shows a fish and a fish stalker in water. (a) Does the stalker see the fish in the general region of point a or point b? (b) Does the fish see the (wild) eyes of the stalker in the general region of point c or point d?

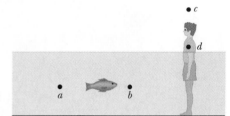

FIGURE 35-25 Question 3.

4. Figure 35-26 shows a coordinate system in front of a flat mirror, with the x axis perpendicular to the mirror. Draw the image of the system in the mirror. (a) Which axis is reversed by the reflection? (b) If you face a mirror, is your image inverted (top for bottom)? Are your left and right reversed (as in common belief)? (c) What then is reversed?

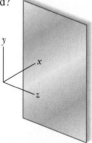

FIGURE 35-26 Question 4.

5. *Putt-putt optics:* Figure 35-27 is an overhead view of a room whose walls are covered with flat mirrors, showing also a sketch of the path of a ray of light from a source S to a target T. The light reflects three times between S and T. The path that produces a ''hit'' with three reflections must reflect off mirrors ab, bc, and cd; it can be found in the following way. Draw the virtual image I_1 of the target in mirror cd. Then draw the virtual image I_2 of I_1 in the dotted extension of mirror bc shown. Then draw the virtual image I_3 of I_2 in the dotted extension of mirror ab. Now aim the ray toward I_3 and reflect it off the three mirrors. Is there a way to hit the target with (a) two reflections and (b) four reflections?

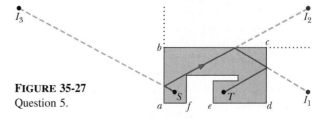

FIGURE 35-27 Question 5.

(a)

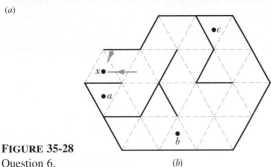

FIGURE 35-28
Question 6.
(b)

6. In the mirror maze of Fig. 35-28a, many "virtual hallways" seem to extend away from you because you see multiple reflections from the mirrors that form the walls of the maze. Those mirrors are placed along some sides of repeated equilateral triangles on the floor. The floor plan for a similar but different maze is shown in Fig. 35-28b; every wall section within this maze is mirrored. If you stand at entrance x, (a) which of the maze monsters a, b, and c hiding in the maze can you see along the virtual hallways extending from entrance x; (b) how many times does each visible monster appear in a hallway; and (c) what is at the far end of a hallway? (*Hint:* The two rays shown are coming down virtual hallways; follow them back into the maze. Do they pass through a triangle with a monster? If so, how many times? For additional analysis, see Jearl Walker, "The Amateur Scientist," *Scientific American,* Vol. 254, pages 120–126, June 1986.

7. A penguin waddles along the central axis of a concave mirror, from the focal point to an effectively infinite distance. (a) How does its image move? (b) Does the height of its image increase continually, decrease continually, or change in some more complicated manner?

8. When a *T. rex* pursues a jeep in the movie *Jurassic Park,* we see a reflected image of the *T. rex* via a side-view mirror, on which is printed the (then darkly humorous) warning: "Objects in mirror are closer than they appear." Is the mirror flat, convex, or concave?

9. Figure 35-29 shows four thin lenses, all of the same material, with sides that either are flat or have a radius of curvature of magnitude 10 cm. Without written calculation, rank the lenses according to the magnitude of the focal length, greatest first.

FIGURE 35-29
Question 9.   (a)   (b)   (c)   (d)

10. An object lies before a thin symmetric converging lens. Does the image distance increase, decrease, or remain the same if we increase (a) the index of refraction n of the lens, (b) the magnitude of the radius of curvature of the two sides, and (c) the index of refraction n_{med} of the surrounding medium, keeping n_{med} less than n?

11. A concave mirror and a converging lens (glass with $n = 1.5$) both have a focal length of 3 cm when in air. When they are in water ($n = 1.33$), are their focal lengths greater than, less than, or equal to 3 cm?

12. A converging lens with index of refraction 1.5 is submerged in three separate liquids with indices of refraction 1.3, 1.5, and 1.7. (a) Rank the liquids (by giving their indices of refraction) according to the magnitude of the lens's focal length f in them, greatest first. (b) What is the sign of f in each liquid?

13. The table details six variations of the basic arrangement of two thin lenses represented in Fig. 35-30. (The points labeled F_1 and F_2 are the focal points of lenses 1 and 2.) An object is distance p_1 to the left of lens 1, as in Fig. 35-16. (a) For which variations can we tell, *without calculation,* whether the final image (that due to lens 2) is to the left or right of lens 2 and whether it has the same orientation as the object? (b) For those "easy" variations, give the image location as "left" or "right" and the orientation as "same" or "inverted."

VARIATION	LENS 1	LENS 2	
1	Converging	Converging	$p_1 < f_1$
2	Converging	Converging	$p_1 > f_1$
3	Diverging	Converging	$p_1 < f_1$
4	Diverging	Converging	$p_1 > f_1$
5	Diverging	Diverging	$p_1 < f_1$
6	Diverging	Diverging	$p_1 > f_1$

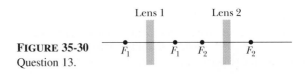

FIGURE 35-30
Question 13.

14. Figure 35-31 shows two situations in which an object sits before two thin lenses whose focal lengths have the same magnitude (one focal point is shown). In each situation, does the final image (due to lens 2) move leftward, move rightward, or remain in place if we move lens 2 toward lens 1?

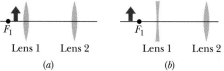

Lens 1   Lens 2   Lens 1   Lens 2

(a)   (b)

FIGURE 35-31   Question 14.

15. Much of the bending of light rays necessary for human vision occurs at the cornea (at the air–eye interface). The cornea has an index of refraction somewhat greater than that of water. (a) When your eye is submerged in a swimming pool, is the bending of light rays at the cornea greater than, less than, or the same as in air? (b) The Central American fish *Anableps anableps* can see simultaneously above and below water because it swims with its eyes partially extending above the water surface. To provide clear sight in both media, is the radius of curvature of the submerged portion of the cornea greater than, less than, or equal to that of the exposed portion?

EXERCISES & PROBLEMS

SECTION 35-2 Plane Mirrors

1E. Figure 35-32 shows an idealized submarine periscope (without submarine) that consists of two parallel plane mirrors set at 45° to the vertical periscope axis and separated by distance *L*. A penguin is sighted at a distance *D* from the top mirror. (a) Is the image seen by a submarine officer peering into the periscope real or virtual? (b) Does it have the same or the opposite orientation as the penguin? (c) Is the size (height) of the image greater than, less than, or the same size as that of the penguin? (d) What is the distance of the image from the bottom mirror?

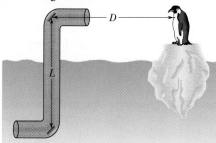

FIGURE 35-32   Exercise 1.

2E. If you move directly toward a plane mirror at speed *v*, at what speed does your image move toward you (a) in your reference frame and (b) in the reference frame of the mirror?

3E. A moth at about eye level is 10 cm in front of a plane mirror; you are behind the moth, 30 cm from the mirror. For what distance must you focus your eyes to see the image of the moth in the mirror; that is, what is the distance between your eyes and the apparent position of the image?

4E. You look through a camera toward an image of a hummingbird in a plane mirror. The camera is 4.30 m in front of the mirror. The bird is at camera level, 5.00 m to your right and 3.30 m from the mirror. For what distance must you focus your camera lens to get a clear photo of the image; that is, what is the distance between the lens and the apparent position of the image?

5E. Light travels from point *A* to point *B* via reflection at point *O* on the surface of a mirror. Without using calculus, show that length *AOB* is a minimum when the angle of incidence θ is equal to the angle of reflection ϕ. (*Hint:* Consider the virtual image of *A* in the mirror.)

6E. Figure 35-33*a* is an overhead view of two vertical plane mirrors with an object *O* placed between them. If you look into the mirrors, you see multiple images of *O*. You can find them by drawing the reflection in each mirror of the angular region between the mirrors, as is done for the left-hand mirror in Fig. 35-33*b*. Then draw the reflection of the reflection. Continue this on the left and on the right until the reflections meet or overlap at the rear of the mirrors. Then you can count the number of images of *O*. (a) If $\theta = 90°$, how many images of *O* would you see? (b) Draw their locations and orientations (as in Fig. 35-33*b*).

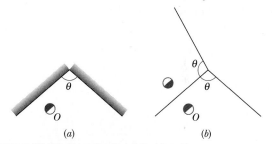

(a)   (b)

FIGURE 35-33   Exercise 6 and Problem 7.

7P. Repeat Exercise 6 for the mirror angle θ equal to (a) 45°, (b) 60°, and (c) 120°. (d) Explain why there are several possible answers for (c).

8P. Figure 35-34 shows an overhead view of a corridor with a plane mirror *M* mounted at one end. A burglar *B* sneaks along the corridor directly toward the center of the mirror. If *d* = 3.0 m, how far from the mirror will she be when the security guard *S* can first see her in the mirror?

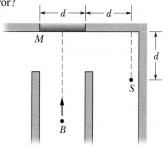

FIGURE 35-34
Problem 8.

9P. Prove that if a plane mirror is rotated through an angle α, the reflected beam is rotated through an angle 2α. Show that this result is reasonable for $\alpha = 45°$.

10P. A point object is 10 cm away from a plane mirror while the eye of an observer (with pupil diameter 5.0 mm) is 20 cm away. Assuming both the eye and the point to be on the same line perpendicular to the mirror surface, find the area of the mirror used in observing the reflection of the point. (*Hint:* Adapt Fig. 35-4.)

11P. You put a point source of light S a distance d in front of a screen A. How is the light intensity at the center of the screen changed if you put a completely reflecting mirror M a distance d behind the source, as in Fig. 35-35? (*Hint:* Use Eq. 34-27.)

FIGURE 35-35
Problem 11.

12P. Figure 35-36 shows a small lightbulb suspended 250 cm above the surface of the water in a swimming pool. The water is 200 cm deep, and the bottom of the pool is a large mirror. How far below the mirror's surface is the image of the bulb? (*Hint:* Construct a diagram of two rays like that of Fig. 35-3, but take into account the bending of light rays by refraction. Assume that the rays are close to a vertical axis through the bulb, and use the small-angle approximation that $\sin\theta \approx \tan\theta \approx \theta$.)

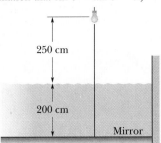

FIGURE 35-36
Problem 12.

13P*. A *corner reflector*, much used in optical, microwave, and other applications, consists of three plane mirrors fastened together to form the corner of a cube. The device has the following property: after three reflections, an incident ray is returned with its direction exactly reversed. Prove this result.

SECTION 35-4 Images from Spherical Mirrors

14E. Equation 35-4 is accurate only if we restrict our attention to rays reflected nearly along the central axis of a mirror, unlike what is drawn (for clarity) in Figs. 35-7b and c. With a ruler, measure r and p in those two parts of Fig. 35-7 and calculate, with Eq. 35-4, the predicted value of i. Then measure i and compare the predicted and measured values.

15E. A concave shaving mirror has a radius of curvature of 35.0 cm. It is positioned so that the (upright) image of a man's face is 2.50 times the size of the face. How far is the mirror from the face?

16P. Fill in Table 35-3, each row of which refers to a different combination of an object and either a plane mirror, a spherical convex mirror, or a spherical concave mirror. Distances are in centimeters. If a number lacks a sign, find the sign. Sketch each combination and draw in enough rays to locate the object and its image.

TABLE 35-3 PROBLEM 16: MIRRORS

TYPE	f	r	i	p	m	REAL IMAGE?	INVERTED IMAGE?
(a) Concave	20			+10			
(b)				+10	+1.0	No	
(c)	+20			+30			
(d)				+60	−0.50		
(e)		−40	−10				
(f)	20				+0.10		
(g) Convex		40	4.0				
(h)				+24	0.50		Yes

17P. A short straight object of length L lies along the central axis of a spherical mirror, a distance p from the mirror. (a) Show that its image in the mirror will have a length L' where

$$L' = L\left(\frac{f}{p-f}\right)^2.$$

(*Hint:* Locate the two ends of the object.) (b) Show that the *longitudinal magnification* $m' (= L'/L)$ is equal to m^2, where m is the lateral magnification.

18P. (a) A luminous point is moving at speed v_O toward a spherical mirror, along the central axis of the mirror. Show that the image of this point is moving at speed

$$v_I = -\left(\frac{r}{2p-r}\right)^2 v_O,$$

where p is the distance of the luminous point from the mirror at any given time. (*Hint:* Start with Eq. 35-4.) Now assume that the mirror is concave, with $r = 15$ cm, and let $v_O = 5.0$ cm/s. Find the speed of the image when (b) $p = 30$ cm (far outside the focal point), (c) $p = 8.0$ cm (just outside the focal point), and (d) $p = 10$ mm (very near the mirror).

SECTION 35-5 Spherical Refracting Surfaces

19P. A beam of parallel light rays from a laser is incident on a solid transparent sphere of index of refraction n (Fig. 35-37). (a) If a point image is produced at the back of the sphere, what is the index of refraction of the sphere? (b) What index of refraction, if any, will produce a point image at the center of the sphere?

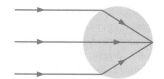

FIGURE 35-37
Problem 19.

20P. Fill in Table 35-4, each row of which refers to a different combination of a point object and a spherical refracting surface separating two media with different indices of refraction. Distances are in centimeters. If a number lacks a sign, find the sign. Sketch each combination and draw in enough rays to locate the object and image.

TABLE 35-4 PROBLEM 20: SPHERICAL REFRACTING SURFACES

	n_1	n_2	p	i	r	INVERTED IMAGE?
(a)	1.0	1.5	+10		+30	
(b)	1.0	1.5	+10	−13		
(c)	1.0	1.5		+600	+30	
(d)	1.0		+20	−20	−20	
(e)	1.5	1.0	+10	−6.0		
(f)	1.5	1.0		−7.5	−30	
(g)	1.5	1.0	+70		+30	
(h)	1.5		+100	+600	−30	

21P. You look downward at a penny that lies at the bottom of a pool of liquid with depth d and index of refraction n (Fig. 35-38). Because you view with two eyes, which intercept different rays of light from the penny, you perceive the penny to be where extensions of the intercepted rays cross, at depth d_a instead of d. Assuming that the intercepted rays in Fig. 35-38 are close to a vertical axis through the penny, show that $d_a = d/n$. (*Hint:* Use the small-angle approximation that $\sin \theta \approx \tan \theta \approx \theta$.)

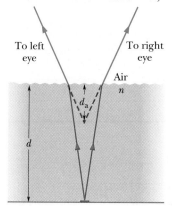

FIGURE 35-38
Problem 21.

22P. A 20-mm-thick layer of water ($n = 1.33$) floats on a 40-mm-thick layer of carbon tetrachloride ($n = 1.46$) in a tank. A penny lies at the bottom of the tank. At what depth below the top

water surface do you perceive the penny? (*Hint:* Use the result and assumptions of Problem 21 and work with a ray diagram of the situation.)

23P*. A goldfish in a spherical fish bowl of radius R is at the level of the center of the bowl and at distance $R/2$ from the glass. What magnification of the fish is produced by the water of the bowl for a viewer looking along a line that includes the fish and the center, from the fish's side of the center? The index of refraction of the water in the bowl is 1.33. Neglect the glass wall of the bowl. Assume the viewer looks with one eye. (*Hint:* Equation 35-5 holds, but Eq. 35-6 does not. You need to work with a ray diagram of the situation and assume that the rays are close to the observer's line of sight.)

SECTION 35-6 Thin Lenses

24E. An object is 20 cm to the left of a thin diverging lens having a 30 cm focal length. What is the image distance i? Find the image position with a ray diagram.

25E. Two coaxial converging lenses, with focal lengths f_1 and f_2, are positioned a distance $f_1 + f_2$ apart, as shown in Fig. 35-39. Arrangements like this are called *beam expanders* and are often used to increase the diameter of a light beam from a laser. (a) If W_1 is the incident beam width, show that the width of the emerging beam is $W_2 = (f_2/f_1)W_1$. (b) Show how a combination of one diverging and one converging lens can also be arranged as a beam expander. Incident rays parallel to the lens axis should exit parallel to the axis.

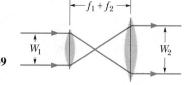

FIGURE 35-39
Exercise 25.

26E. Calculate the ratio of the intensity of the beam emerging from the beam expander of Exercise 25 to the intensity of the incident beam.

27E. A double-convex lens is to be made of glass with an index of refraction of 1.5. One surface is to have twice the radius of curvature of the other and the focal length is to be 60 mm. What are the radii?

28E. You produce an image of the Sun on a screen, using a thin lens whose focal length is 20.0 cm. What is the diameter of the image? (See Appendix C for needed data on the Sun.)

29E. A lens is made of glass having an index of refraction of 1.5. One side of the lens is flat, and the other convex with a radius of curvature of 20 cm. (a) Find the focal length of the lens. (b) If an object is placed 40 cm in front of the lens, where will the image be located?

30E. Using the lens maker's formula (Eq. 35-10), decide which of the thin lenses in Fig. 35-40 are converging and which are diverging for incident light rays that are parallel to the central axis of the lens.

FIGURE 35-40
Exercise 30.   (a)   (b)   (c)   (d)

31E. The formula

$$\frac{1}{p} + \frac{1}{i} = \frac{1}{f}$$

is called the *Gaussian* form of the thin-lens formula. Another form of this formula, the *Newtonian* form, is obtained by considering the distance x from the object to the first focal point and the distance x' from the second focal point to the image. Show that

$$xx' = f^2.$$

32E. A movie camera with a (single) lens of focal length 75 mm takes a picture of a 180 cm high person standing 27 m away. What is the height of the image of the person on the film?

33P. You have a supply of flat glass disks ($n = 1.5$) and a lens-grinding machine that can be set to grind radii of curvature of either 40 or 60 cm. You are asked to prepare a set of six lenses like those shown in Fig. 35-41. What will be the focal length of each lens? Will the lens form a real or a virtual image of the Sun? (*Note:* Where you have a choice of radii of curvature, select the smaller one.)

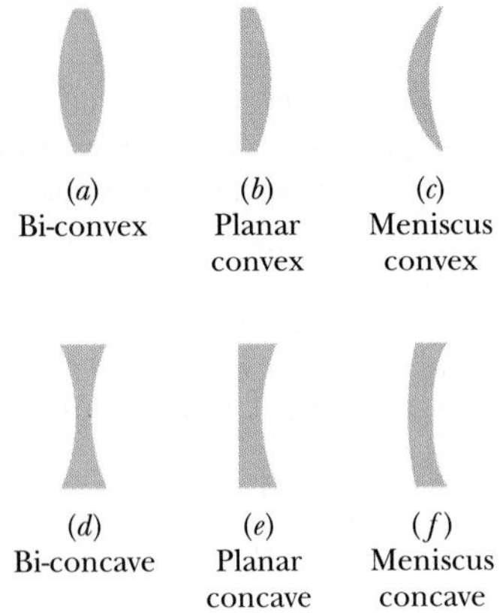

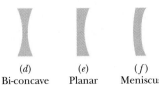

(a)   (b)   (c)
Bi-convex   Planar   Meniscus
        convex   convex

(d)   (e)   (f)

FIGURE 35-41
Problem 33.
      Bi-concave   Planar   Meniscus
                concave   concave

34P. To the extent possible, fill in Table 35-5, each row of which refers to a different combination of an object and a thin lens. Distances are in centimeters. For the type of lens, use C for converging and D for diverging. If a number (except for the index of refraction) lacks a sign, find the sign. Sketch each combination and draw in enough rays to locate the object and image.

35P. A converging lens with a focal length of $+20$ cm is located 10 cm to the left of a diverging lens having a focal length of -15 cm. If an object is located 40 cm to the left of the converging lens, locate and describe completely the final image formed by the diverging lens.

36P. An object is placed 1.0 m in front of a converging lens, of focal length 0.50 m, which is 2.0 m in front of a plane mirror. (a) Where is the final image, measured from the lens, that would be seen by an eye looking toward the mirror through the lens (and just past the object)? (b) Is the final image real or virtual? (c) Is the orientation of the final image the same as the object or inverted? (d) What is the lateral magnification?

37P. In Fig. 35-42, an object is placed a distance in front of a converging lens equal to twice the focal length f_1 of the lens. On the other side of the lens is a concave mirror of focal length f_2 separated from the lens by a distance $2(f_1 + f_2)$. (a) Find the location, type, orientation, and lateral magnification of the final image, as seen by an eye looking toward the mirror through the lens and just past (to one side of) the object. (b) Draw a ray diagram to locate the image.

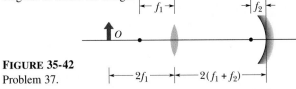

FIGURE 35-42
Problem 37.

38P. In Fig. 35-43, a real inverted image I of an object O is formed by a certain lens (not shown); the object–image separation is $d = 40.0$ cm, measured along the central axis of the lens.

TABLE 35-5   **PROBLEM 34: THIN LENSES**

	TYPE	f	r_1	r_2	i	p	n	m	REAL IMAGE?	INVERTED IMAGE?
(a)	C	10				$+20$				
(b)		$+10$				$+5.0$				
(c)		10				$+5.0$		>1.0		
(d)		10				$+5.0$		<1.0		
(e)			$+30$	-30		$+10$	1.5			
(f)			-30	$+30$		$+10$	1.5			
(g)			-30	-60		$+10$	1.5			
(h)						$+10$		0.50		No
(i)						$+10$		-0.50		

The image is just half the size of the object. (a) What kind of lens must be used to produce this image? (b) How far from the object must the lens be placed? (c) What is the focal length of the lens?

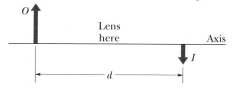

FIGURE 35-43 Problem 38.

39P. An object is 20 cm to the left of a lens with a focal length of +10 cm. A second lens of focal length +12.5 cm is 30 cm to the right of the first lens. (a) Find the location and relative size of the final image. (b) Verify your conclusions by drawing the lens system to scale and constructing a ray diagram. (c) Is the final image real or virtual? (d) Is it inverted?

40P. Two thin lenses of focal lengths f_1 and f_2 are in contact. Show that they are equivalent to a single thin lens with a focal length given by

$$f = \frac{f_1 f_2}{f_1 + f_2}.$$

41P. The *power P* of a lens is defined as $P = 1/f$, where f is the focal length. The unit of power is the *diopter,* where 1 diopter = 1 m^{-1}. (a) Why is this a reasonable definition to use for lens power? (b) Show that the net power of two lenses in contact is given by $P = P_1 + P_2$, where P_1 and P_2 are the powers of the two lenses. (*Hint:* See Problem 40.)

42P. An illuminated slide is held 44 cm from a screen. How far from the slide must a lens of focal length 11 cm be placed to form an image of the slide's picture on the screen?

43P. Show that the distance between an object and its real image formed by a thin converging lens is always greater than or equal to four times the focal length of the lens.

44P. A luminous object and a screen are a fixed distance D apart. (a) Show that a converging lens of focal length f, placed between object and screen, will form a real image on the screen for two lens positions that are separated by a distance

$$d = \sqrt{D(D - 4f)}.$$

(b) Show that the ratio of the two image sizes for these two lens positions is

$$\left(\frac{D - d}{D + d}\right)^2.$$

45P. A narrow beam of parallel light rays is incident on a glass sphere from the left, directed toward the center of the sphere. (The sphere is a lens but certainly not a *thin* lens.) Approximate the angle of incidence of the rays as 0°, and assume that the index of refraction of the glass is $n < 2.0$. Find the image distance i (from the right side of the sphere) in terms of n and the radius r of the sphere. (*Hint:* Apply Eq. 35-8 to locate the image produced by refraction at the left side of the sphere; then use that image as the object for refraction at the right side of the sphere to locate the final image. In the second refraction, is the object distance p positive or negative?)

SECTION 35-7 Optical Instruments

46E. In a microscope of the type shown in Fig. 35-18, the focal length of the objective is 4.00 cm, and that of the eyepiece is 8.00 cm. The distance between the lenses is 25.0 cm. (a) What is the tube length s? (b) If image I in Fig. 35-18 is to be just inside focal point F_1', how far from the objective should the object be? (c) What then is the lateral magnification m of the objective? (d) What then is the angular magnification m_θ of the eyepiece? (e) What then is the overall magnification M of the microscope?

47E. If the angular magnification of an astronomical telescope is 36 and the diameter of the objective is 75 mm, what is the minimum diameter of the eyepiece required to collect all the light entering the objective from a distant point source located on the axis of the instrument?

48P. A simple magnifying lens of focal length f is placed near the eye of someone whose near point P_n is 25 cm from the eye. An object is positioned so that its image in the magnifying lens appears at P_n. (a) What is the angular magnification of the lens? (b) What is the angular magnification if the object is moved so that its image appears at infinity? (c) Evaluate the angular magnifications of (a) and (b) for $f = 10$ cm. (Viewing an image at P_n requires effort by muscles in the eye, whereas for many people viewing an image at infinity requires no effort.)

49P. (a) Show that if the object O in Fig. 35-17c is moved from focal point F_1 toward the eye, the image moves in from infinity and the angle θ' (and thus the angular magnification m_θ) increases. (b) If you continue this process, at what image location will m_θ have its maximum usable value? (You can then still increase m_θ, but the image will no longer be clear.) (c) Show that the maximum usable value of m_θ is $1 + (25$ cm$)/f$. (d) Show that in this situation the angular magnification is equal to the lateral magnification.

50P. Figure 35-44a shows the basic structure of a human eye. Light refracts into the eye through the cornea and is then further redirected by a lens whose shape (and thus ability to focus the light) is controlled by muscles. We can treat the cornea and eye lens as a single effective thin lens (Fig. 35-44b). If the muscles are relaxed, a "normal" eye focuses parallel light rays from a distant object O to a point on the retina at the back of the eye, where processing of the visual information begins. As an object is brought close to the eye, the muscles change the shape of the lens so that rays form an inverted real image on the retina (Fig. 35-44c). (a) Suppose the "relaxed" focal length f of the effective thin lens of the eye is 2.50 cm. If an object is located at a distance $p = 40.0$ cm, what focal length f' of the effective thin lens is required for the object to be seen clearly? (b) Do the eye muscles increase or decrease the radii of curvature of the eye lens to produce focal length f'?

51P. In an eye that is *farsighted*, the eye focuses parallel rays so that the image would form behind the retina, as in Fig. 35-45a. In an eye that is *nearsighted*, the image is formed in front of the retina, as in Fig. 35-45b. (a) How would you design a corrective lens for each eye defect? Make a ray diagram for each case. (b) If you need eyeglasses only for reading, are you nearsighted or far-

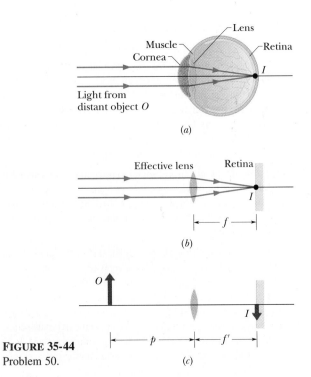

FIGURE 35-44
Problem 50.

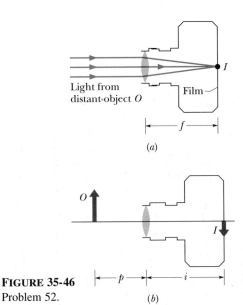

FIGURE 35-46
Problem 52.

sighted? (c) What is the function of bifocal glasses, in which the upper and lower parts have different focal lengths?

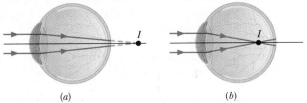

FIGURE 35-45 Problem 51.

52P. Figure 35-46*a* shows the basic structure of a camera. A lens can be moved forward or back to produce an image on film at the back of the camera. For a certain camera, with the distance between the lens and the film set at $f = 5.0$ cm, parallel light rays from a distant object O converge to a point image on the film. The object is now brought closer, to a distance of $p = 100$ cm, and the lens–film distance is adjusted so that an inverted real image forms on the film (Fig. 35-46*b*). (a) What is the lens–film distance (the image distance i) now? (b) By how much was the lens–film distance changed?

53P. In a certain compound microscope, the object is 10.0 mm from the objective. The lenses are 300 mm apart and the intermediate image is 50.0 mm from the eyepiece. What overall magnification is produced?

Electronic Computation

54. The equation $1/p + 1/i = 2/r$ for spherical mirrors is an approximation that is valid if the image is formed by rays that make only small angles with the central axis. In reality, many of the angles are large, which smears out the image a little. You can use a computer to find out how much. Refer to Fig. 35-20 and consider a ray that leaves a point source (the object) on the central axis and that makes an angle α with that axis.

First find the point of intersection of the ray with the mirror. If the coordinates of this point are x and y and the origin is placed at the center of curvature, then $y = (x + p - r)\tan \alpha$ and $x^2 + y^2 = r^2$, where p is the object distance and r is the mirror's radius of curvature. Use $\tan \beta = y/x$ to find the angle β at the point of intersection, and then use $\alpha + \gamma = 2\beta$ to find the value of γ. Finally use $\tan \gamma = y/(x + i - r)$ to find the image distance i.

(a) Suppose $r = 12$ cm and $p = 20$ cm. For each of the following values of α, find the position of the image, that is, the position of the point where the reflected ray crosses the central axis: 0.500, 0.100, 0.0100 rad. Compare the results with that obtained with the equation $1/p + 1/i = 2/r$. (b) Repeat the calculations for $p = 4.00$ cm.

36
Interference

At first glance, the top surface of the <u>Morpho</u> butterfly's wing is simply a beautiful blue-green. But there is something strange about the color, for it almost glimmers, unlike the colors of most objects. And if you change your perspective, or if the wing moves, the tint of the color changes. The wing is said to be iridescent, and the blue-green we see hides the wing's "true" dull brown color that appears on the bottom surface. What, then, is so different about the top surface that gives us this arresting display?

36-1 INTERFERENCE

Sunlight, as the rainbow shows us, is a composite of all the colors of the visible spectrum. The colors reveal themselves in the rainbow because the incident wavelengths are bent through different angles as they pass through raindrops that produce the bow. However, soap bubbles and oil slicks can also show striking colors, produced not by refraction but by constructive and destructive **interference** of light. The interfering waves combine either to enhance or to suppress certain colors in the spectrum of the incident sunlight. Interference of light waves is thus a superposition phenomenon like those we discussed in Chapter 17.

This selective enhancement or suppression of wavelengths has many applications. When light encounters an ordinary glass surface, for example, about 4% of the incident energy is reflected, thus weakening the transmitted beam by that amount. This unwanted loss of light can be a real problem in optical systems with many components. A thin, transparent "interference film," deposited on the glass surface, can reduce the amount of reflected light (and thus enhance the transmitted light) by destructive interference. The bluish cast of a camera lens reveals the presence of such a coating. Interference coatings can also be used to enhance—rather than reduce—the ability of a surface to reflect light.

To understand interference, we must go beyond the restrictions of geometrical optics and employ the full power of wave optics. In fact, as you will see, the existence of interference phenomena is perhaps our most convincing evidence that light is a wave—because interference cannot be explained other than with waves.

36-2 LIGHT AS A WAVE

The first person to advance a convincing wave theory for light was Dutch physicist Christian Huygens, in 1678. While much less comprehensive than the later electromagnetic theory of Maxwell, Huygens' theory was simpler mathematically and remains useful today. Its great advantages are that it accounts for the laws of reflection and refraction in terms of waves and gives physical meaning to the index of refraction.

Huygens' wave theory is based on a geometrical construction that allows us to tell where a given wavefront will be at any time in the future if we know its present position. This construction is based on **Huygens' principle,** which is:

> All points on a wavefront serve as point sources of spherical secondary wavelets. After a time t, the new position of the wavefront will be that of a surface tangent to these secondary wavelets.

Here is a simple example. At the left in Fig. 36-1, the present location of a wavefront of a plane wave traveling to the right in vacuum is represented by plane ab, perpendicular to the page. Where will the wavefront be at time Δt later? We let several points on plane ab (the dots) serve as sources of spherical secondary wavelets that are emitted at $t = 0$. At time Δt, the radius of all these spherical wavelets will have grown to $c\,\Delta t$, where c is the speed of light in vacuum. We draw plane de tangent to these wavelets at time Δt. This plane represents the wavefront of the plane wave at time Δt; it is parallel to plane ab and a perpendicular distance $c\,\Delta t$ from it.

The Law of Refraction

We now use Huygens' principle to derive the law of refraction, Eq. 34-44 (Snell's law). Figure 36-2 shows three stages in the refraction of several wavefronts at a plane interface between air (medium 1) and glass (medium 2). We arbitrarily choose the wavefronts in the incident beam to be separated by λ_1, the wavelength in medium 1. Let the speed of light in air be v_1 and that in glass be v_2. We assume that $v_2 < v_1$, which happens to be true.

Angle θ_1 in Fig. 36-2a is the angle between the wavefront and the interface; this is the same as the angle between the *normal* to the wavefront (that is, the incident ray) and the *normal* to the interface; thus θ_1 is the angle of incidence. As the wave moves into the glass (Fig. 36-2b), the time ($= \lambda_1/v_1$) for a Huygens wavelet to expand from

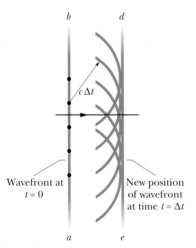

FIGURE 36-1 The propagation of a plane wave in vacuum, as portrayed by Huygens' principle.

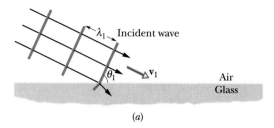

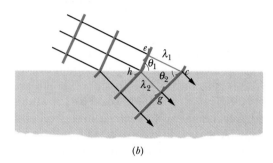

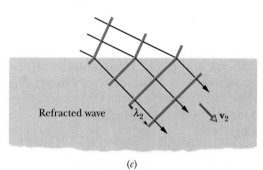

(c)

FIGURE 36-2 The refraction of a plane wave at an air–glass interface, as portrayed by Huygens' principle. The wavelength in glass is smaller than that in air. For simplicity, the reflected wave is not shown.

point e to include point c will equal the time ($= \lambda_2/v_2$) for a wavelet in the glass to expand at the reduced speed v_2 from h to include g. By equating these times, we obtain the relation

$$\frac{\lambda_1}{\lambda_2} = \frac{v_1}{v_2}, \qquad (36\text{-}1)$$

which shows that the wavelengths of light in two media are proportional to the speeds of light in those media.

By Huygens' principle, the refracted wavefront must be tangent to an arc of radius λ_2 centered on h, say at point g. The refracted wavefront must also be tangent to an arc of radius λ_1 centered on e, say at c. Then the refracted wavefront must be oriented as shown. Note that θ_2, the angle between the refracted wavefront and the interface, is actually the angle of refraction.

For the right triangles hce and hcg in Fig. 36-2b we

may write

$$\sin \theta_1 = \frac{\lambda_1}{hc} \qquad \text{(for triangle } hce\text{)}$$

and $\qquad \sin \theta_2 = \frac{\lambda_2}{hc} \qquad \text{(for triangle } hcg\text{)}.$

Dividing the first of these two equations by the second and using Eq. 36-1, we find

$$\frac{\sin \theta_1}{\sin \theta_2} = \frac{\lambda_1}{\lambda_2} = \frac{v_1}{v_2}. \qquad (36\text{-}2)$$

We can define an **index of refraction** for each medium as the ratio of the speed of light in vacuum to the speed of light v in the medium. Thus

$$n = \frac{c}{v} \qquad \text{(index of refraction).} \qquad (36\text{-}3)$$

In particular, for our two media, we have

$$n_1 = \frac{c}{v_1} \quad \text{and} \quad n_2 = \frac{c}{v_2}. \qquad (36\text{-}4)$$

If we combine Eqs. 36-2 and 36-4 we find

$$\frac{\sin \theta_1}{\sin \theta_2} = \frac{c/n_1}{c/n_2} = \frac{n_2}{n_1} \qquad (36\text{-}5)$$

or

$$n_1 \sin \theta_1 = n_2 \sin \theta_2 \qquad \begin{array}{l}\text{(law of}\\\text{refraction),}\end{array} \qquad (36\text{-}6)$$

as introduced in Chapter 34.

CHECKPOINT **1:** The figure shows a monochromatic ray of light traveling across parallel interfaces, from an original material a, through layers of material b and c, and then back into material a. Rank the materials according to the speed of light in them, greatest first.

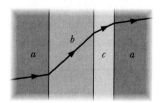

Wavelength and Index of Refraction

We have now seen that the wavelength of light changes when the speed of the light changes, as happens when light crosses an interface from one medium into another. Further, the speed of light in any medium depends on the index of refraction of the medium, according to Eq. 36-3. Thus the wavelength of light in any medium depends on the index of refraction of the medium. Let a certain monochromatic light have wavelength λ and speed c in vacuum and wavelength λ_n and speed v in a medium with an index of refraction n. Then we can rewrite Eq. 36-1 as

$$\lambda_n = \lambda \frac{v}{c}. \qquad (36\text{-}7)$$

Using Eq. 36-3 to substitute $1/n$ for v/c then yields

$$\lambda_n = \frac{\lambda}{n}. \qquad (36\text{-}8)$$

This equation relates the wavelength of light in any medium to its wavelength in vacuum. It tells us that the larger the index of refraction of a medium, the smaller the wavelength of light in that medium.

This fact is important in certain situations involving the interference of light waves. For example, in Fig. 36-3, the *waves of the rays* (that is, the waves represented by the rays) have identical wavelengths λ and are initially in phase in air ($n \approx 1$). One of the waves travels through medium 1 of index of refraction n_1 and length L. The other travels through medium 2 of index of refraction n_2 and the same length L. Because the wavelength of the light differs in the two media, the two waves may no longer be in phase when they leave these media.

> The phase difference between two light waves can change if the waves travel through different materials having different indices of refraction.

As we shall discuss soon, this change in the phase difference can determine the interference of the light waves if they reach some common point. To find their new phase difference in terms of wavelengths, we first count the number N_1 of wavelengths there are in the length L of medium 1. From Eq. 36-8, the wavelength in medium 1 is $\lambda_{n1} = \lambda/n_1$. So

$$N_1 = \frac{L}{\lambda_{n1}} = \frac{Ln_1}{\lambda}. \qquad (36\text{-}9)$$

Similarly, we count the number N_2 of wavelengths there are in the length L of medium 2, where the wavelength is

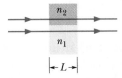

FIGURE 36-3 Two light rays travel through two media having different indices of refraction.

$\lambda_{n2} = \lambda/n_2$:

$$N_2 = \frac{L}{\lambda_{n2}} = \frac{Ln_2}{\lambda}. \qquad (36\text{-}10)$$

To find the new phase difference between the waves, we subtract the smaller of N_1 and N_2 from the larger. Assuming $n_2 > n_1$, we would obtain

$$N_2 - N_1 = \frac{Ln_2}{\lambda} - \frac{Ln_1}{\lambda} = \frac{L}{\lambda}(n_2 - n_1). \qquad (36\text{-}11)$$

Suppose Eq. 36-11 tells us that the waves now have a phase difference of 45.6 wavelengths. That is equivalent to taking the initially in-phase waves and shifting one of them by 45.6 wavelengths. However, a shift of an integer number of wavelengths (such as 45) would put the waves back in phase. So it is only the decimal fraction (here, 0.6) that is important. A phase difference of 45.6 wavelengths is equivalent to a phase difference of 0.6 wavelength.

A phase difference of 0.5 wavelength puts the waves exactly out of phase. If the waves were to reach some common point, they would then undergo fully destructive interference, producing darkness at that point. With a phase difference of 0.0 or 1.0 wavelength, they would, instead, undergo fully constructive interference, resulting in brightness at the common point. Our phase difference of 0.6 wavelength is an intermediate situation, but closer to destructive interference, and the waves would produce a dimly illuminated crossing point.

We can also express phase difference in terms of radians and degrees, as we have done already. A phase difference of one wavelength is equivalent to phase differences of 2π rad and 360°.

SAMPLE PROBLEM 36-1

In Fig. 36-3, the two light waves that are represented by the rays have wavelength 550.0 nm before entering media 1 and 2. Medium 1 is now just air, and medium 2 is a transparent plastic layer of index of refraction 1.600 and thickness 2.600 μm.

(a) What is the phase difference of the emerging waves, in wavelengths?

SOLUTION: From Eq. 36-11, with $n_1 = 1.000$, $n_2 = 1.600$, $L = 2.600$ μm, and $\lambda = 550.0$ nm, we have

$$N_2 - N_1 = \frac{L}{\lambda}(n_2 - n_1)$$

$$= \frac{2.600 \times 10^{-6}\text{ m}}{5.500 \times 10^{-7}\text{ m}}(1.600 - 1.000)$$

$$= 2.84, \qquad \text{(Answer)}$$

which is equivalent to a phase difference of 0.84 wavelength.

(b) If the rays of the waves were angled slightly so that the waves reached the same point on a distant viewing screen, what type of interference would the waves produce at that point?

SOLUTION: The effective phase difference of 0.84 wavelength is an intermediate situation, but closer to fully constructive interference (1.0) than to fully destructive interference (0.5).

(c) What is the phase difference in radians and in degrees?

SOLUTION: In radians,

$$(0.84)(2\pi \text{ rad}) = 5.3 \text{ rad}. \qquad \text{(Answer)}$$

In degrees,

$$(0.84)(360°) = 302° \approx 300°. \qquad \text{(Answer)}$$

CHECKPOINT **2:** The light waves of the rays in Fig. 36-3 have the same wavelength and are initially in phase. (a) If 7.60 wavelengths fit within the length of the top layer and 5.50 wavelengths fit within that of the bottom layer, which layer has the greater index of refraction? (b) If the rays are angled slightly so that they meet at the same point on a distant screen, will the interference there result in brightness, bright intermediate illumination, dark intermediate illumination, or darkness?

36-3 DIFFRACTION

In the next section we shall discuss the experiment that first proved that light is a wave. To prepare for that discussion, we must introduce the idea of **diffraction** of waves, a phenomenon that we explore much more fully in Chapter 37. Its essence is this: if a wave encounters a barrier that has an opening of dimensions similar to the wavelength, the part of the wave that passes through the opening will flare out —will *diffract*—into the region beyond the barrier. The flaring out is consistent with the spreading of the wavelets in the Huygens construction of Fig. 36-1. Diffraction occurs for waves of all types, not just light waves; Fig. 36-4 shows the diffraction of water waves traveling across the surface of water in a shallow tank.

Figure 36-5*a* shows the situation schematically for an incident plane wave of wavelength λ encountering a slit that has width $a = 6.0\lambda$ and extends into and out of the page. The wave flares out on the far side of the slit. Figures 36-5*b* (with $a = 3.0\lambda$) and 36-5*c* ($a = 1.5\lambda$) illustrate the main feature of diffraction: the narrower the slit, the greater the diffraction.

Diffraction limits geometrical optics, in which we represent an electromagnetic wave with a ray. If we actu-

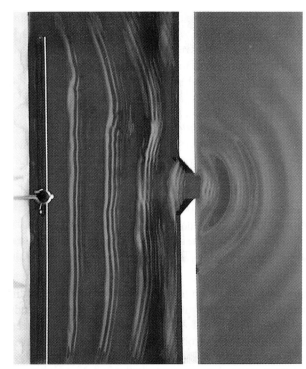

FIGURE 36-4 The diffraction of water waves in a ripple tank. Waves moving from left to right flare out through an opening in a barrier along the water surface.

ally try to form a ray by sending light through a narrow slit, or through a series of narrow slits, diffraction will always defeat our effort because it always causes the light to spread. Indeed, the narrower we make the slits (in the hope of producing a narrower beam), the greater the spreading is. Thus, geometrical optics holds only when slits or other apertures that might be located in the path of light do not have dimensions comparable to or smaller than the wavelength of the light.

36-4 YOUNG'S INTERFERENCE EXPERIMENT

In 1801 Thomas Young experimentally proved that light is a wave, contrary to what most other scientists then thought. He did so by demonstrating that light undergoes interference, as do water waves, sound waves, and waves of all other types. In addition, he was able to measure the average wavelength of sunlight; his value, 570 nm, is impressively close to the modern accepted value of 555 nm. We shall here examine Young's historic experiment as an example of the interference of light waves.

Figure 36-6 gives the basic arrangement of Young's experiment. Light from a distant monochromatic source illuminates slit S_0 in screen A. The emerging light then

FIGURE 36-9 A plot of Eq. 36-21, showing the intensity of a double-slit interference pattern as a function of the phase difference between the waves from the two slits. I_0 is the (uniform) intensity that would appear on the screen if one slit were covered. The average intensity of the fringe pattern is $2I_0$, and the *maximum* intensity (for coherent light) is $4I_0$.

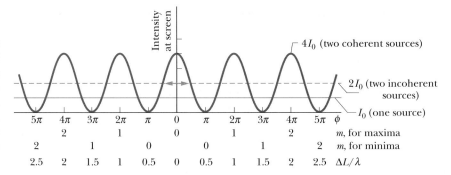

which is just Eq. 36-16, the expression derived earlier for the locations of the fringe minima.

Figure 36-9, which is a plot of Eq. 36-21, shows the intensity pattern for double-slit interference as a function of the phase difference ϕ at the screen. The horizontal solid line is I_0, the (uniform) intensity on the screen when one of the slits is covered up. Note in Eq. 36-21 and the graph that the intensity I (which is always positive) varies from zero at the fringe minima to $4I_0$ at the fringe maxima.

If the waves from the two sources (slits) were *incoherent*, so that no enduring phase relation existed between them, there would be no fringe pattern and the intensity would have the uniform value $2I_0$ for all points on the screen; the horizontal dashed line in Fig. 36-9 shows this uniform value.

Interference cannot create or destroy energy but merely redistributes it over the screen. Thus the *average* intensity on the screen must be the same $2I_0$ regardless of whether the sources are coherent. This follows at once from Eq. 36-21; if we substitute $\frac{1}{2}$, the average value of the cosine-squared function, this equation reduces to $\bar{I} = 2I_0$.

Proof of Eqs. 36-21 and 36-22

We shall combine the electric field components E_1 and E_2, given by Eqs. 36-19 and 36-20, respectively, by the method of phasors discussed in Section 17-10. In Fig. 36-10a, the waves with components E_1 and E_2 are represented by phasors of magnitude E_0 that rotate around the origin at angular speed ω. The values of E_1 and E_2 at any time are the projections of the corresponding phasors onto the vertical axis. Figure 36-10a shows the phasors and their projections at an arbitrary time t. Consistent with Eqs. 36-19 and 36-20, the phasor for E_1 has a rotation angle ωt and the phasor for E_2 has a rotation angle $\omega t + \phi$.

To combine the field components E_1 and E_2 on a phasor diagram, we add them vectorially, as shown in Fig. 36-10b. The magnitude of the vector sum is the amplitude E of the resultant wave, and that wave has a certain phase constant β. To find the amplitude E in Fig. 36-10b, we first

note that the two angles marked β are equal because they are opposite equal-length sides of a triangle. From the theorem (for triangles) that an exterior angle (ϕ) is equal to the sum of the two opposite interior angles ($\beta + \beta$), we see that $\beta = \frac{1}{2}\phi$. Thus we have

$$E = 2(E_0 \cos \beta) = 2E_0 \cos \tfrac{1}{2}\phi. \qquad (36\text{-}26)$$

If we square each side of this relation we obtain

$$E^2 = 4E_0^2 \cos^2 \tfrac{1}{2}\phi. \qquad (36\text{-}27)$$

From Eq. 34-24, we know that the intensity of an electromagnetic wave is proportional to the square of its amplitude. So the waves we are combining in Fig. 36-10b, whose amplitudes are E_0, have an intensity I_0 that is proportional to E_0^2. And the resultant wave, with amplitude E, has an intensity I that is proportional to E^2. Thus,

$$\frac{I}{I_0} = \frac{E^2}{E_0^2}.$$

Substituting Eq. 36-27 into this and rearranging then yield

$$I = 4I_0 \cos^2 \tfrac{1}{2}\phi,$$

which is Eq. 36-21, which we set out to prove.

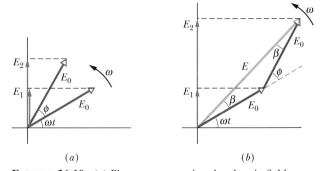

FIGURE 36-10 (*a*) Phasors representing the electric field components of the waves given by Eqs. 36-19 and 36-20. They both have magnitude E_0 and rotate with angular speed ω. (*b*) Vector addition of the two phasors gives the phasor representing the resultant wave, with magnitude E and phase constant β.

It remains to prove Eq. 36-22, which relates the phase difference ϕ between the waves arriving at any point P on the screen of Fig. 36-8 to the angle θ that serves as a locator of that point.

The phase difference ϕ in Eq. 36-20 is associated with the path difference S_1b in Fig. 36-8. If S_1b is $\frac{1}{2}\lambda$, then ϕ is π; if S_1b is λ, then ϕ is 2π, and so on. This suggests

$$\begin{pmatrix} \text{phase} \\ \text{difference} \end{pmatrix} = \frac{2\pi}{\lambda} \begin{pmatrix} \text{path length} \\ \text{difference} \end{pmatrix}. \quad (36\text{-}28)$$

The path difference S_1b in Fig. 36-8b is just $d \sin \theta$, so Eq. 36-28 becomes

$$\phi = \frac{2\pi d}{\lambda} \sin \theta,$$

which is just Eq. 36-22, the other equation that we set out to prove.

Combining More than Two Waves

In a more general case, we might want to find the resultant of more than two sinusoidally varying waves. The general procedure is this:

1. Construct a series of phasors representing the functions to be added. Draw them end to end, maintaining the proper phase relations between adjacent phasors.

2. Construct the vector sum of this array. The length of this vector sum gives the amplitude of the resultant phasor. The angle between the vector sum and the first phasor is the phase of the resultant with respect to this first phasor. The projection of this vector-sum phasor on the vertical axis gives the time variation of the resultant wave.

SAMPLE PROBLEM 36-3

Find the resultant wave $E(t)$ of the following waves:

$$E_1 = E_0 \sin \omega t,$$

$$E_2 = E_0 \sin(\omega t + 60°),$$

$$E_3 = E_0 \sin(\omega t - 30°).$$

SOLUTION: The resultant wave is

$$E(t) = E_1(t) + E_2(t) + E_3(t).$$

In using the method of phasors to find this sum, we are free to evaluate the phasors at any time t. To simplify the problem we choose $t = 0$, for which the phasors representing the three waves are shown in Fig. 36-11. We now treat the addition of the phasors as we would any other addition of vectors. The sum of the horizontal components of E_1, E_2, and E_3 is

$$\sum E_h = E_0 \cos 0 + E_0 \cos 60° + E_0 \cos(-30°)$$

$$= E_0 + 0.500E_0 + 0.866E_0 = 2.37E_0.$$

The sum of the vertical components, which is the value of E at $t = 0$, is

$$\sum E_v = E_0 \sin 0 + E_0 \sin 60° + E_0 \sin(-30°)$$

$$= 0 + 0.866E_0 - 0.500E_0 = 0.366E_0.$$

The resultant wave $E(t)$ has an amplitude E_R of

$$E_R = \sqrt{(2.37E_0)^2 + (0.366E_0)^2} = 2.4E_0,$$

and a phase angle β relative to phasor E_1 of

$$\beta = \tan^{-1}\left(\frac{0.366E_0}{2.37E_0}\right) = 8.8°.$$

We can now write, for the resultant wave $E(t)$,

$$E = E_R \sin(\omega t + \beta)$$

$$= 2.4E_0 \sin(\omega t + 8.8°). \quad \text{(Answer)}$$

Be careful to interpret the angle β correctly in Fig. 36-11: it is the constant angle between E_R and E_1 as the four phasors rotate as a single unit around the origin. The angle between E_R and the horizontal axis does not remain equal to β.

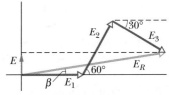

FIGURE 36-11 Sample Problem 36-3. Three phasors E_1, E_2, and E_3, shown at time $t = 0$, combine to give resultant phasor E_R.

CHECKPOINT **4:** Each of four pairs of light waves arrives at a certain point on a screen. The waves have the same wavelength. At the arrival point, their amplitudes and phase differences are (a) $2E_0$, $6E_0$, and π rad; (b) $3E_0$, $5E_0$, and π rad; (c) $9E_0$, $7E_0$, and 3π rad; (d) $2E_0$, $2E_0$, and 0 rad. Rank the four pairs according to the intensity of the light at those points, greatest first. (*Hint:* Draw phasors.)

36-7 INTERFERENCE FROM THIN FILMS

The colors we see when sunlight illuminates a soap bubble or an oil slick are caused by the interference of light waves reflected from the front and back surfaces of a thin transparent film. The thickness of the soap or oil film is typically of the order of magnitude of the wavelength of the (visible) light involved. (We shall not consider greater thicknesses, which spoil the coherence of the light needed to produce colors by interference; we shall discuss lesser thicknesses shortly.)

Figure 36-12 shows a thin transparent film of uniform thickness L and index of refraction n_2, illuminated by bright light of wavelength λ from a distant point source. For now, we assume that air lies on both sides of the film and thus that $n_1 = n_3$ in Fig. 36-12. For simplicity, we also assume that the light rays are almost perpendicular to the film ($\theta \approx 0$). We are interested in whether the film is bright or dark to an observer viewing it almost perpendicularly. (Since the film is brightly illuminated, how could it possibly be dark? You will see.)

The incident light, represented by ray i, intercepts the front (left) surface of the film at point a and undergoes both reflection and refraction there. The reflected ray r_1 is intercepted by the observer's eye. The refracted light crosses the film to point b on the back surface, where it undergoes both reflection and refraction. The light reflected at b crosses back through the film to point c, where it undergoes both reflection and refraction. The light refracted at c, represented by ray r_2, is intercepted by the observer's eye.

If the light waves of rays r_1 and r_2 are exactly in phase at the eye, they produce an interference maximum, and region ac on the film is bright to the observer. If they are exactly out of phase, they produce an interference minimum, and region ac is dark to the observer, *even though it is illuminated*. And if there is some intermediate phase difference, there are intermediate interference and intermediate brightness.

So the key to what the observer sees is the phase difference between the waves of rays r_1 and r_2. Both rays are

derived from the same ray i, but the path involved in producing r_2 involves light traveling twice across the film (a to b, and then b to c), whereas the path involved in producing r_1 involves no travel through the film. Because θ is about zero, we approximate the path length difference between the waves of r_1 and r_2 as $2L$. However, to find the phase difference between the waves, we cannot just find the number of wavelengths λ that is equivalent to a path length difference of $2L$. This simple approach is impossible for two reasons: (1) the path length difference occurs in a medium other than air, and (2) reflections are involved, which can change the phase.

> The phase difference between two waves can change if one or both are reflected.

Before we continue our discussion of interference from thin films, we must discuss changes in phase that are caused by reflections.

Reflection Phase Shifts

Refraction at an interface never causes a phase change. But reflection can, depending on the indices of refraction on the two sides of the interface. Figure 36-13 shows what happens when reflection causes a phase change, using pulses on a denser string (along which pulse travel is relatively slow) and a lighter string (along which pulse travel is relatively fast).

When a pulse traveling slowly along the denser string in Fig. 36-13a reaches the interface with the lighter string, the pulse is partially transmitted and partially reflected, with no change in orientation. For light, this situation corresponds to the incident wave traveling in the medium of greater index of refraction n (recall that greater n means slower speed). In that case, the wave that is reflected at the interface does not undergo a change in phase; that is, the *reflection phase shift* is zero.

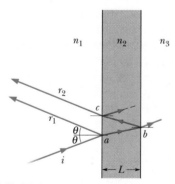

FIGURE 36-12 Light waves, represented with ray i, are incident on a thin film of thickness L and index of refraction n_2. Rays r_1 and r_2 represent light waves reflected by the front and back surfaces of the film. (All three rays are actually nearly perpendicular to the film.) The interference of the waves of r_1 and r_2 with each other depends on their phase difference. The index of refraction n_1 of the medium at the left can differ from the index of refraction n_3 of the medium at the right, but for now we assume that both media are air, with $n_1 = n_3 = 1.0$, which is less than n_2.

FIGURE 36-13 Phase changes when a pulse is reflected at the interface between two stretched strings of different linear densities. The wave speed is greater in the lighter string. (a) The incident pulse is in the denser string. (b) The incident pulse is in the lighter string. Only here is there a phase change.

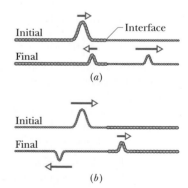

When a pulse traveling more quickly along the lighter string in Fig. 36-13b reaches the interface with the denser string, the transmitted pulse has the same orientation as the incident pulse, but the reflected pulse is inverted. For a sinusoidal wave, such an inversion involves a phase change of π rad, or half a wavelength. For light, this situation corresponds to the incident wave traveling in the medium of lesser index of refraction (with greater speed). In that case, the wave that is reflected at the interface undergoes a phase shift of π rad, or half a wavelength. We can summarize these results for light in terms of the index of refraction of the medium off which (or from which) the light reflects:

Reflection	Reflection phase shift
Off lower index	0
Off higher index	0.5 wavelength

This might be remembered as ''higher means half.''

Equations for Thin-Film Interference

In this chapter we have now seen three ways in which the phase difference between two waves can change:

1. by reflection
2. by the waves traveling along paths of different lengths
3. by the waves traveling through media of different indices of refraction

When light reflects from a thin film, producing the waves of rays r_1 and r_2 in Fig. 36-12, all three ways are involved. Let us consider them one by one.

We first reexamine the two reflections in Fig. 36-12. At point a on the front interface, the incident wave (in air) reflects from the medium having the higher of the two indices of refraction, so the wave of reflected ray r_1 has its phase shifted by 0.5 wavelength. At point b on the back interface, the incident wave reflects from the medium (air) having the lower of the two indices of refraction, so the wave reflected there is not shifted in phase by the reflection, and thus neither is the portion of it that exits the film as ray r_2. We can organize this information with the first line in Table 36-1. It tells us that, so far, as a result of the reflection phase shifts, the waves of r_1 and r_2 have a phase difference of 0.5 wavelength and thus are exactly out of phase.

Now we must consider the path length difference $2L$ that occurs because the wave of ray r_2 crosses the film twice. (This difference $2L$ is shown on the second line in Table 36-1.) If the waves of r_1 and r_2 are to be exactly in phase so that they produce fully constructive interference,

TABLE 36-1 AN ORGANIZING TABLE FOR THIN-FILM INTERFERENCE IN AIR[a]

	r_1	r_2
Reflection phase shifts	0.5 wavelength	0
Path length difference	$2L$	
Index in which path length difference occurs	n_2	
In phase[a]:	$2L = \dfrac{\text{odd number}}{2} \times \dfrac{\lambda}{n_2}$	
Out of phase[a]:	$2L = \text{integer} \times \dfrac{\lambda}{n_2}$	

[a]Valid for $n_2 > n_1$ and $n_2 > n_3$.

the path length $2L$ must cause an additional phase difference of 0.5, 1.5, 2.5, . . . wavelengths. Only then will the net phase difference be an integer number of wavelengths. Thus, for a bright film, we must have

$$2L = \frac{\text{odd number}}{2} \times \text{wavelength}$$
$$\text{(in-phase waves).} \quad (36\text{-}29)$$

The wavelength we need here is the wavelength λ_{n2} of the light in the medium containing path length $2L$, that is, in the medium with index of refraction n_2. So, we can rewrite Eq. 36-29 as

$$2L = \frac{\text{odd number}}{2} \times \lambda_{n2} \quad \text{(in-phase waves).} \quad (36\text{-}30)$$

If, instead, the waves are to be exactly out of phase so that there is fully destructive interference, the path length $2L$ must cause either no additional phase difference or a phase difference of 1, 2, 3, . . . wavelengths. Only then will the net phase difference be an odd number of half-wavelengths. So, for a dark film, we must have

$$2L = \text{integer} \times \text{wavelength}, \quad (36\text{-}31)$$

where, again, the wavelength is the wavelength λ_{n2} in the medium containing $2L$. So, this time we have

$$2L = \text{integer} \times \lambda_{n2} \quad \text{(out-of-phase waves).} \quad (36\text{-}32)$$

Now recalling that the wave of ray r_2 traveled through a medium of index of refraction n_2 whereas the wave of ray r_1 did not, we can use Eq. 36-8 ($\lambda_n = \lambda/n$) to write the wavelength of the wave inside the film as

$$\lambda_{n2} = \frac{\lambda}{n_2}, \quad (36\text{-}33)$$

where λ is the wavelength of the incident light in vacuum (and approximately also in air). Substituting Eq. 36-33 into

Eq. 36-30 and replacing "odd number/2" with $(m + \frac{1}{2})$ give us

$$2L = (m + \tfrac{1}{2}) \frac{\lambda}{n_2}, \quad \text{for } m = 0, 1, 2, \ldots$$
$$\text{(maxima—bright film in air).} \quad (36\text{-}34)$$

Similarly, with m replacing "integer," Eq. 36-32 yields

$$2L = m \frac{\lambda}{n_2}, \quad \text{for } m = 0, 1, 2, \ldots$$
$$\text{(minima—dark film in air).} \quad (36\text{-}35)$$

For a given film thickness L, Eqs. 36-34 and 36-35 tell us the wavelengths of light for which the film appears bright and dark, respectively, one wavelength for each value of m. Intermediate wavelengths give intermediate brightnesses. For a given wavelength λ, Eqs. 36-34 and 36-35 tell us the thicknesses of the films that appear bright and dark in that light, respectively, one thickness for each value of m. Intermediate thicknesses give intermediate brightnesses.

A special situation arises when a film is so thin that L is much less than λ, say, $L < 0.1\lambda$. Then the path length difference $2L$ can be neglected, and the phase difference between r_1 and r_2 is due *only* to reflection phase shifts. If the film of Fig. 36-12, where the reflections cause a phase difference of 0.5 wavelength, has thickness $L < 0.1\lambda$, then r_1 and r_2 are exactly out of phase, and thus the film is dark, regardless of the wavelength and even the intensity of the light that illuminates it. This special situation corresponds to $m = 0$ in Eq. 36-35. We shall count any $L < 0.1\lambda$ as being the least thickness that makes the film of Fig. 36-12 dark. The next greater thickness that makes the film dark is that corresponding to $m = 1$.

Figure 36-14 shows a vertical soap film whose thickness increases from top to bottom because the weight of the film has caused it to slump. Bright white light illuminates the film. However, the top portion is so thin that it is dark. In the (somewhat thicker) middle we see fringes, or bands, whose color depends primarily on the wavelength at which reflected light undergoes fully constructive interference for a particular thickness. Toward the (thickest) bottom of the film the fringes become progressively narrower and the colors begin to overlap and fade.

FIGURE 36-14 The reflection of light from a soapy water film spanning a vertical loop. The top portion is so thin that the light reflected there undergoes destructive interference, making that portion dark. Colored interference fringes, or bands, decorate the rest of the film but are marred by circulation of liquid within the film as the liquid is gradually pulled downward by gravitation.

PROBLEM SOLVING TACTICS

TACTIC 1: *Thin-Film Equations*
Some students believe that Eq. 36-34 gives the maxima and Eq. 36-35 gives the minima for *all* thin-film situations. This is not true. These relations were derived only for the situation in which $n_2 > n_1$ and $n_2 > n_3$ in Fig. 36-12.

The appropriate equations for other relative values of the indices of refraction can be derived by following the reasoning of this section and constructing new versions of Table 36-1. In each case you will end up with Eqs. 36-34 and 36-35, but sometimes Eq. 36-34 will give the minima and Eq. 36-35 will give the maxima—the opposite of what we found here. Which equation gives which depends on whether the reflections at the two interfaces give the same reflection phase shift.

CHECKPOINT **5:** The figure shows four situations in which light reflects perpendicularly from a thin film (as in Fig. 36-12), with the indices of refraction as given. (a) For which situations does reflection cause a zero phase difference for the two reflected rays? (b) For which situations will the film be dark if the path length difference $2L$ causes a phase difference of 0.5 wavelength?

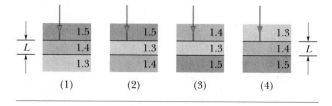

SAMPLE PROBLEM 36-4

White light, with a uniform intensity across the visible wavelength range of 400–690 nm, is perpendicularly incident on a water film, of index of refraction $n_2 = 1.33$ and thickness $L = 320$ nm, that is suspended in air. At what wavelength λ is the light reflected by the film brightest to an observer?

SOLUTION: This situation is like that of Fig. 36-12, for which Eq. 36-34 gives the interference maxima. Solving for λ and inserting the given data, we obtain

$$\lambda = \frac{2n_2L}{m + \frac{1}{2}} = \frac{(2)(1.33)(320\text{ nm})}{m + \frac{1}{2}} = \frac{851\text{ nm}}{m + \frac{1}{2}}.$$

For $m = 0$, this gives us $\lambda = 1700$ nm, which is in the infrared region. For $m = 1$, we find $\lambda = 567$ nm, which is yellow-green light, near the middle of the visible spectrum. For $m = 2$, $\lambda = 340$ nm, which is in the ultraviolet region. So the wavelength at which the light seen by the observer is brightest is

$$\lambda = 567\text{ nm.} \qquad \text{(Answer)}$$

Solving for L and inserting the given data, we obtain

$$L = \frac{\lambda}{4n_2} = \frac{550\text{ nm}}{(4)(1.38)} = 99.6\text{ nm.} \qquad \text{(Answer)}$$

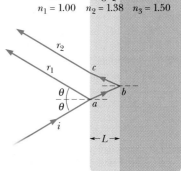

FIGURE 36-15 Sample Problem 36-5. Unwanted reflections from glass can be suppressed (at a chosen wavelength) by coating the glass with a thin transparent film of magnesium fluoride of a properly chosen thickness.

SAMPLE PROBLEM 36-5

A glass lens is coated on one side with a thin film of magnesium fluoride (MgF_2) to reduce reflection from the lens surface (Fig. 36-15). The index of refraction of MgF_2 is 1.38; that of the glass is 1.50. What is the least coating thickness that eliminates (via interference) the reflections at the middle of the visible spectrum ($\lambda = 550$ nm)? Assume that the light is approximately perpendicular to the lens surface.

SOLUTION: Figure 36-15 differs from Fig. 36-12 in that now $n_3 > n_2 > n_1$. This means there is now a reflection phase shift of 0.5 wavelength associated with the reflections at *both* front and back interfaces of the thin film. Constructing a table like Table 36-1, we fill in 0.5 and 0.5 for the first line. For the second and third lines, the path length difference is still $2L$ and it still occurs in a medium (here MgF_2) having index of refraction n_2.

The reflections alone tend to put the waves of r_1 and r_2 in phase. For these rays to be out of phase so that the reflections from the lens are eliminated, the path length difference $2L$ within the film must be

$$2L = \frac{\text{odd number}}{2} \times \text{wavelength}$$

$$= (m + \tfrac{1}{2})\lambda_{n2}, \qquad \text{for } m = 0, 1, 2, \ldots.$$

Substituting λ/n_2 for λ_{n2} yields

$$2L = (m + \tfrac{1}{2})\frac{\lambda}{n_2}, \qquad \text{for } m = 0, 1, 2, \ldots.$$

We want the least thickness for the coating, that is, the smallest L. Thus we choose $m = 0$, the smallest value of m.

SAMPLE PROBLEM 36-6

Figure 36-16a shows a transparent plastic block with a thin wedge of the plastic removed at the right. A broad beam of red light, with wavelength $\lambda = 632.8$ nm, is directed directly downward through the top of the block (at an incidence angle of 0°). Some of the light is reflected back up from the top and bottom surfaces of the wedge, which acts as a thin film (of air) with a thickness that varies uniformly and gradually from L_L at the left-hand end to L_R at the right-hand end. An observer

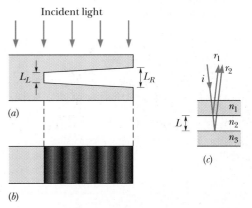

FIGURE 36-16 Sample Problem 36-6. (a) Red light is incident on a thin air-filled wedge in the side of a transparent plastic block. The thickness of the wedge is L_L at the left end and L_R at the right end. (b) The view from above the block: an interference pattern of six dark fringes and five bright red fringes lies over the region of the wedge. (c) A representation of the incident ray i, reflected rays r_1 and r_2, and thickness L of the wedge anywhere along the length of the wedge.

looking down on the block sees an interference pattern consisting of six dark fringes and five bright red fringes along the wedge. What is the change in thickness $\Delta L\, (= L_R - L_L)$ along the wedge?

SOLUTION: This thin-film problem differs from preceding problems because the thickness of the film varies. It is that variation that produces the observed variation between dark and bright fringes along the wedge. Because the observer sees more dark fringes than bright fringes, we can assume that a dark fringe is produced at both the left and right ends of the film. Thus the interference pattern is that shown in Fig. 36-16b, which we can use to determine the change in thickness ΔL of the wedge.

We can represent the reflection of light by the top and bottom surfaces of the wedge anywhere along its length with Fig. 36-16c, at a place where we assume that the wedge has thickness L. From what we know about phase shifts by reflection, we see that the reflection phase shift for ray r_1 is zero and that for ray r_2 is 0.5 wavelength. Constructing a table like Table 36-1, we fill in 0 and 0.5 for the first line. For the second and third lines, the path length difference is still $2L$ and it still occurs in a medium (here air) with index of refraction n_2. Thus, for fully destructive interference we find that

$$2L = \text{integer} \times \frac{\lambda}{n_2} = m\,\frac{\lambda}{n_2}. \qquad (36\text{-}36)$$

We can apply this equation at any point along the wedge where a dark fringe is observed. The least value of the integer m is associated with the least thickness of the wedge where a dark fringe is observed. And progressively greater values of m are associated with progressively greater thicknesses of the wedge where a dark fringe is observed.

A dark fringe happens to be observed at the left end of the wedge, where the thickness is least. Applying Eq. 36-36 to that end, substituting L_L for L, and then solving for L_L, we have

$$L_L = \frac{m_L \lambda}{2n_2}, \qquad (36\text{-}37)$$

where m_L is the integer associated with the dark fringe at the left end and n_2 is the index of refraction of the material inside the wedge (air).

We can also apply Eq. 36-36 to the right end of the wedge, where another dark fringe is seen. There the thickness is L_R, and the integer associated with L_R and this dark fringe is $m_L + 5$ (because the fringe is the fifth one from the fringe at the left-hand end). Substituting L_R for L and $m_L + 5$ for m into Eq. 36-36 and solving for L_R yield

$$L_R = \frac{(m_L + 5)\lambda}{2n_2}. \qquad (36\text{-}38)$$

Subtracting Eq. 36-37 from Eq. 36-38 then gives us the change in thickness ΔL of the wedge:

$$\Delta L = L_R - L_L = \frac{(m_L + 5)\lambda}{2n_2} - \frac{m_L \lambda}{2n_2} = \frac{5}{2}\frac{\lambda}{n_2}.$$

Substituting 632.8×10^{-9} m for λ and 1.00 for n_2 into this equation, we find

$$\Delta L = \frac{5}{2}\frac{632.8 \times 10^{-9}\ \text{m}}{1.00}$$

$$= 1.58 \times 10^{-6}\ \text{m.} \qquad \text{(Answer)}$$

SAMPLE PROBLEM 36-7

The iridescence seen in the top surface of *Morpho* butterfly wings is due to constructive interference of the light reflected by thin terraces of transparent cuticle-like material. The terraces extend outward, parallel to the wings, from a central structure that is approximately perpendicular to the wing. Cross sections of the central structure and terraces are shown in the electron micrograph of Fig. 36-17a. The terraces have index of refraction $n = 1.53$ and thickness $D_t = 63.5$ nm; they are separated (by air) by $D_a = 127$ nm. If the incident light is perpendicular to the terraces (see Fig. 36-17b, where the angle of the incident light is exaggerated), at what wavelength of visible light do the reflections from the terraces have an interference maximum?

SOLUTION: Let us first consider rays r_1 and r_2 in Fig. 36-17b, which involve reflections at points a and b. This situation is just like that of Fig. 36-12, and so Eq. 36-34 gives the interference maxima. Solving Eq. 36-34 for λ gives us

$$\lambda = \frac{2n_2 L}{m + \frac{1}{2}}.$$

Substituting $D_t\, (= 63.5$ nm$)$ for L and $n\, (= 1.53)$ for n_2, we have

$$\lambda = \frac{2nD_t}{m + \frac{1}{2}} = \frac{(2)(1.53)(63.5\ \text{nm})}{m + \frac{1}{2}} = \frac{194\ \text{nm}}{m + \frac{1}{2}}.$$

For $m = 0$, we find an interference maximum at $\lambda = 388$ nm, which is in the ultraviolet region. For all larger values of m, λ is even smaller, farther into the ultraviolet. So rays r_1 and r_2 do not produce the bright blue-green color of the *Morpho*.

Let us next consider rays r_1 and r_3 in Fig. 36-17b. The wave producing the latter passes through a terrace and then through air to the next terrace, where it reflects at point d. Then it travels upward, resulting in ray r_3. The path length difference between the waves leading to rays r_1 and r_3 is $2D_t + 2D_a$. This situation differs considerably from that of Fig. 36-12, and Eq. 36-34 does not apply. To find a new equation for interference maxima for this new situation, we first consider the reflections involved and then count the wavelengths along path length difference $2D_t + 2D_a$.

The reflections at points a and d both introduce a phase change of half a wavelength. So the reflections alone tend to put the waves of rays r_1 and r_3 in phase. Thus for these waves actually to end up in phase, the number of wavelengths along the path length difference $2D_t + 2D_a$ must be an integer. The wavelength within the terrace is $\lambda_n = \lambda/n$. So the number of incident light is not exactly perpendicular to the terraces but travels along a slanted path, the paths taken by the waves represented by r_1 and r_3 change, and so does the wavelength of maximum interference. Thus as the wing moves in your view, the wavelength at which the wing is brightest changes slightly, producing iridescence of the wing.

36-8 MICHELSON'S INTERFEROMETER

An **interferometer** is a device that can be used to measure lengths or changes in length with great accuracy by means of interference fringes. We describe the form originally devised and built by A. A. Michelson in 1881. Consider light that leaves point P on extended source S (Fig. 36-18) and encounters a *beam splitter M*. This is a mirror with the following property: it transmits half the incident light, reflecting the rest. In the figure we have assumed, for convenience, that this mirror possesses negligible thickness. At M the light thus divides into two waves. One proceeds by transmission toward mirror M_1; the other proceeds by reflection toward M_2. The waves are reflected at each of these mirrors and are sent back along their directions of incidence, each wave eventually entering the telescope T. What the observer sees is a pattern of curved or approximately straight interference fringes; the latter resemble the stripes on a zebra.

The path length difference for the two waves when they recombine is $2d_2 - 2d_1$, and anything that changes this path difference will cause a change in the phase between these two waves at the eye. As an example, if mirror M_2 is moved by a distance $\frac{1}{2}\lambda$, the path length difference is changed by λ and the fringe pattern is shifted by one fringe (as if each dark stripe on a zebra had moved to where the

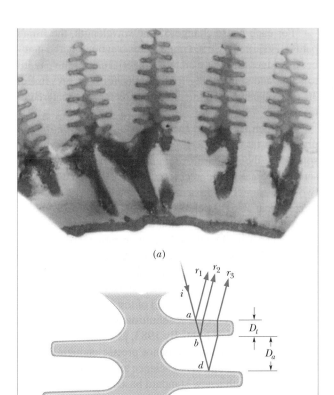

(a)

(b)

FIGURE 36-17 Sample Problem 36-7. (*a*) An electron micrograph shows the cross section of terrace structures of cuticle material that stick up from the top surface of a *Morpho* wing. (*b*) Light waves reflecting at points a and b on a terrace, represented by rays r_1 and r_2, interfere at the eye of an observer. The wave of ray r_1 also interferes with the wave that reflects at point d and is represented by ray r_3.

wavelengths in length $2D_t$ is

$$N_t = \frac{2D_t}{\lambda_n} = \frac{2D_t n}{\lambda}.$$

Similarly, the number of wavelengths in length $2D_a$ is

$$N_a = \frac{2D_a}{\lambda}.$$

For the waves of rays r_1 and r_3 to be in phase, we need $N_t + N_a$ to be equal to an integer m. Thus for an interference maximum,

$$\frac{2D_t n}{\lambda} + \frac{2D_a}{\lambda} = m, \qquad \text{for } m = 1, 2, 3, \ldots.$$

Solving for λ and substituting the given data, we obtain

$$\lambda = \frac{(2)(63.5 \text{ nm})(1.53) + (2)(127 \text{ nm})}{m} = \frac{448 \text{ nm}}{m}.$$

For $m = 1$, we find

$$\lambda = 448 \text{ nm}. \qquad \text{(Answer)}$$

This wavelength corresponds to the bright blue-green color of the top surface of a *Morpho* wing. Further, when the

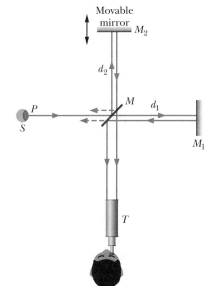

FIGURE 36-18 Michelson's interferometer, showing the path of light originating at point P of an extended source S. Mirror M splits the light into two beams, which reflect from mirrors M_1 and M_2 back to M and then to telescope T. In the telescope an observer sees a pattern of interference fringes.

6. Is there an interference maximum, a minimum, an intermediate state closer to a maximum, or an intermediate state closer to a minimum at point P in Fig. 36-8 if the path length difference of the two rays is (a) 2.2λ, (b) 3.5λ, (c) 1.8λ, and (d) 1.0λ? For each situation, give the value of m associated with the maximum or minimum involved.

7. (a) If you move from one bright fringe in a two-slit interference pattern to the next one farther out, (a) does the path length difference ΔL increase or decrease and (b) by how much does it change, in wavelengths λ?

8. Does the spacing between fringes in a two-slit interference pattern increase, decrease, or stay the same if (a) the slit separation is increased, (b) the color of the light is switched from red to blue, and (c) the whole apparatus is submerged in cooking sherry? (d) If the slits are illuminated with white light, then at any side maximum, does the blue component or the red component peak closer to the central maximum?

9. In Fig. 36-23, a thin, transparent plastic layer has been placed over the lower slit in a double-slit experiment. Does this cause the central maximum (the fringe where waves arrive with a phase difference of zero wavelengths) to move up or down the screen? (*Hint:* Is the wavelength in the plastic greater than or less than that in air?)

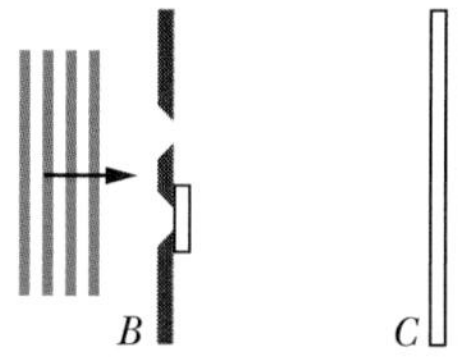

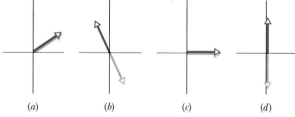

FIGURE 36-23 Question 9.

10. Figure 36-24 shows, at different times, the phasors representing the two light waves arriving at four different points on the viewing screen in a double-slit interference experiment. Assuming all eight phasors have the same length, rank the points according to the intensity of the light there, greatest first.

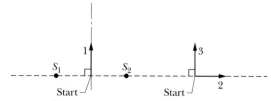

(a) (b) (c) (d)

FIGURE 36-24 Question 10.

11. Figure 36-25 shows two sources S_1 and S_2 that emit radio waves of wavelength λ in all directions. The sources are exactly in phase and are separated by a distance equal to 1.5λ. The vertical broken line is the perpendicular bisector of the distance be-

FIGURE 36-25 Question 11.

tween the sources. (a) If we start at the indicated start point and travel along path 1, does the interference produce a maximum all along the path, a minimum all along the path, or alternating maxima and minima? Repeat for (b) path 2 and (c) path 3.

12. Whole milk is a liquid suspension of fat and other particles. If you hold a spoon partially filled with milk in bright sunlight, you will see fleeting points of color near the perimeter of the milk. What causes them?

13. Figure 36-26 shows two rays of light encountering interfaces, where they reflect and refract. Which of the resulting waves are shifted in phase at the interface?

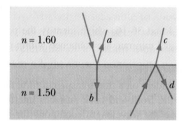

FIGURE 36-26 Question 13.

14. Suppose that the equation $2L = (m + \frac{1}{2})\lambda/n_2$ gives the maxima for interference by a certain thin film. (a) For a given film thickness, does $m = 2$ correspond to the maximum due to the second longest wavelength, the second shortest wavelength, the third longest wavelength, or the third shortest wavelength? (b) For a given wavelength, what value of m corresponds to the third least thickness giving a maximum?

15. Figure 36-27a shows the cross section of a vertical thin film whose width increases downward owing to its weight. Figure 36-27b is a face-on view of the film, showing four bright interference fringes that result when the film is illuminated with a perpendicular beam of red light. Points in the cross section corresponding to the bright fringes are labeled. In terms of the wavelength of the light inside the film, what is the difference in film thickness between (a) points a and b and (b) points b and d?

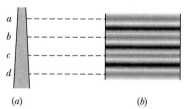

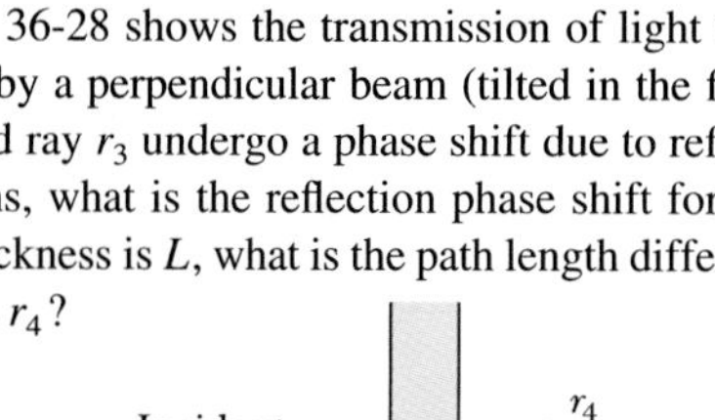

(a) (b)

FIGURE 36-27 Question 15.

16. Figure 36-28 shows the transmission of light through a thin film in air by a perpendicular beam (tilted in the figure for clarity). (a) Did ray r_3 undergo a phase shift due to reflection? (b) In wavelengths, what is the reflection phase shift for ray r_4? (c) If the film thickness is L, what is the path length difference between rays r_3 and r_4?

Incident light

r_4
r_3

FIGURE 36-28 Question 16.

17. Sunlight illuminates a thin film of oil that floats on water, which has a greater index of refraction than the oil. The edge of the film has thickness $L < 0.1\lambda$. Is the edge dark (like the corresponding thin region of the soap film in Fig. 36-14) or bright?

18. The eyes of some animals contain reflectors that send light to receptors where the light is absorbed. In the scallop, the reflector consists of many thin transparent layers alternating between high and low indices of refraction. With the proper layer thicknesses, the combined reflections from the interfaces end up in phase with one another, thereby giving a much brighter reflection than a single biological surface or layer could give. Figure 36-29 shows such an arrangement of alternating layers, along with the reflec-

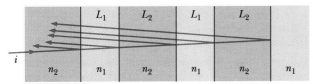

FIGURE 36-29 Question 18.

tions due to a single perpendicularly incident ray i. In terms of the indices of refraction n_1 and n_2 and the wavelength λ of visible light, should the thicknesses be (a) $L_1 = \lambda/4n_1$ and $L_2 = \lambda/4n_2$ or (b) $L_1 = \lambda/2n_1$ and $\lambda/2n_2$?

19. Figure 36-30 shows four situations in which light of wavelength λ is incident perpendicularly on a very thin layer. The indicated indices of refraction are $n_1 = 1.33$ and $n_2 = 1.50$. In each situation the thin layer has thickness $L < 0.1\lambda$. In which situations will the light reflected by the thin layer be approximately eliminated by interference?

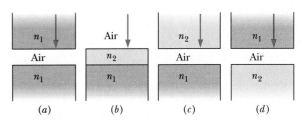

FIGURE 36-30 Question 19.

EXERCISES & PROBLEMS

SECTION 36-2 Light as a Wave

1E. The wavelength of yellow sodium light in air is 589 nm. (a) What is its frequency? (b) What is its wavelength in glass whose index of refraction is 1.52? (c) From the results of (a) and (b) find its speed in this glass.

2E. How much faster, in meters per second, does light travel in sapphire than in diamond? See Table 34-1.

3E. Derive the law of reflection using Huygens' principle.

4E. The speed of yellow light (from a sodium lamp) in a certain liquid is measured to be 1.92×10^8 m/s. What is the index of refraction of this liquid for the light?

5E. What is the speed in fused quartz of light of wavelength 550 nm? (See Fig. 34-19.)

6E. When an electron moves through a medium at a speed exceeding the speed of light in that medium, the electron radiates electromagnetic energy (the *Cerenkov effect*). What minimum speed must an electron have in a liquid of refractive index 1.54 in order to radiate?

7E. A laser beam travels along the axis of a straight section of pipeline, 1 mi long. The pipe normally contains air at standard temperature and pressure (see Table 34-1), but it may also be evacuated. In which case would the travel time for the beam be greater, and by how much?

8P. One end of a stick is pushed through water at speed v, which is greater than the speed u of water waves. Applying Huygens' construction to the water waves produced by the stick, show that a conical wavefront is set up and that its half-angle θ (see Fig. 18-22) is given by

$$\sin\theta = u/v.$$

This is familiar as the bow wave of a ship and the shock wave caused by an object moving through air with a speed exceeding that of sound.

9P. Ocean waves moving at a speed of 4.0 m/s are approaching a beach at an angle of 30° to the normal, as shown in Fig. 36-31. Suppose the water depth changes abruptly at a certain distance from the beach and the wave speed there drops to 3.0 m/s. Close to the beach, what is the angle θ between the direction of wave motion and the normal? (Assume the same law of refraction as for light.) Explain why most waves come in normal to a shore even though at large distances they approach at a variety of angles.

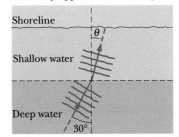

FIGURE 36-31
Problem 9.

10P. In Fig. 36-32, light travels from point A to point B, through two regions having indices of refraction n_1 and n_2. Show that the path that requires the least travel time from A to B is the path for which θ_1 and θ_2 in the figure satisfy Eq. 36-6.

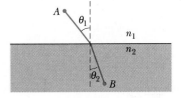

FIGURE 36-32
Problem 10.

11P. In Fig. 36-33, two pulses of light are sent through layers of plastic with the indices of refraction indicated and with thicknesses of either L or $2L$ as shown. (a) Which pulse travels through the plastic in less time? (b) In terms of L/c, what is the difference in the traversal times of the pulses?

	$\leftarrow L \rightarrow$	$\leftarrow L \rightarrow$	$\leftarrow L \rightarrow$	$\leftarrow L \rightarrow$
Pulse 2	1.55	1.70	1.60	1.45
Pulse 1	1.59		1.65	1.50

FIGURE 36-33
Problem 11.

12P. In Fig. 36-3, assume two waves of light in air, of wavelength 400 nm, are initially in phase. One travels through a glass layer of index of refraction $n_1 = 1.60$ and thickness L. The other travels through an equally thick plastic layer of index of refraction $n_2 = 1.50$. (a) What is the (least) value of L if the waves are to end up with a phase difference of 5.65 rad? (b) If the waves arrive at some common point after emerging, what type of interference do they undergo?

13P. Suppose the two waves in Fig. 36-3 have wavelength 500 nm in air. In wavelengths, what is their phase difference after traversing media 1 and 2 if (a) $n_1 = 1.50$, $n_2 = 1.60$, and $L = 8.50$ μm; (b) $n_1 = 1.62$, $n_2 = 1.72$, and $L = 8.50$ μm; and (c) $n_1 = 1.59$, $n_2 = 1.79$, and $L = 3.25$ μm? (d) Suppose that in each of these three situations the waves arrive at a common point after emerging. Rank the situations according to the brightness the waves produce at the common point.

14P. In Fig. 36-3, assume the two light waves, of wavelength 620 nm in air, are initially out of phase by π rad. The indices of refraction of the media are $n_1 = 1.45$ and $n_2 = 1.65$. (a) What is the least thickness L that will put the waves exactly in phase once they pass through the two media? (b) What is the next greater L that will do this?

15P. Two waves of light in air, of wavelength 600.0 nm, are initially in phase. They then travel through plastic layers as shown in Fig. 36-34, with $L_1 = 4.00$ μm, $L_2 = 3.50$ μm, $n_1 = 1.40$, and $n_2 = 1.60$. (a) In wavelengths, what is their phase difference after they both have emerged from the layers? (b) If the waves later arrive at some common point, what type of interference do they undergo?

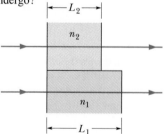

FIGURE 36-34 Problem 15.

SECTION 36-4 Young's Interference Experiment

16E. Monochromatic green light, of wavelength 550 nm, illuminates two parallel narrow slits 7.70 μm apart. Calculate the angular deviation (θ in Fig. 36-8) of the third-order (for $m = 3$) bright fringe (a) in radians and (b) in degrees.

17E. What is the phase difference between the waves from the two slits arriving at the mth dark fringe in a Young's double-slit experiment?

18E. If the slit separation d in Young's experiment is doubled, how must the distance D of the viewing screen be changed to maintain the same fringe spacing?

19E. Suppose Young's experiment is performed with blue-green light of wavelength 500 nm. The slits are 1.20 mm apart, and the viewing screen is 5.40 m from the slits. How far apart are the bright fringes?

20E. Find the slit separation of a double-slit arrangement that will produce interference fringes 0.018 rad apart on a distant screen. Assume sodium light ($\lambda = 589$ nm).

21E. A double-slit arrangement produces interference fringes for sodium light ($\lambda = 589$ nm) that have an angular separation of 3.50×10^{-3} rad. For what wavelength would the angular separation be 10.0% greater?

22E. In a double-slit arrangement the slits are separated by a distance equal to 100 times the wavelength of the light passing through the slits. (a) What is the angular separation in radians between the central maximum and an adjacent maximum? (b) What is the distance between these maxima on a screen 50.0 cm from the slits?

23E. In a double-slit experiment (Fig. 36-8), $\lambda = 546$ nm, $d = 0.10$ mm, and $D = 20$ cm. On a viewing screen, what is the distance between the fifth maximum and the seventh minimum from the central maximum?

24E. A double-slit arrangement produces interference fringes for sodium light ($\lambda = 589$ nm) that are 0.20° apart. What is the angular fringe separation if the entire arrangement is immersed in water ($n = 1.33$)?

25E. Two radio-frequency point sources separated by 2.0 m are radiating in phase with $\lambda = 0.50$ m. A detector moves in a circular path around the two sources in a plane containing them. Without written calculation, find how many maxima it detects.

26E. Source A and B emit long-range radio waves of wavelength 400 m, with the phase of the emission from A ahead of that from source B by 90°. The distance r_A from A to a detector is greater than the corresponding distance r_B by 100 m. What is the phase difference at the detector?

27P. In a double-slit experiment the distance between slits is 5.0 mm and the slits are 1.0 m from the screen. Two interference patterns can be seen on the screen: one due to light with wavelength 480 nm, and the other due to light with wavelength 600 nm. What is the separation on the screen between the third-order ($m = 3$) bright fringes of the two different patterns?

28P. If the distance between the first and tenth minima of a double-slit pattern is 18 mm and the slits are separated by 10.15 mm with the screen 50 cm from the slits, what is the wavelength of the light used?

29P. In Fig. 36-35, A and B are identical radiators of waves that are in phase and of the same wavelength λ. The radiators are separated by distance $d = 3.00\lambda$. Find the greatest distance from A, along the x axis, for which fully destructive interference occurs. Express this distance in wavelengths.

FIGURE 36-35 Problems 29 and 39.

30P. Laser light of wavelength 632.8 nm passes through a double-slit arrangement at the front of a lecture room, reflects off a mirror 20.0 m away at the back of the room, and then produces an interference pattern on a screen at the front of the room. The distance between adjacent bright fringes is 10.0 cm. (a) What is the slit separation? (b) What happens to the pattern when the lecturer places a thin cellophane sheet over one slit, thereby increasing by 2.50 the number of wavelengths along the path that includes the cellophane?

31P. Sodium light ($\lambda = 589$ nm) illuminates two slits separated by $d = 2.0$ mm. The slit–screen distance D is 40 mm. What percentage error is made by using Eq. 36-14 to locate the $m = 10$ bright fringe on the screen rather than using the exact path length difference?

32P. Two point sources, S_1 and S_2 in Fig. 36-36, emit waves in phase and at the same frequency. Show that all curves (such as that given) over which the phase difference for rays r_1 and r_2 is a constant are hyperbolas. (*Hint:* A constant phase difference implies a constant difference in length between r_1 and r_2.)

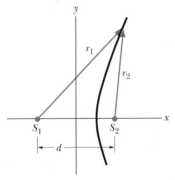

FIGURE 36-36 Problem 32.

33P. A thin flake of mica ($n = 1.58$) is used to cover one slit of a double-slit arrangement. The central point on the screen is now occupied by what had been the seventh bright side fringe ($m = 7$) before the mica was used. If $\lambda = 550$ nm, what is the thickness of the mica? (*Hint:* Consider the wavelength within the mica.)

34P. One slit of a double-slit arrangement is covered by a thin glass plate of refractive index 1.4, and the other by a thin glass plate of refractive index 1.7. The point on the screen at which the central maximum fell before the glass plates were inserted is now occupied by what had been the $m = 5$ bright fringe. Assuming that $\lambda = 480$ nm and that the plates have the same thickness t, find t.

SECTION 36-6 Intensity in Double-Slit Interference

35E. Find the sum y of the following quantities:

$$y_1 = 10 \sin \omega t \quad \text{and} \quad y_2 = 8.0 \sin(\omega t + 30°).$$

36E. Two waves of the same frequency have amplitudes 1.00 and 2.00. They interfere at a point where their phase difference is 60.0°. What is the resultant amplitude?

37E. Light of wavelength 600 nm is incident normally on two parallel narrow slits separated by 0.60 mm. Sketch the intensity pattern observed on a distant screen as a function of angle θ for the range of values $0 \le \theta \le 0.0040$ rad.

38E. Add the following quantities using the phasor method:

$$y_1 = 10 \sin \omega t$$
$$y_2 = 15 \sin(\omega t + 30°)$$
$$y_3 = 5.0 \sin(\omega t - 45°)$$

39P. A and B in Fig. 36-35 are point sources of electromagnetic waves of wavelength 1.00 m. They are in phase and separated by $d = 4.00$ m, and they emit at the same power. (a) If a detector is moved to the right along the x axis from point A, at what distances from A are the first three interference maxima detected? (b) Is the intensity of the nearest minimum exactly zero? (*Hint:* Does the intensity of a wave from a point source remain constant with an increase in distance from the source?)

40P. The double horizontal arrow in Fig. 36-9 marks the points on the intensity curve where the intensity of the central fringe is half the maximum intensity. Show that the angular separation $\Delta\theta$ between the corresponding points on the screen is

$$\Delta\theta = \frac{\lambda}{2d}$$

if θ in Fig. 36-8 is small enough so that $\sin \theta \approx \theta$.

41P*. Suppose that one of the slits of a double-slit arrangement is wider than the other, so that the amplitude of the light reaching the central part of the screen from one slit, acting alone, is twice that from the other slit, acting alone. Derive an expression for the light intensity I at the screen in terms of θ, corresponding to Eqs. 36-21 and 36-22.

SECTION 36-7 Interference from Thin Films

42E. In Fig. 36-37, light wave W_1 reflects once from a reflecting surface while light wave W_2 reflects twice from that surface and once from a reflecting sliver at distance L from the mirror. The waves are initially in phase and have a wavelength of 620 nm. Neglect the slight tilt of the rays. (a) For what least value of L are the reflected waves exactly out of phase? (b) How far must the sliver be moved to put the waves exactly out of phase again?

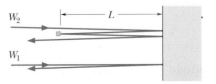

FIGURE 36-37
Exercise 42.

43E. Suppose the light waves of Exercise 42 are initially exactly out of phase. Find an expression for the values of L in terms of the wavelength λ for the situations in which the reflected waves are exactly in phase.

44E. Bright light of wavelength 585 nm is incident perpendicularly on a soap film ($n = 1.33$) of thickness 1.21 μm, suspended in air. Is the light reflected by the two surfaces of the film closer to interfering fully destructively or fully constructively?

45E. Light of wavelength 624 nm is incident perpendicularly on a soap film (with $n = 1.33$) suspended in air. What are the smallest two thicknesses of the film for which the reflections from the film undergo fully constructive interference?

46E. A lens with index of refraction greater than 1.30 is coated with a thin transparent film of index of refraction 1.30 to eliminate by interference the reflection of red light at wavelength 680 nm that is incident perpendicularly on the lens. What minimum film thickness is needed?

47E. A camera lens with index of refraction greater than 1.30 is coated with a thin transparent film of index of refraction 1.25 to eliminate by interference the reflection of light at wavelength λ that is incident perpendicularly on the lens. In terms of λ, what minimum film thickness is needed?

48E. A thin film suspended in air is 0.410 μm thick and illuminated with white light that is incident perpendicularly on its surface. The index of refraction of the film is 1.50. At what wavelengths will visible light reflected from the two surfaces of the film undergo fully constructive interference?

49E. The rhinestones in costume jewelry are glass with index of refraction 1.50. To make them more reflective, they are often coated with a layer of silicon monoxide of index of refraction 2.00. What is the minimum coating thickness needed to ensure that light of wavelength 560 nm and of perpendicular incidence will be reflected from the two surfaces of the coating with fully constructive interference?

50E. We wish to coat flat glass ($n = 1.50$) with a transparent material ($n = 1.25$) so that reflection of light at wavelength 600 nm is eliminated by interference. What minimum thickness can the coating have to do this?

51P. In Fig. 36-38, light of wavelength 600 nm is incident perpendicularly on five sections of a transparent structure suspended in air. The structure has index of refraction 1.50. The thickness of each section is given in terms of $L = 4.00$ μm. For which sections will the light that is reflected from the top and bottom surfaces of that section undergo fully constructive interference?

52P. In Fig. 36-39, light is incident perpendicularly on four thin layers of thickness L. The indices of refraction of the thin layers and of the media above and below these layers are given. Let λ represent the wavelength of the light in air, and n_2 represent the index of refraction of the thin layer in each situation. Consider only the transmission of light that undergoes no reflection or two reflections, as in Fig. 36-39a. For which of the situations does the expression

$$\lambda = \frac{2Ln_2}{m}, \qquad \text{for } m = 0, 1, 2, \ldots,$$

give the wavelengths of the transmitted light that undergoes fully constructive interference?

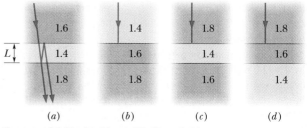

FIGURE 36-39 Problems 52, 53, and 56.

53P. A disabled tanker leaks kerosene ($n = 1.20$) into the Persian Gulf, creating a large slick on top of the water ($n = 1.30$). (a) If you are looking straight down from an airplane, while the Sun is overhead, at a region of the slick where its thickness is 460 nm, for which wavelength(s) of visible light is the reflection brightest because of constructive interference? (b) If you are scuba diving directly under this same region of the slick, for which wavelength(s) of visible light is the transmitted intensity strongest? (*Hint:* Use Fig. 36-39a with appropriate indices of refraction.)

54P. A plane wave of monochromatic light is incident normally on a uniformly thin film of oil that covers a glass plate. The wavelength of the source can be varied continuously. Fully destructive interference of the reflected light is observed for wavelengths of 500 and 700 nm and for no wavelengths in between. If the index of refraction of the oil is 1.30 and that of the glass is 1.50, find the thickness of the oil film.

55P. The reflection of perpendicularly incident white light by a soap film in air has an interference maximum at 600 nm and a minimum at 450 nm, with no minimum in between. If $n = 1.33$ for the film, what is the film thickness, assumed uniform?

56P. A sheet of glass having an index of refraction of 1.40 is to be coated with a film of material having a refractive index of 1.55 such that green light with a wavelength of 525 nm is preferentially transmitted via constructive interference. (a) What is the minimum thickness of the film that will achieve the result? (*Hint:* Use Fig. 36-39a with appropriate indices of refraction.) (b) Why are other parts of the visible spectrum not also preferentially transmitted? (c) Will the transmission of any colors be sharply reduced? If so, which colors?

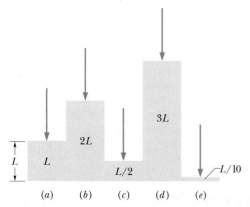

FIGURE 36-38 Problem 51.

57P. A plane monochromatic light wave in air is perpendicularly incident on a thin film of oil that covers a glass plate. The wavelength of the source may be varied continuously. Fully destructive interference of the reflected light is observed for wavelengths of 500 and 700 nm and for no wavelength in between. The index of refraction of glass is 1.50. Show that the index of refraction of the oil must be less than 1.50.

58P. A thin film of acetone ($n = 1.25$) coats a thick glass plate ($n = 1.50$). White light is incident normal to the film. In the reflections, fully destructive interference occurs at 600 nm and fully constructive interference at 700 nm. Calculate the thickness of the acetone film.

59P. Suppose that in Fig. 36-12 the light is not incident perpendicularly on the thin film but at an angle $\theta_i > 0$. Find an equation like Eqs. 36-34 and 36-35 that gives the interference maxima for the waves of rays r_1 and r_2. The wavelength is λ, the film thickness is L, and $n_2 > n_1 = n_3 = 1.0$.

60P. From a medium of index of refraction n_1, monochromatic light of wavelength λ is incident normally on a thin film of uniform thickness L (where $L > 0.1\lambda$) and index of refraction n_2. The light transmitted by the film travels into a medium with index of refraction n_3. Find expressions for the minimum film thickness (in terms of λ and the indices of refraction) for the following cases: (a) minimum light is reflected (hence maximum light is transmitted) with $n_1 < n_2 > n_3$; (b) minimum light is reflected (hence maximum light is transmitted) with $n_1 < n_2 < n_3$; and (c) maximum light is reflected (hence minimum light is transmitted) with $n_1 < n_2 < n_3$.

61P. In Sample Problem 36-5 assume that the coating eliminates the reflection of light of wavelength 550 nm at normal incidence. Calculate the factor by which reflection is diminished by the coating at 450 and 650 nm.

62P. In Fig. 36-40, a broad beam of light of wavelength 683 nm is sent directly downward through the top plate of a pair of glass plates. The plates are 120 mm long, touch at the left end, and are separated by a wire of diameter 0.048 mm at the right end. The air between the plates acts as a thin film. How many bright fringes will be seen by an observer looking down through the top plate?

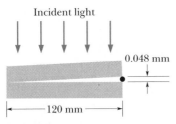

Incident light

0.048 mm

120 mm

FIGURE 36-40 Problems 62 and 63.

63P. In Fig. 36-40, white light is sent directly downward through the top plate of a pair of glass plates. The plates touch at the left end and are separated by wire (of diameter 0.048 mm) at the right end; the air between the plates acts as a thin film. An observer looking down through the top plate sees bright and dark fringes due to that film. (a) Is a dark fringe or a bright fringe seen at the left end? (b) To the right of that end, fully destructive interference occurs at different locations for different wave-

lengths of the light. Does it occur first for the red end or the blue end of the visible spectrum?

64P. In Fig. 36-41a, a broad beam of light of wavelength 600 nm is sent directly downward through a glass plate ($n = 1.5$) that, with a plastic plate ($n = 1.2$), forms a thin wedge of air which acts as a thin film. An observer looking down through the top plate sees the fringe pattern shown in Fig. 36-41b, with dark fringes centered on ends A and B. (a) What is the thickness of the wedge at B? (b) How many dark fringes will the observer see if the air between the plates is replaced with water ($n = 1.33$)?

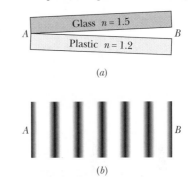

Glass $n = 1.5$

Plastic $n = 1.2$

A B

(a)

A B

(b)

FIGURE 36-41 Problem 64.

65P. A broad beam of light of wavelength 630 nm is incident at 90° on a thin, wedge-shaped film with index of refraction 1.50. An observer intercepting the light transmitted by the film sees 10 bright and 9 dark fringes along the length of the film. By how much does the film thickness change over this length?

66P. Two glass plates are held together at one end to form a wedge of air that acts as a thin film. A broad beam of light of wavelength 480 nm is directed through the plates, perpendicular to the first plate. An observer intercepting light reflected from the plates sees on the plates an interference pattern that is due to the wedge of air. How much thicker is the wedge at the sixteenth bright fringe than it is at the sixth bright fringe, counting from where the plates touch?

67P. A broad beam of monochromatic light is directed perpendicularly through two glass plates that are held together at one end, creating a wedge of air between them. An observer intercepting light reflected from the wedge of air, which acts as a thin film, sees 4001 dark fringes along the length of the wedge. When the air between the plates is evacuated, only 4000 dark fringes are seen. Calculate the index of refraction of air from these data.

68P. Figure 36-42a shows a lens with radius of curvature R lying on a plane glass plate and illuminated from above by light with wavelength λ. Figure 36-42b shows that circular interference fringes (called *Newton's rings*) appear, associated with the variable thickness d of the air film between the lens and the plate. Find the radii r of the interference maxima assuming $r/R \ll 1$.

69P. In a Newton's rings experiment (see Problem 68), the radius of curvature R of the lens is 5.0 m and its diameter is 20 mm. (a) How many bright rings are produced? Assume that $\lambda = 589$ nm. (b) How many bright rings would be produced if the arrangement were immersed in water ($n = 1.33$)?

70P. A Newton's rings apparatus is to be used to determine the radius of curvature of a lens (see Fig. 36-42 and Problem 68). The radii of the nth and $(n + 20)$th bright rings are measured and found to be 0.162 and 0.368 cm, respectively, in light of wavelength 546 nm. Calculate the radius of curvature of the lower surface of the lens.

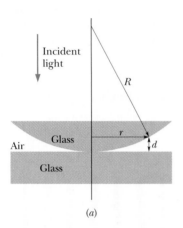

(a)

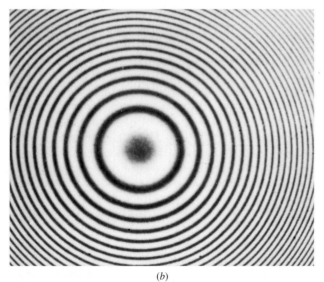

(b)

FIGURE 36-42 Problems 68 through 71.

71P. (a) Use the result of Problem 68 to show that, in a Newton's rings experiment, the difference in radius between adjacent bright rings (maxima) is given by

$$\Delta r = r_{m+1} - r_m \approx \tfrac{1}{2}\sqrt{\lambda R/m},$$

assuming $m \gg 1$. (b) Now show that the *area* between adjacent bright rings is given by

$$A = \pi\lambda R,$$

assuming $m \gg 1$. Note that this area is independent of m.

72P. In Fig. 36-43, a microwave transmitter at height a above the water level of a wide lake transmits microwaves of wavelength λ toward a receiver on the opposite shore, a distance x above the water level. The microwaves reflecting from the water

interfere with the microwaves arriving directly from the transmitter. Assuming that the lake width D is much larger than a and x, and that $\lambda \geq a$, at what values of x is the signal at the receiver maximum? (*Hint:* Does the reflection cause a phase change?)

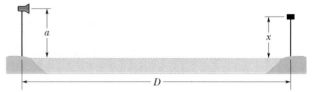

FIGURE 36-43 Problem 72.

SECTION 36-8 Michelson's Interferometer

73E. If mirror M_2 in a Michelson interferometer is moved through 0.233 mm, a shift of 792 fringes occurs. What is the wavelength of the light producing the fringe pattern?

74E. A thin film with index of refraction $n = 1.40$ is placed in one arm of a Michelson interferometer, perpendicular to the optical path. If this causes a shift of 7.0 fringes of the pattern produced by light of wavelength 589 nm, what is the film thickness?

75P. An airtight chamber 5.0 cm long with glass windows is placed in one arm of a Michelson interferometer as indicated in Fig. 36-44. Light of wavelength $\lambda = 500$ nm is used. When the air has been completely evacuated from the chamber, there has been a shift of 60 fringes. From these data, find the index of refraction of air at atmospheric pressure.

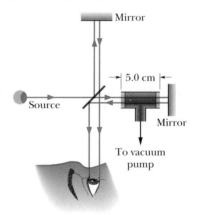

FIGURE 36-44 Problem 75.

76P. The element sodium can emit light at two wavelengths, $\lambda_1 = 589.10$ nm and $\lambda_2 = 589.59$ nm. If light from sodium is used in a Michelson interferometer, through what distance must one mirror be moved to cause the fringe pattern for one wavelength to shift 1.00 fringe more than the pattern for the other wavelength?

77P. Write an expression for the intensity observed in a Michelson interferometer (Fig. 36-18) as a function of the position of the movable mirror. Measure the position of the mirror from the point at which $d_1 = d_2$.

78P. By the late 1800s, most scientists believed that light (any electromagnetic wave) required a medium in which to travel, that it could not travel through vacuum. One reason for this belief was that any other type of wave known to the scientists requires a medium. For example, sound waves can travel through air, water, or ground but not through vacuum. Thus, reasoned the scientists, when light travels from the Sun or any other star to Earth, it cannot be traveling through vacuum; instead, it must be traveling through a medium that fills all of space and through which Earth slips. Presumably, light has a certain speed c through this medium, which was called *aether* (or *ether*).

In 1887 Michelson and Edward Morely used a version of Michelson's interferometer to test for the effects of aether on the travel of light within the device. Specifically, the motion of the device through aether as Earth moves around the Sun should affect the interference pattern produced by the device. Scientists assumed that the Sun is approximately stationary in aether; hence the speed of the interferometer through aether should be Earth's speed v about the Sun.

Figure 36-45a shows the basic arrangement of mirrors in the 1887 experiment. The mirrors were mounted on a heavy slab that was suspended on a pool of mercury so that the slab could be rotated smoothly about a vertical axis. Michelson and Morely wanted to monitor the interference pattern as they rotated the slab, changing the orientation of the interferometer arms relative to the motion through aether. A fringe shift in the interference pattern during the rotation would clearly signal the presence of aether.

Figure 36-45b, an overhead view of the equipment, shows the path of the light. To improve the possibility of fringe shift, the light was reflected several times along the arms of the interferometer, instead of only once along each arm as indicated in the basic interferometer of Fig. 36-18. This repeated reflection increased the effective length of each arm to about 10 m. In spite of the added complexity, the interferometer of Figs. 36-45a and b functions just like the simpler interferometer of Fig. 36-18; so we can use Fig. 36-18 in our discussion here by merely taking the arm lengths d_1 and d_2 to be 10 m each.

Let us assume that there is aether through which light has speed c. Figure 36-45c shows a side view of the arm of length d_1 from the aether reference frame as the interferometer moves rightward through it with velocity **v**. (For simplicity, the beam splitter M of Fig. 36-18 is drawn parallel to the mirror M_1 at the far end of the arm.) Figure 36-45d shows the arm just as a particular portion of the light (represented by a dot) begins its travel along the arm. We shall follow this light to find the path length along the arm.

As the light moves at speed c rightward through aether and toward mirror M_1, that mirror moves rightward at speed v. Figure 36-45e shows the positions of M and M_1 when the light reaches M_1, reflecting there. The light now moves leftward through aether at speed c while M moves rightward. Figure 36-45f shows the positions of M and M_1 when the light has returned to M. (a) Show that the total time of travel for this light, from M to M_1 and then back to M, is

$$t_1 = \frac{2cd_1}{c^2 - v^2}$$

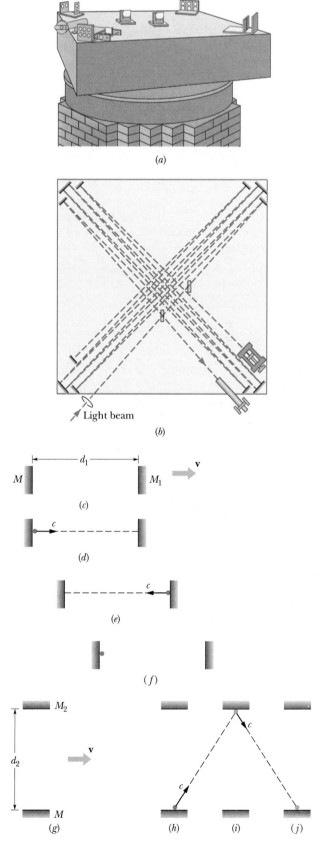

FIGURE 36-45 Problem 78.

and thus that the path length L_1 traveled by the light along this arm is

$$L_1 = ct_1 = \frac{2c^2 d_1}{c^2 - v^2}.$$

Figure 36-45g shows a view of the arm of length d_2; that arm also moves rightward with velocity $\mathbf{v}$ through the aether. For simplicity, the beam splitter M of Fig. 36-18 is now drawn parallel to the mirror M_2 at the far end of this arm. Figure 36-45h shows the arm just as a particular portion of the light (the dot) begins its travel along the arm. Because the arm moves rightward during the flight of the light, the path of the light is angled rightward toward the position that M_2 will have when the light reaches that mirror (Fig. 36-45i). The reflection of the light from M_2 sends the light angled rightward toward the position that M will have when the light returns to it (Fig. 36-45j). (b) Show that the total time of travel for the light, from M to M_2 and then back to M, is

$$t_2 = \frac{2d_2}{\sqrt{c^2 - v^2}}$$

and thus that the path length L_2 traveled by the light along this arm is

$$L_2 = ct_2 = \frac{2cd_2}{\sqrt{c^2 - v^2}}.$$

Substitute d for d_1 and d_2 in the expressions for L_1 and L_2. Then expand the two expressions by using the binomial expansion (given in Appendix E and explained in the Problem Solving Tactic on page 145); retain the first two terms in each expansion.

(c) Show that path length L_1 is greater than path length L_2 and that their difference ΔL is

$$\Delta L = \frac{dv^2}{c^2}.$$

(d) Next show that the phase difference between the light traveling along L_1 and that along L_2 is

$$\frac{\Delta L}{\lambda} = \frac{dv^2}{\lambda c^2},$$

where λ is the wavelength of the light. This phase difference determines the fringe pattern produced by the light arriving at the telescope in the interferometer.

Now rotate the interferometer by $90°$ so that the arm of length d_2 is along the direction of motion through the aether and the arm of length d_1 is perpendicular to that direction. (e) Show that the shift in the fringe pattern due to the rotation is

$$\text{shift} = \frac{2dv^2}{\lambda c^2}.$$

(f) Evaluate the shift, setting $c = 3.0 \times 10^8$ m/s, $d = 10$ m, and $\lambda = 500$ nm and using data about Earth given in Appendix C.

This expected fringe shift would have been easily observable. However, Michelson and Morely observed no fringe shift, which cast grave doubt on the existence of aether. In fact, the idea of aether soon disappeared. Moreover, the null result of Michelson and Morely led, at least indirectly, to Einstein's special theory of relativity.

Georges Seurat painted <u>Sunday Afternoon on the Island of La Grande Jatte</u> using not brush strokes in the usual sense, but rather a myriad of small colored dots, in a style of painting now known as pointillism. You can see the dots if you stand close enough to the painting, but as you move away from it, they eventually blend and cannot be distinguished. Moreover, the color that you see at any given place on the painting changes as you move away—which is why Seurat painted with the dots. What causes this change in color?

37-1 DIFFRACTION AND THE WAVE THEORY OF LIGHT

In Chapter 36 we defined diffraction rather loosely as the flaring of light as it emerges from a narrow slit. More than just flaring occurs, however, because the light produces an interference pattern called a **diffraction pattern.** For example, when monochromatic light from a distant source (or a laser) passes through a narrow slit and is then intercepted by a viewing screen, the light produces on the screen a diffraction pattern like that in Fig. 37-1. This pattern consists of a broad and intense (very bright) central maximum and a number of narrower and less intense maxima (called **secondary** or **side** maxima) to both sides. In between the maxima are minima.

Such a pattern would be totally unexpected in geometrical optics: if light traveled in straight lines as rays, then the slit would merely allow some of those rays through and they would form a sharp, bright rendition of the slit on the viewing screen. As in Chapter 36, we again must conclude that geometrical optics is only an approximation.

Diffraction of light is not limited to situations of light passing through a narrow opening (such as a slit or pinhole). It also occurs when light passes an edge, such as the edges of the razor blade in Fig. 37-2. Note the lines of maxima and minima that run approximately parallel to the edges, both on the inside of the blade and on the outside. As the light passes, say, the vertical edge at the left, it flares left and right and undergoes interference, producing the pattern along the left edge. The right-most portion of that pattern actually lies within what would have been the shadow of the blade if geometrical optics prevailed.

You encounter a common example of diffraction when you look at a clear blue sky and see tiny specks and hairlike structures floating in your view. These *floaters,* as they are called, are produced when light passes the edges of tiny bits of vitreous humor (the transparent material fill-

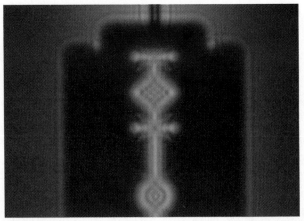

FIGURE 37-2 The diffraction pattern of a razor blade in monochromatic light. Note the lines of alternating maximum and minimum intensity.

ing most of the eyeball). These bits have broken off from the main section and now float in a water layer just in front of the retina where light is detected. What you are seeing when a floater is in your field of vision is the diffraction pattern produced by one of these floating bits. If you sight through a pinhole in an otherwise opaque sheet so as to make the light entering your eye approximately a plane wave, you might be able to distinguish individual maxima and minima in the patterns.

The Fresnel Bright Spot

Diffraction finds a ready explanation in the wave theory of light. However, this theory, originally advanced by Huygens and used 123 years later by Young to explain double-slit interference, was very slow in being adopted, largely because it ran counter to Newton's theory that light was a stream of particles.

Newton's view was the prevailing view in French scientific circles of the early nineteenth century, when Augustin Fresnel was a young military engineer. Fresnel, who believed in the wave theory of light, submitted a paper to the French Academy of Sciences describing his experiments and his wave-theory explanations of them.

In 1819, the Academy, dominated by the supporters of Newton and thinking to challenge the wave point of view, organized a prize competition for an essay on the subject of diffraction. Fresnel won. The Newtonians, however, were neither converted nor silenced. One of them, S. D. Poisson, pointed out the "strange result" that if Fresnel's theories were correct, then light waves should flare into the shadow region of a sphere as they pass the edge of the sphere, producing a bright spot at the center of the shadow. The prize committee arranged a test of the famous mathemati-

FIGURE 37-1 This diffraction pattern appeared on a viewing screen when light that had passed through a narrow horizontal slit reached the screen. The diffraction process causes light to flare out perpendicular to the long sides of the slit. The process also produces an interference pattern consisting of a broad central maximum, less intense and narrower secondary (or side) maxima, and minima.

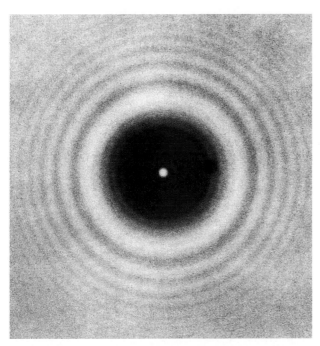

FIGURE 37-3 The diffraction pattern of a disk. Note the concentric diffraction rings and the Fresnel bright spot at the center of the pattern. This experiment is essentially identical to that arranged by the committee testing Fresnel's theories, because both the sphere they used and the disk used here have a cross section with a circular edge.

cian's prediction and discovered (see Fig. 37-3) that the predicted *Fresnel bright spot,* as we call it today, was indeed there! Nothing builds confidence in a theory so much as having one of its unexpected and counterintuitive predictions verified by experiment.

37-2 DIFFRACTION BY A SINGLE SLIT: LOCATING THE MINIMA

Let us now consider how plane waves of light of wavelength λ are diffracted by a single long narrow slit of width a in an otherwise opaque screen B, as shown in cross section in Fig. 37-4a. (In that figure, the slit's length extends into and out of the page.) When the diffracted light reaches viewing screen C, waves from different points within the slit undergo interference and produce a diffraction pattern of bright and dark fringes (interference maxima and minima) on the screen. To locate the fringes, we shall use a procedure somewhat similar to the one we used to locate the fringes in a two-slit interference pattern. However, diffraction is more mathematically challenging, and here we shall be able to find equations for only the dark fringes.

Before we do that, however, we can justify the central

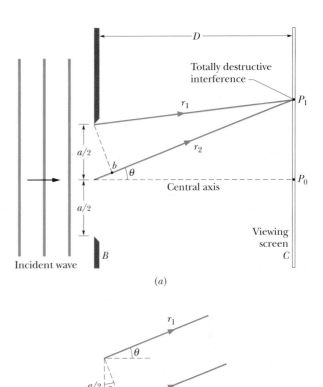

FIGURE 37-4 (*a*) Waves from the top points of two zones of width $a/2$ undergo totally destructive interference at point P_1 on viewing screen C. (*b*) For $D \gg a$, we can approximate rays r_1 and r_2 as being parallel, at angle θ to the central axis.

bright fringe seen in Fig. 37-1 by noting that the waves from all points in the slit travel about the same distance to reach the center of the pattern and thus are in phase there. As for the other bright fringes, we can say only that they are approximately halfway between adjacent dark fringes.

To find the dark fringes, we shall use a clever (and simplifying) strategy that involves pairing up all the rays coming through the slit and then finding what conditions cause the waves of the rays in each pair to cancel each other. Figure 37-4a shows how we apply this strategy to locate the first dark fringe, at point P_1. First, we mentally divide the slit into two *zones* of equal widths $a/2$. Then we extend to P_1 a light ray r_1 from the top point of the top zone and a light ray r_2 from the top point of the bottom zone. A central axis is drawn from the center of the slit to screen C, and P_1 is located at an angle θ to that axis.

The waves of the pair of rays r_1 and r_2 are in phase within the slit because they originate from the same wavefront passing through the slit. However, to produce the first dark fringe they must be out of phase by $\lambda/2$ when they reach P_1; this phase difference is due to their path length difference, with the wave of r_2 traveling a longer path to reach P_1 than the wave of r_1. To display this path length difference, we find a point b on ray r_2 such that the path length from b to P_1 matches the path length of ray r_1. Then the path length difference between the two rays is the distance from the center of the slit to b.

When viewing screen C is near screen B, as in Fig. 37-4a, the diffraction pattern on C is difficult to describe mathematically. However, we can simplify the mathematics considerably if we arrange for the screen separation D to be much larger than the slit width a. Then we can approximate rays r_1 and r_2 as being parallel, at angle θ to the central axis (Fig. 37-4b). We can also approximate the triangle formed by point b, the top point of the slit, and the center of the slit as being a right triangle, and one of the angles inside that triangle as being θ. The path length difference between rays r_1 and r_2 (which is still the distance from the center of the slit to point b) is then equal to $(a/2) \sin \theta$.

We can repeat this analysis for any other pair of rays originating at corresponding points in the two zones (say, at the midpoints of the zones) and extending to point P_1. Each such pair of rays has the same path length difference $(a/2) \sin \theta$. Setting this common path length difference equal to $\lambda/2$, we have

$$\frac{a}{2} \sin \theta = \frac{\lambda}{2},$$

which gives us

$$a \sin \theta = \lambda \quad \text{(first minimum)}. \quad (37\text{-}1)$$

Given slit width a and wavelength λ, Eq. 37-1 tells us the angle θ of the first dark fringe above and (by symmetry) below the central axis.

Note that if we begin with $a > \lambda$ and then narrow the slit while holding the wavelength constant, the angle for the first dark fringe increases; that is, the extent of the diffraction (the extent of the flaring) is *greater* for a *narrower* slit. For $a = \lambda$, the angle of the first dark fringes is 90°. Since these dark fringes mark the two edges of the central bright fringe, that bright fringe must cover the entire viewing screen.

We find the second dark fringes above and below the central axis as we found the first dark fringes, except that we now divide the slit into *four* zones of equal widths $a/4$, as shown in Fig. 37-5a. We then extend rays r_1, r_2, r_3, and r_4 from the top points of the zones to point P_2, the location of the second dark fringe above the central axis. To produce that fringe, the path length difference between r_1 and r_2, that between r_2 and r_3, and that between r_3 and r_4 must each be equal to $\lambda/2$.

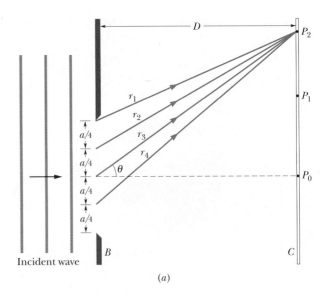

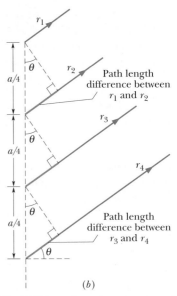

FIGURE 37-5 (a) Waves from the top points of four zones of width $a/4$ undergo totally destructive interference at point P_2. (b) For $D \gg a$, we can approximate rays r_1, r_2, r_3, and r_4 as being parallel, at angle θ to the central axis.

For $D \gg a$, we can approximate these four rays as being parallel, at angle θ to the central axis. To display their path length differences, we extend a perpendicular line through each adjacent pair of rays, as shown in Fig. 37-5b, to form a series of right triangles, each of which has a path length difference as one side. We see from the top triangle that the path length difference between r_1 and r_2 is $(a/4) \sin \theta$. Similarly, from the bottom triangle, the path length difference between r_3 and r_4 is also $(a/4) \sin \theta$. In fact, the path length difference between the members of any pair of rays that originate at corresponding points in two adjacent zones is $(a/4) \sin \theta$. Since in each such case the path length difference is equal to $\lambda/2$, we have

$$\frac{a}{4} \sin \theta = \frac{\lambda}{2},$$

which gives us

$$a \sin \theta = 2\lambda \quad \text{(second minimum).} \quad (37\text{-}2)$$

We could now continue to locate dark fringes in the diffraction pattern by splitting up the slit into more zones of equal width. We would always choose an even number of zones so that the zones (and their waves) could be paired as we have been doing. We would find that the dark fringes can be located with the following general equation:

$$a \sin \theta = m\lambda, \quad \text{for } m = 1, 2, 3, \ldots$$
$$\text{(minima—dark fringes).} \quad (37\text{-}3)$$

You can remember this result in the following way. Draw a triangle like the one in Fig. 37-4b, but for the full slit width a, and note that the path length difference between the top and bottom rays from the slit equals $a \sin \theta$. So, Equation 37-3 says:

In a single-slit diffraction experiment, dark fringes are produced where the path length differences ($a \sin \theta$) between the top and bottom rays are equal to λ, 2λ, 3λ, ...

This may seem to be wrong, because the waves of those two particular rays will be exactly in phase with each other. However, they each will still be part of a pair of waves that are exactly out of phase with each other; thus, *each* will be canceled by some other wave.

Equations 37-1, 37-2, and 37-3 are derived for the case of $D \gg a$. However, they also apply if we place a converging lens between the slit and the viewing screen and then move the screen in so that it coincides with the focal plane of the lens. The rays that now arrive at any point on the screen are *exactly* parallel (rather than approximately) when they leave the slit—they are like the initially parallel rays of Fig. 35-13a that are directed to a point by a lens.

CHECKPOINT 1: We produce a diffraction pattern on a viewing screen by using a long narrow slit illuminated by blue light. Does the pattern expand away from the bright center or contract toward it if we (a) switch to yellow light or (b) decrease the slit width?

SAMPLE PROBLEM 37-1

A slit of width a is illuminated by white light.

(a) For what value of a will the first minimum for red light of $\lambda = 650$ nm be at $\theta = 15°$?

SOLUTION: At the first minimum, $m = 1$ in Eq. 37-3. Solving for a, we then find

$$a = \frac{m\lambda}{\sin \theta} = \frac{(1)(650 \text{ nm})}{\sin 15°}$$
$$= 2511 \text{ nm} \approx 2.5 \text{ } \mu\text{m.} \quad \text{(Answer)}$$

For the incident light to flare out that much ($\pm 15°$) the slit has to be very fine indeed, amounting to about four times the wavelength. Note that a fine human hair may be about $100 \text{ } \mu\text{m}$ in diameter.

(b) What is the wavelength λ' of the light whose first side diffraction maximum is at 15°, thus coinciding with the first minimum for the red light?

SOLUTION: This maximum is about halfway between the first and second minima produced with wavelength λ'. We can find it without too much error by putting $m = 1.5$ in Eq. 37-3, obtaining

$$a \sin \theta = 1.5\lambda'.$$

Solving for λ' and substituting known data yield

$$\lambda' = \frac{a \sin \theta}{1.5} = \frac{(2511 \text{ nm})(\sin 15°)}{1.5}$$
$$= 430 \text{ nm.} \quad \text{(Answer)}$$

Light of this wavelength is violet. The first side maximum for light of wavelength 430 nm will always coincide with the first minimum for light of wavelength 650 nm, no matter what the slit width. If the slit is relatively narrow, the angle θ at which this overlap occurs will be relatively large, and conversely.

37-3 INTENSITY IN SINGLE-SLIT DIFFRACTION, QUALITATIVELY

In Section 37-2 we saw how to find the positions of the maxima and the minima in a single-slit diffraction pattern. Now we turn to a more general problem: find an expression for the intensity I of the pattern as a function of θ, the angular position of a point on a viewing screen.

To do this, we divide the slit of Fig. 37-4a into N zones of equal widths Δx so small that we can assume that each zone acts as a source of Huygens' wavelets. We wish to superimpose the wavelets arriving at an arbitrary point P on the viewing screen, at angle θ to the central axis, so that we can determine the amplitude E_θ of the resultant wave at P. The intensity of the light at P is then proportional to the square of the amplitude.

To find E_θ, we need the phase relationships among the arriving wavelets. The phase difference between wavelets from adjacent zones is given by

$$\begin{pmatrix} \text{phase} \\ \text{difference} \end{pmatrix} = \left(\frac{2\pi}{\lambda} \right) \begin{pmatrix} \text{path length} \\ \text{difference} \end{pmatrix}.$$

For point P at angle θ, the path length difference between wavelets from adjacent zones is $\Delta x \sin \theta$. So the phase difference $\Delta\phi$ between wavelets from adjacent zones is

$$\Delta\phi = \left(\frac{2\pi}{\lambda} \right) (\Delta x \sin \theta). \quad (37\text{-}4)$$

We assume that the wavelets arriving at P all have the same amplitude ΔE. To find the amplitude E_θ of the resultant wave at P, we add the amplitudes ΔE via phasors. To do this, we construct a diagram of N phasors, one corresponding to the wavelet from each zone in the slit.

For point P_0 at $\theta = 0$ on the central axis of Fig. 37-4a, Eq. 37-4 tells us that the phase difference $\Delta\phi$ between the wavelets is zero. That is, the wavelets all arrive in phase. Figure 37-6a is the corresponding phasor diagram; adjacent phasors represent wavelets from adjacent zones and are arranged head to tail. Because there is zero phase difference between the wavelets, there is zero angle between each pair of adjacent phasors. The amplitude E_θ of the net wave at P_0 is the vector sum of these phasors. This arrangement of the phasors turns out to be the one that gives the greatest value for the wave amplitude E_θ. We call this value E_m; that is, E_m is the value of E_θ for $\theta = 0$.

We next consider a point P that is at a small angle θ to the central axis. Equation 37-4 now tells us that the phase difference $\Delta\phi$ between wavelets from adjacent zones is no longer zero. Figure 37-6b shows the corresponding phasor diagram; as before, the phasors are arranged head to tail, but now there is an angle $\Delta\phi$ between adjacent phasors. The amplitude E_θ at this new point is still the vector sum of the phasors, but it is smaller than that in Fig. 37-6a, which means that the intensity of the light is less at this new point P than at P_0.

If we continue to increase θ, the angle $\Delta\phi$ between adjacent phasors increases, and eventually the chain of phasors curls completely around so that the head of the last phasor reaches the tail of the first phasor (Fig. 37-6c). The amplitude E_θ is now zero, which means that the intensity of the light is also zero. We have reached the first minimum, or dark fringe, in the diffraction pattern. The first and last phasors now have a phase difference of 2π rad, which means that the path length difference between the top and bottom rays through the slit equals one wavelength. Recall that this is the condition we determined for the first diffraction minimum.

As we continue to increase θ, the angle $\Delta\phi$ between adjacent phasors increases, the chain of phasors begins to wrap back on itself, and the resulting coil begins to shrink. Amplitude E_θ now grows larger until it reaches a maximum value in the arrangement shown in Fig. 37-6d. This arrangement corresponds to the first side maximum in the diffraction pattern.

If we increase θ a bit more, the resulting shrinkage of the coil decreases E_θ, which means that the intensity also decreases. When θ is increased enough, the head of the last phasor again meets the tail of the first phasor. We have then reached the second minimum.

We could continue this qualitative method of determining the maxima and minima of the diffraction pattern but, instead, we shall now turn to a quantitative method.

$\mathbf{C}$HECKPOINT **2:** The figures represent, in smoother form (with more phasors) than Fig. 37-6, the phasor diagrams for points on opposite sides of a certain diffraction maximum. (a) Which maximum is it? (b) What is the approximate value of m (in Eq. 37-3) that corresponds to this maximum?

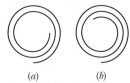

(a) (b)

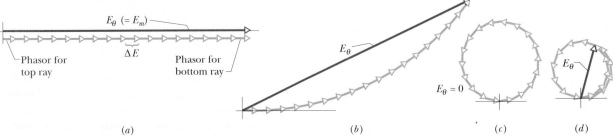

(a)

(b) (c) (d)

FIGURE 37-6 Phasor diagrams for $N = 18$ phasors, corresponding to 18 zones in a single slit. Resultant amplitudes E_θ are shown for (a) the central maximum at $\theta = 0$, (b) a point on the screen lying at a small angle θ to the central axis, (c) the first minimum, and (d) the first side maximum.

37-4 INTENSITY IN SINGLE-SLIT DIFFRACTION, QUANTITATIVELY

Equation 37-3 tells us how to locate the minima of the single-slit diffraction pattern on screen C of Fig. 37-4a as a function of the angle θ in that figure. Here we wish to derive an expression for the intensity I of the pattern as a function of θ. We state, and shall prove below, that the intensity is given by

$$I = I_m \left(\frac{\sin \alpha}{\alpha} \right)^2, \qquad (37\text{-}5)$$

where

$$\alpha = \tfrac{1}{2}\phi = \frac{\pi a}{\lambda} \sin \theta. \qquad (37\text{-}6)$$

The symbol α is just a convenient connection between the angle θ that locates a point on the viewing screen and the light intensity I at that point. I_m is the greatest value of the intensities I_θ in the pattern, and it occurs at the central maximum (where $\theta = 0$). And ϕ is the phase difference (in radians) between the top and bottom rays from the slit.

Study of Eq. 37-5 shows that intensity minima will occur when

$$\alpha = m\pi, \qquad \text{for } m = 1, 2, 3, \ldots . \qquad (37\text{-}7)$$

If we put this result into Eq. 37-6 we find

$$m\pi = \frac{\pi a}{\lambda} \sin \theta, \qquad \text{for } m = 1, 2, 3, \ldots$$

or $\quad a \sin \theta = m\lambda, \qquad \text{for } m = 1, 2, 3, \ldots$
$$\text{(minima—dark fringes)}, \qquad (37\text{-}8)$$

which is exactly Eq. 37-3, the expression that we derived earlier for the location of the minima.

Figure 37-7 shows plots of the intensity of a single-slit diffraction pattern, calculated with Eqs. 37-5 and 37-6 for three slit widths: $a = \lambda$, $a = 5\lambda$, and $a = 10\lambda$. Note that as the slit width increases (relative to the wavelength), the width of the *central diffraction maximum* (the central hill-like region of the graphs) decreases; that is, the light undergoes less flaring by the slit. The secondary maxima also decrease in width (and become weaker). In the limit of slit width a being much greater than wavelength λ, the secondary maxima due to the slit disappear; we then no longer have single-slit diffraction (but we still have diffraction due to the edges of the wide slit, like that produced by the edges of the razor blade in Fig. 37-2).

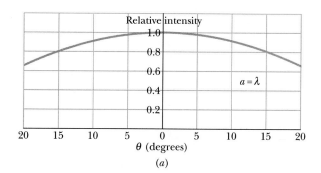

(a)

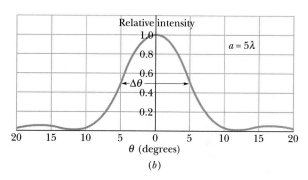

(b)

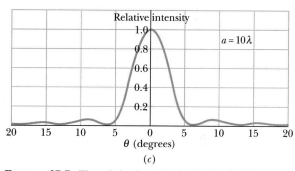

(c)

FIGURE 37-7 The relative intensity in single-slit diffraction for three values of the ratio a/λ. The wider the slit, the narrower is the central diffraction maximum.

Proof of Eqs. 37-5 and 37-6

The arc of phasors in Fig. 37-8 represents the wavelets that reach an arbitrary point P on the viewing screen of Fig. 37-4, corresponding to a particular small angle θ. The amplitude E_θ of the resultant wave at P is the vector sum of these phasors. If we divide the slit of Fig. 37-4 into infinitesimal zones of width Δx, the arc of phasors in Fig. 37-8 approaches the arc of a circle; we call its radius R as indicated in that figure. The length of the arc must be E_m, the amplitude at the center of the diffraction pattern, because if we straightened out the arc we would have the phasor arrangement of Fig. 37-6a (shown lightly in Fig. 37-8).

The angle ϕ in the lower part of Fig. 37-8 is the difference in phase between the infinitesimal vectors at the left

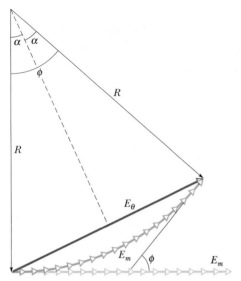

FIGURE 37-8 A construction used to calculate the intensity in single-slit diffraction. The situation shown corresponds to that of Fig. 37-6b.

and right ends of arc E_m. From the geometry, ϕ is also the angle between the two radii marked R in Fig. 37-8. The dashed line in that figure then forms two congruent triangles with angle $\frac{1}{2}\phi$. From either triangle we can write

$$\sin \frac{1}{2}\phi = \frac{E_\theta}{2R}. \qquad (37\text{-}9)$$

In radian measure, ϕ is (with E_m considered to be a circular arc)

$$\phi = \frac{E_m}{R}.$$

Solving this equation for R and substituting in Eq. 37-9 yield, after some manipulation,

$$E_\theta = \frac{E_m}{\frac{1}{2}\phi} \sin \frac{1}{2}\phi. \qquad (37\text{-}10)$$

In Section 34-4 we saw that the intensity of an electromagnetic wave is proportional to the square of the amplitude of its electric field. Here, this means that the maximum intensity I_m (at the center of the diffraction pattern) is proportional to E_m^2 and the intensity I at angle θ is proportional to E_θ^2. Thus, we may write

$$\frac{I}{I_m} = \frac{E_\theta^2}{E_m^2}. \qquad (37\text{-}11)$$

Substituting for E_θ with Eq. 37-10 and then substituting $\alpha = \frac{1}{2}\phi$, we are led to the following expression for the

intensity as a function of θ:

$$I = I_m \left(\frac{\sin \alpha}{\alpha} \right)^2.$$

This is exactly Eq. 37-5, one of the two equations we set out to prove.

The second equation we wish to prove relates α to θ. The phase difference ϕ between the rays from the top and bottom of the entire slit may be related to a path length difference with Eq. 37-4; it tells us that

$$\phi = \left(\frac{2\pi}{\lambda} \right) (a \sin \theta),$$

where a is the sum of the widths Δx of the infinitesimal strips. But $\phi = 2\alpha$, so this equation reduces to Eq. 37-6.

SAMPLE PROBLEM 37-2

Find the intensities of the first three secondary maxima (side maxima) in the single-slit diffraction pattern of Fig. 37-1, measured relative to the intensity of the central maximum.

SOLUTION: The secondary maxima lie approximately halfway between the minima, which are given by Eq. 37-7 ($\alpha = m\pi$). The secondary maxima are then given (approximately) by

$$\alpha = (m + \tfrac{1}{2})\pi, \qquad \text{for } m = 1, 2, 3, \ldots,$$

with α in radian measure. If we substitute this result into Eq. 37-5 we obtain

$$\frac{I}{I_m} = \left(\frac{\sin \alpha}{\alpha} \right)^2 = \left(\frac{\sin(m + \frac{1}{2})\pi}{(m + \frac{1}{2})\pi} \right)^2,$$

$$\text{for } m = 1, 2, 3, \ldots .$$

The first of the secondary maxima occurs for $m = 1$, its relative intensity being

$$\frac{I_1}{I_m} = \left(\frac{\sin(1 + \frac{1}{2})\pi}{(1 + \frac{1}{2})\pi} \right)^2 = \left(\frac{\sin 1.5\pi}{1.5\pi} \right)^2$$

$$= 4.50 \times 10^{-2} \approx 4.5\%. \qquad \text{(Answer)}$$

For $m = 2$ and $m = 3$ we find that

$$\frac{I_2}{I_m} = 1.6\% \quad \text{and} \quad \frac{I_3}{I_m} = 0.83\%. \qquad \text{(Answer)}$$

Successive secondary maxima decrease rapidly in intensity. The pattern of Fig. 37-1 was deliberately overexposed to reveal them.

CHECKPOINT 3: Two wavelengths, 650 and 430 nm, are used separately in a single-slit diffraction experiment. The figure shows the results as graphs of intensity I versus angle θ for the two diffraction patterns. If both wavelengths are then used simultaneously, what color will be seen in the combined diffraction pattern at (a) angle A and (c) angle B?

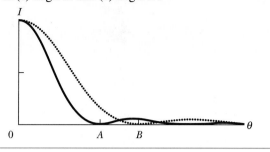

37-5 DIFFRACTION BY A CIRCULAR APERTURE

Here we consider diffraction by a circular aperture, that is, a circular opening such as a circular lens, through which light can pass. Figure 37-9 shows the image of a distant point source of light (a star, for instance) formed on photographic film placed in the focal plane of a converging lens. This image is not a point, as the geometrical optics treatment would suggest, but a circular disk surrounded by several progressively fainter secondary rings. Comparison

with Fig. 37-1 leaves little doubt that we are dealing with a diffraction phenomenon. Here, however, the aperture is a circle of diameter d rather than a rectangular slit.

The analysis of such patterns is complex. It shows, however, that the first minimum for the diffraction pattern of a circular aperture of diameter d is given by

$$\sin \theta = 1.22 \frac{\lambda}{d} \qquad \text{(first minimum; circular aperture).} \qquad (37\text{-}12)$$

Compare this with Eq. 37-1,

$$\sin \theta = \frac{\lambda}{a} \qquad \text{(first minimum; single slit),} \qquad (37\text{-}13)$$

which locates the first minimum for a long narrow slit of width a. The main difference is the factor 1.22, which enters because of the circular shape of the aperture.

Resolvability

The fact that lens images are diffraction patterns is important when we wish to *resolve* (distinguish) two distant point objects whose angular separation is small. Figure 37-10 shows the visual appearances and corresponding intensity patterns for two distant point objects (stars, say) with small angular separations. In Figure 37-10*a*, the objects are not resolved because of diffraction; that is, their diffraction patterns overlap so much that the two objects cannot be distinguished from a single point object. In Fig. 37-10*b* the objects are barely resolved, and in Fig. 37-10*c* they are fully resolved.

In Fig. 37-10*b* the angular separation of the two point sources is such that the central maximum of the diffraction

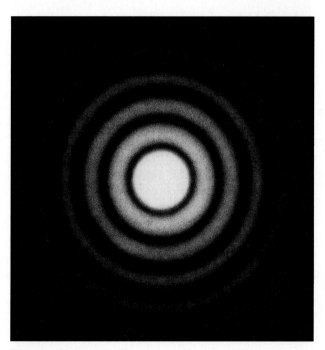

FIGURE 37-9 The diffraction pattern of a circular aperture. Note the central maximum and the circular secondary maxima. The figure has been overexposed to bring out these secondary maxima, which are much less intense than the central maximum.

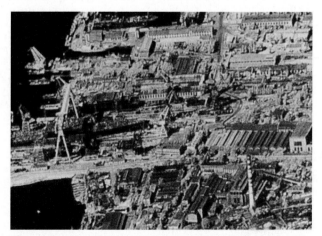

The construction of a Soviet aircraft carrier can be seen in this image made by a spy satellite and published in 1984. The image has been "cleaned" by a computer to remove diffraction effects and to improve resolution. Today, images from spy satellites can resolve much smaller details than shown here.

FIGURE 37-10 Above, the images of two point sources (stars), formed by a converging lens. Below, representations of the image intensities. In (a) the angular separation of the sources is too small for them to be distinguished; in (b) they can be marginally distinguished, and in (c) they are clearly distinguished. Rayleigh's criterion is just satisfied in (b), with the central maximum of one diffraction pattern coinciding with the first minimum of the other.

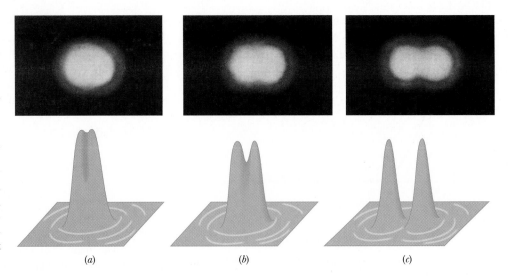

(a)   (b)   (c)

pattern of one source is centered on the first minimum of the diffraction pattern of the other, a condition called **Rayleigh's criterion** for resolvability. From Eq. 37-12, two objects that are barely resolvable by this criterion must have an angular separation θ_R of

$$\theta_R = \sin^{-1} \frac{1.22\lambda}{d}.$$

Since the angles involved are small, we can replace $\sin \theta_R$ with θ_R expressed in radians:

$$\theta_R = 1.22 \frac{\lambda}{d} \quad \text{(Rayleigh's criterion).} \quad (37\text{-}14)$$

Rayleigh's criterion for resolvability is only an approximation, because resolvability depends on many factors, such as the relative brightness of the sources and their surroundings, turbulence in the air between the sources and the observer, and the functioning of the observer's visual system. However, for the sake of calculations here, we shall take Eq. 37-14 as being a precise criterion: if the angular separation θ between the sources is greater than θ_R, we can resolve the sources; if it is less, we cannot.

When we wish to use a lens to resolve objects of small angular separation, it is desirable to make the diffraction pattern as small as possible. According to Eq. 37-14, this can be done either by increasing the lens diameter or by using light of a shorter wavelength.

For this reason ultraviolet light is often used with microscopes; because of its shorter wavelength, it permits finer detail to be examined than would be possible for the same microscope operated with visible light. In Chapter 40 of the extended version of this text, we show that beams of electrons behave like waves under some circumstances. In an *electron microscope* such beams may have an effective

wavelength that is 10^{-5} of the wavelength of visible light. They permit the detailed examination of tiny structures, like that in Fig. 37-11, that would be blurred by diffraction if viewed with an optical microscope.

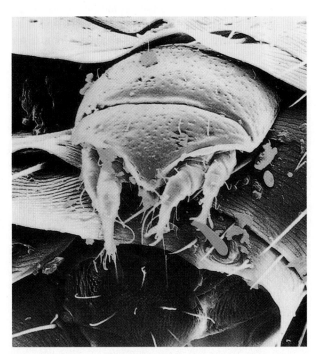

FIGURE 37-11 A false-color scanning electron micrograph of a mite that is on the back of a hedgehog flea.

SAMPLE PROBLEM 37-3

A circular converging lens, with diameter $d = 32$ mm and focal length $f = 24$ cm, forms images of distant point objects in the focal plane of the lens. Light of wavelength $\lambda = 550$ nm is used.

(a) Considering diffraction by the lens, what angular separation must two such objects have to satisfy Rayleigh's criterion for resolvability?

SOLUTION: Figure 37-12 shows two distant point objects P_1 and P_2, the lens, and a viewing screen in the focal plane of the lens. It also shows, at the right, plots of the light intensity I versus position on the screen for the central maxima of the images formed by the lens. From the perspective of the lens, the angular separation θ_o of the objects equals the angular separation θ_i of the images. So, if the images are to satisfy Rayleigh's criterion for resolvability, the angular separations on both sides of the lens must be given by Eq. 37-14 (assuming small angles). Substituting the given data, we obtain from Eq. 37-14

$$\theta_o = \theta_i = \theta_R = 1.22 \frac{\lambda}{d}$$

$$= \frac{(1.22)(550 \times 10^{-9} \text{ m})}{32 \times 10^{-3} \text{ m}} = 2.1 \times 10^{-5} \text{ rad}. \quad \text{(Answer)}$$

At this angular separation, each central maximum in the two intensity curves of Fig. 37-12 is centered on the first minimum of the other curve.

(b) What is the separation Δx of the centers of the *images* in the focal plane? (That is, what is the separation of the *central peaks* in the two curves?)

SOLUTION: From either triangle between the lens and the screen in Fig. 37-12, we see that $\tan \theta_i/2 = \Delta x/2f$. Rearranging this and making the approximation $\tan \theta \approx \theta$, we find

$$\Delta x = f\theta_i, \quad (37\text{-}15)$$

where θ_i is in radian measure. Substituting known data then yields

$$\Delta x = (0.24 \text{ m})(2.1 \times 10^{-5} \text{ rad}) = 5.0 \ \mu\text{m}. \quad \text{(Answer)}$$

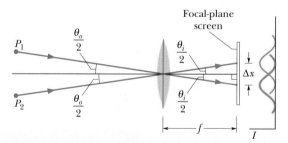

FIGURE 37-12 Sample Problem 37-3. Light from two distant point objects P_1 and P_2 passes through a converging lens and forms images on a viewing screen in the focal plane of the lens. Only one representative ray from each object is shown. The images are not points but diffraction patterns, with intensities approximately as plotted at the right. The angular separation of the objects is θ_o and that of the images is θ_i; the central maxima of the images have a separation Δx.

SAMPLE PROBLEM 37-4

Approximate the colored dots in Seurat's *Sunday Afternoon on the Island of La Grande Jatte* as closely spaced circles with center-to-center separations $D = 2.0$ mm (Fig. 37-13). If the diameter of the pupil of your eye is $d = 1.5$ mm, what is the minimum viewing distance from which you cannot distinguish any dots?

SOLUTION: Consider any two adjacent dots that you can distinguish when you are close to the painting. As you move away, you can distinguish the dots until their angular separation θ (in your view) decreases to that given by Rayleigh's criterion (Eq. 37-14):

$$\theta_R = 1.22 \frac{\lambda}{d}. \quad (37\text{-}16)$$

Because the angular separation is then small, we can approximate $\sin \theta$ as θ and then write

$$\theta = \frac{D}{L}, \quad (37\text{-}17)$$

in which L is your distance from the dots.

Setting θ of Eq. 37-17 equal to θ_R of Eq. 37-16 and solving for L, we obtain

$$L = \frac{Dd}{1.22\lambda}. \quad (37\text{-}18)$$

Equation 37-18 tells us that L is larger for smaller λ. Thus, as you move away from the painting, adjacent red dots (corresponding to a long wavelength) become indistinguishable before adjacent blue dots do. So to find the least distance L at which *no* colored dots are distinguishable, we substitute $\lambda = 400$ nm (blue or violet light) and the given data into Eq. 37-18, finding

$$L = \frac{(2.0 \times 10^{-3} \text{ m})(1.5 \times 10^{-3} \text{ m})}{(1.22)(400 \times 10^{-9} \text{ m})} = 6.1 \text{ m}. \quad \text{(Answer)}$$

At this or a greater distance, the colors of all adjacent dots blend together. The color you then perceive at any given spot on the painting is a blended color that may not actually exist there. In other words, Seurat uses the viewer's eyes to create the colors of his art.

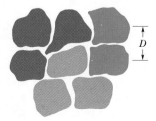

FIGURE 37-13 Sample Problem 37-4. Representation of dots on a Seurat painting.

CHECKPOINT **4:** Suppose that you can barely resolve two red dots, owing to diffraction by the pupil of your eye. If we increase the general illumination around you so that the pupil decreases in diameter, does the resolvability of the dots improve or diminish? Consider only diffraction. (You might experiment to check your answer.)

37-6 DIFFRACTION BY A DOUBLE SLIT

In the double-slit experiments of Chapter 36, we implicitly assumed that the slits were narrow compared to the wavelength of the light illuminating them; that is, $a \ll \lambda$. For such narrow slits, the central maximum of the diffraction pattern of either slit covers the entire viewing screen. Moreover, the interference of light from the two slits produces bright fringes that all have approximately the same intensity (Fig. 36-9).

In practice with visible light, however, the condition $a \ll \lambda$ is often not met. For relatively wide slits, the interference of light from two slits produces bright fringes that do not all have the same intensity. In fact, their intensity is modified by the diffraction of the light through each slit.

As an example, the intensity plot of Fig. 37-14a suggests the double-slit fringe pattern that would occur if the slits were infinitely narrow (and thus $a \ll \lambda$); all the bright interference fringes would have the same intensity. The intensity plot of Fig. 37-14b is that of a single actual slit; the diffraction pattern has a broad central maximum and weaker secondary maxima at $\pm 17°$. The plot of Fig. 37-14c suggests the resulting interference pattern for two actual slits. The plot was constructed by using the curve of Fig. 37-14b as an *envelope* on the intensity plot in Fig. 37-14a. The positions of the fringes are not changed; only the intensity is affected.

Figure 37-15a shows an actual pattern in which both double-slit interference and diffraction are evident. If one slit is covered, the single-slit diffraction pattern of Fig.

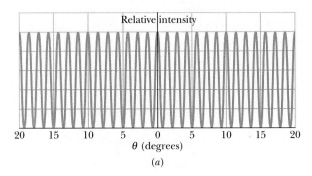

(a)

FIGURE 37-14 (*a*) The intensity pattern to be expected in a double-slit interference experiment with vanishingly narrow slits. (*b*) The intensity pattern in diffraction by a typical slit of width *a* (not vanishingly narrow). (*c*) The intensity pattern to be expected for two slits of width *a*. The curve of (*b*) acts as an envelope, limiting the intensity of the double-slit fringes in (*a*). Note that the first minima of the diffraction pattern of (*b*) eliminate the double-slit fringes that would occur near 12° in (*c*).

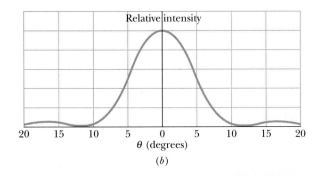

(b)

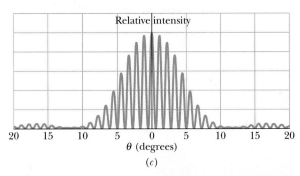

(c)

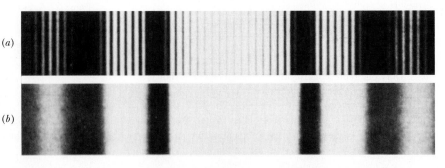

(a)

(b)

FIGURE 37-15   (*a*) Interference fringes for a double-slit system; compare with Fig. 37-14c. (*b*) The diffraction pattern of a single slit; compare with Fig. 37-14b.

37-15*b* results. Note the correspondence between Figs. 37-15*a* and 37-14*c*, and between Figs. 37-15*b* and 37-14*b*. In comparing these figures bear in mind that Fig. 37-15 has been deliberately overexposed to bring out the faint secondary maxima and that two secondary maxima (rather than one) are shown.

With diffraction effects taken into account, the intensity of a double-slit interference pattern is given by

$$I = I_m \, (\cos^2 \beta) \left(\frac{\sin \alpha}{\alpha} \right)^2 \qquad \begin{array}{c} \text{(double} \\ \text{slit),} \end{array} \quad (37\text{-}19)$$

in which

$$\beta = \left(\frac{\pi d}{\lambda} \right) \sin \theta \qquad (37\text{-}20)$$

and

$$\alpha = \left(\frac{\pi a}{\lambda} \right) \sin \theta. \qquad (37\text{-}21)$$

Here *d* is the distance between the centers of the slits, and *a* is the slit width. Note carefully that the right side of Eq. 37-19 is the product of I_m and two factors. (1) The *interference factor* $\cos^2 \beta$ is due to the interference between two slits with slit separation *d* (as given by Eqs. 36-21 and 36-22). (2) The *diffraction factor* $[(\sin \alpha)/\alpha]^2$ is due to diffraction by a single slit of width *a* (as given by Eqs. 37-5 and 37-6).

Let us check these factors. If we let $a \rightarrow 0$ in Eq. 37-21, for example, then $\alpha \rightarrow 0$ and $(\sin \alpha)/\alpha \rightarrow 1$. Equation 37-19 then reduces, as it must, to an equation describing the interference pattern for a pair of vanishingly narrow slits with slit separation *d*. Similarly, putting $d = 0$ in Eq. 37-20 is equivalent physically to causing the two slits to merge into a single slit of width *a*. Then Eq. 37-20 yields $\beta = 0$ and $\cos^2 \beta = 1$. In this case Eq. 37-19 reduces, as it must, to an equation describing the diffraction pattern for a single slit of width *a*.

The double-slit pattern described by Eq. 37-19 and displayed in Fig. 37-15*a* combines interference and diffraction in an intimate way. Both are superposition effects, in that they result from the combining of waves with different phases at a given point. If the combining waves originate from a finite (and usually small) number of elementary coherent sources—as in a double-slit experiment with $a \ll \lambda$—we call the process *interference*. If the combining waves originate in a single wavefront—as in a single-slit experiment—we call the process *diffraction*. This distinction between interference and diffraction (which is somewhat arbitrary and not always adhered to) is a convenient one, but we should not forget that both are superposition effects and usually both are present simultaneously (as in Fig. 37-15*a*).

SAMPLE PROBLEM 37-5

In a double-slit experiment, the wavelength λ of the light source is 405 nm, the slit separation *d* is 19.44 μm, and the slit width *a* is 4.050 μm.

(a) How many bright fringes are within the central peak of the diffraction envelope?

SOLUTION: The limits of the central diffraction peak are the first diffraction minima, each of which is located at the angle θ given by Eq. 37-3 with $m = 1$:

$$a \sin \theta = \lambda. \qquad (37\text{-}22)$$

The locations of the bright fringes of the double-slit interference pattern are given by Eq. 36-14:

$$d \sin \theta = m\lambda, \quad \text{for } m = 0, 1, 2, \ldots . \quad (37\text{-}23)$$

We can locate the first diffraction minimum within the double-slit fringe pattern by dividing Eq. 37-23 by Eq. 37-22 and solving for *m*. By doing so and then substituting the given data, we obtain

$$m = \frac{d}{a} = \frac{19.44 \ \mu\text{m}}{4.050 \ \mu\text{m}} = 4.8.$$

This tells us that the first diffraction minimum occurs just before the bright fringe for $m = 5$ in Eq. 37-23. So, within the central diffraction peak we have the central bright fringe ($m = 0$) and four bright fringes (up to $m = 4$) on each side of it. Thus a total of nine bright fringes of the double-slit interference pattern are within the central peak of the diffraction envelope. The bright fringes to one side of the central bright fringe are shown in Fig. 37-16.

(b) How many bright fringes are within either of the first side peaks of the diffraction envelope?

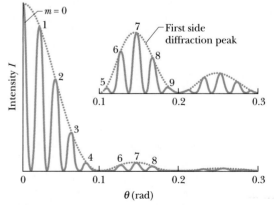

FIGURE 37-16 Sample Problem 37-5. One side of the intensity pattern for a two-slit interference experiment; the diffraction envelope is indicated by dots. The insert shows (vertically expanded) the intensity pattern for the first and second side diffraction peaks.

SOLUTION: The outer limits of the first side diffraction peaks are the second diffraction minima, each of which is at the angle θ given by Eq. 37-3 with $m = 2$:

$$a \sin \theta = 2\lambda. \qquad (37\text{-}24)$$

Dividing Eq. 37-23 by Eq. 37-24, we find

$$m = \frac{2d}{a} = \frac{(2)(19.44 \ \mu m)}{4.050 \ \mu m} = 9.6.$$

This tells us that the second diffraction minimum occurs just before the bright fringe for $m = 10$ in Eq. 37-23. So, within the first side diffraction peak we have the fringes from $m = 5$ to $m = 9$ and thus a total of five bright fringes of the double-slit interference pattern (shown in the insert of Fig. 37-16). However, if the $m = 5$ bright fringe, which is almost eliminated by the first diffraction minimum, is considered too dim to count, then only four bright fringes are in the first side diffraction peak.

$\mathbb{C}$HECKPOINT **5:** If we increase the wavelength of the light source in Sample Problem 37-5 to 550 nm, do (a) the width of the central diffraction peak and (b) the number of bright fringes within that peak increase, decrease, or remain the same?

37-7 DIFFRACTION GRATINGS

One of the most useful tools in the study of light and of objects that emit and absorb light is the **diffraction grating.** Somewhat like the double-slit arrangement of Fig. 36-8, this device has a much greater number N of slits, often called *rulings,* perhaps as many as several thousand per millimeter. An idealized grating consisting of only five slits is represented in Fig. 37-17. When monochromatic light is sent through the slits, it forms narrow interference fringes that can be analyzed to determine the wavelength of the light. (Diffraction gratings can also be opaque surfaces with narrow parallel grooves arranged like the slits in Fig. 37-17. Light then scatters back from the grooves to form interference fringes rather than being transmitted through open slits.)

With monochromatic light incident on a diffraction grating, if we gradually increase the number of slits from two to a large number N, the intensity pattern changes from the typical double-slit pattern of Fig. 37-14c to a much more complicated pattern and then eventually to a simple pattern like that shown in Fig. 37-18a. The maxima are now very narrow (and so are called *lines*); they are separated by relatively wide dark regions. What you would see on a viewing screen using monochromatic red light from, say, a helium–neon laser, is shown in Fig. 37-18b.

We use a familiar procedure to find the locations of the bright lines on the viewing screen. We first assume that the screen is far enough from the grating that the rays reaching a particular point P on the screen are approximately parallel when they leave the grating (Fig. 37-19). Then we apply to each pair of adjacent rulings the same reasoning we used for double-slit interference. The separation d between rulings is called the *grating spacing.* (If N rulings occupy a total width w, then $d = w/N$.) The path length difference between adjacent rays is again $d \sin \theta$ (Fig. 37-19), where θ is the angle from the central axis of the grating (and of the pattern) to point P. A line is located at P if the path length difference between adjacent rays is an integer number of wavelengths, that is, if

$$d \sin \theta = m\lambda, \qquad \text{for } m = 0, 1, 2, \ldots \\ \text{(maxima—lines)}, \qquad (37\text{-}25)$$

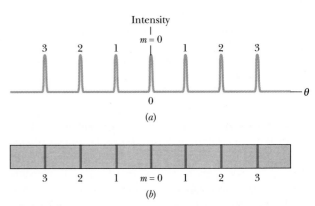

(a)

(b)

FIGURE 37-18 (a) The intensity pattern produced by a diffraction grating with a great many rulings consists of narrow peaks that are labeled with an order number m. (b) The corresponding bright fringes seen on the screen are called lines and are also labeled with m. Lines of the zeroth, first, second, and third orders are shown.

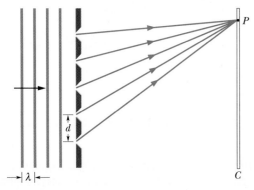

FIGURE 37-17 An idealized diffraction grating, consisting of only five rulings, that produces an interference pattern on a distant viewing screen C.

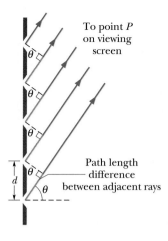

FIGURE 37-19 The rays from the rulings in a diffraction grating to a distant point P are approximately parallel. The path length difference between each two adjacent rays is $d \sin \theta$, where θ is measured as shown. (The rulings extend into and out of the page.)

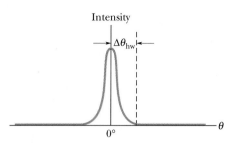

FIGURE 37-20 The half-width $\Delta \theta_{hw}$ of the central line is measured from the center of that line to the adjacent minimum on a plot of I versus θ like Fig. 37-18a.

where λ is the wavelength of the light. Each integer m represents a different line; hence these integers can be used to label the lines, as in Fig. 37-18. The integers are then called the *order numbers,* and the lines are called the zeroth-order line (the central line, with $m = 0$), the first-order line, the second-order line, and so on.

If we rewrite Eq. 37-25 as $\theta = \sin^{-1}(m\lambda/d)$ we see that, for a given diffraction grating, the angle from the central axis to any line (say, the third-order line) depends on the wavelength of the light being used. Thus, when light of an unknown wavelength is sent through a diffraction grating, measurements of the angles to the higher order lines can be used in Eq. 37-25 to determine the wavelength. Even light of several unknown wavelengths can be distinguished and identified in this way. We cannot do that with the double-slit arrangement of Section 36-4, even though the same equation and wavelength dependence apply there. In double-slit interference, the bright fringes due to different wavelengths overlap too much to be distinguished.

Width of the Lines

A grating's ability to resolve (separate) lines of different wavelengths depends on the width of the lines. We shall here derive an expression for the *half-width* of the central line (the line for which $m = 0$) and then state an expression for the half-widths of the higher order lines. We measure the half-width of the central line as the angle $\Delta \theta_{hw}$ from the center of the line at $\theta = 0$ outward to where the line effectively ends and darkness effectively begins with the first minimum (Fig. 37-20). At such a minimum, the N rays from the N slits of the grating cancel one another. (The actual width of the central line is, of course $2\Delta \theta_{hw}$, but line widths are usually compared via half-widths.)

In Section 37-2 we were also concerned with the cancellation of a great many rays, there due to diffraction through a single slit. We obtained Eq. 37-3 which, owing to the similarity of the two situations, we can use to find the first minimum here. It tells us that the first minimum

occurs where the path length difference between the top and bottom rays equals λ. For single-slit diffraction, this difference is $a \sin \theta$. For a grating of N rulings, each separated from the next by distance d, the distance between the top and bottom rulings is Nd (Fig. 37-21). So, the path length difference between the top and bottom rays here is $Nd \sin \Delta \theta_{hw}$. Thus, the first minimum occurs where

$$Nd \sin \Delta \theta_{hw} = \lambda. \qquad (37\text{-}26)$$

Because $\Delta \theta_{hw}$ is small, $\sin \Delta \theta_{hw} = \Delta \theta_{hw}$ (in radian measure). Substituting this in Eq. 37-26 gives the half-width of the central line as

$$\Delta \theta_{hw} = \frac{\lambda}{Nd} \quad \text{(half-width of central line).} \qquad (37\text{-}27)$$

We state without proof that the half-width of any other line depends on its location relative to the central axis and is

$$\Delta \theta_{hw} = \frac{\lambda}{Nd \cos \theta} \quad \text{(half-width of line at } \theta). \qquad (37\text{-}28)$$

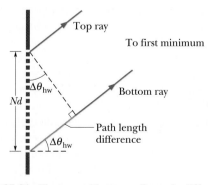

FIGURE 37-21 The top and bottom rulings of a diffraction grating of N rulings are separated by distance Nd. The top and bottom rays passing through these rulings have a path length difference of $Nd \sin \Delta \theta_{hw}$, if $\Delta \theta_{hw}$ is measured as shown. (The angle here is greatly exaggerated for clarity.)

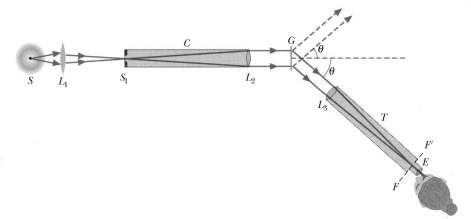

FIGURE 37-22 A simple type of grating spectroscope used to analyze the wavelengths of the light emitted by source S.

Note that for light of a given wavelength λ and a given ruling separation d, the widths of the lines decrease with an increase in the number N of rulings. Thus, of two diffraction gratings, the grating with the larger value of N is better able to distinguish between wavelengths because its diffraction lines are narrower and so produce less overlap.

An Application of Diffraction Gratings

Diffraction gratings are widely used to determine the wavelengths that are emitted by sources of light ranging from lamps to stars. Figure 37-22 shows a simple *grating spectroscope* in which a grating is used for this purpose. Light from source S is focused by lens L_1 on a slit S_1 placed in the focal plane of lens L_2. The light emerging from tube C (called a *collimator*) is a plane wave and is incident perpendicularly on grating G, where it is diffracted into a diffraction pattern, with the $m = 0$ order diffracted at angle $\theta = 0$ along the central axis of the grating.

We can view the diffraction pattern that would appear on a viewing screen at any angle θ simply by orienting telescope T in Fig. 37-22 to that angle. Lens L_3 of the telescope then focuses the light diffracted at angle θ (and at slightly smaller and larger angles) onto a focal plane FF' within the telescope. When we look through eyepiece E, we see a magnified view of this focused image.

By changing the angle θ of the telescope, we can examine the entire diffraction pattern. For any order number other than $m = 0$, the original light is spread out according to wavelength (or color) so that we can determine, with Eq. 37-25, just what wavelengths are being emitted by the source. If the source emits a broad band of wavelengths, what we see as we rotate the telescope through the angles corresponding to an order m is a broad band of color, with the shorter wavelength end at a smaller angle θ than the longer wavelength end. If the source emits discrete wavelengths, what we see are discrete vertical lines of color corresponding to those wavelengths.

For example, the light emitted by a hydrogen lamp, which contains hydrogen gas, has four discrete wavelengths in the visible range. If our eyes intercept this light directly, it appears to be white. If, instead, we view it through a grating spectroscope, we can distinguish, in several orders, the lines of the four colors corresponding to these visible wavelengths. (Such lines are called *emission lines.*) Four orders are represented in Fig. 37-23. In the central order ($m = 0$), the lines corresponding to all four wavelengths are superimposed, giving a single white line at $\theta = 0$. The colors are separated in the higher orders.

The third order is not shown in Fig. 37-23 for the sake of clarity; it actually overlaps the second and fourth orders. The fourth-order red line is missing because it is not formed by the grating used here (which is the grating of Sample Problem 37-6). That is, when we attempt to solve Eq. 37-25 for the angle θ for the red wavelength when $m = 4$, we find that $\sin \theta$ is greater than unity, which is not possible. The fourth order is then said to be *incomplete* for this grating; it might not be incomplete for a grating with greater spacing d, which will spread the lines less than in Fig. 37-23. Figure 37-24 is a photograph of the visible emission lines produced by cadmium.

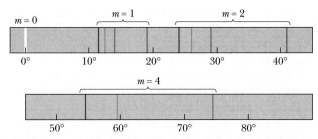

FIGURE 37-23 The zeroth, first, second, and fourth orders of the visible emission lines from hydrogen. Note that the lines are farther apart at greater angles. (They are also dimmer and wider, although that is not shown here.)

FIGURE 37-24 The visible emission lines of cadmium, as seen through a grating spectroscope.

SAMPLE PROBLEM 37-6

A diffraction grating has 1.26×10^4 rulings uniformly spaced over width $w = 25.4$ mm. It is illuminated at normal incidence by blue light of wavelength 450 nm.

(a) At what angles to the central axis do the second-order maxima occur?

SOLUTION: The grating spacing d is

$$d = \frac{w}{N} = \frac{25.4 \times 10^{-3} \text{ m}}{1.26 \times 10^4}$$
$$= 2.016 \times 10^{-6} \text{ m} = 2016 \text{ nm}.$$

The second-order maxima correspond to $m = 2$ in Eq. 37-25. For $\lambda = 450$ nm, we thus have

$$\theta = \sin^{-1} \frac{m\lambda}{d} = \sin^{-1} \frac{(2)(450 \text{ nm})}{2016 \text{ nm}}$$
$$= 26.51° \approx 26.5°. \quad \text{(Answer)}$$

(b) What is the half-width of the second-order line?

SOLUTION: From Eq. 37-28,

$$\Delta\theta_{hw} = \frac{\lambda}{Nd \cos\theta} = \frac{450 \text{ nm}}{(1.26 \times 10^4)(2016 \text{ nm})(\cos 26.51°)}$$
$$= 1.98 \times 10^{-5} \text{ rad}. \quad \text{(Answer)}$$

CHECKPOINT 6: The figure shows lines of different orders produced by a diffraction grating in monochromatic red light. (a) Is the center of the pattern to the left or right? (b) If we switch to monochromatic green light, will the half-widths of the lines then produced in the same orders be greater than, less than, or the same as the half-widths of the lines shown?

37-8 GRATINGS: DISPERSION AND RESOLVING POWER (OPTIONAL)

Dispersion

To be useful in distinguishing wavelengths that are close to each other (as in a grating spectroscope), a grating must spread apart the diffraction lines associated with the

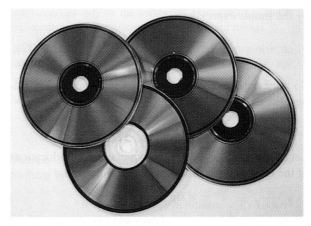

The fine rulings, each 0.5 μm wide, on a compact disc function as a diffraction grating. When a small source of white light illuminates a disc, the diffracted light forms colored "lanes" that are the composite of the diffraction patterns from the rulings.

various wavelengths. This spreading, called **dispersion,** is defined as

$$D = \frac{\Delta\theta}{\Delta\lambda} \quad \text{(dispersion defined).} \quad (37\text{-}29)$$

Here $\Delta\theta$ is the angular separation of two lines whose wavelengths differ by $\Delta\lambda$. The greater D is, the greater is the distance between two emission lines whose wavelengths differ by $\Delta\lambda$. We show below that the dispersion of a grating at angle θ is given by

$$D = \frac{m}{d \cos\theta} \quad \text{(dispersion of a grating).} \quad (37\text{-}30)$$

Thus to achieve high dispersion, we must use a grating of small grating spacing (small d) and work in high orders (large m). Note that the dispersion does not depend on the number of rulings. The SI unit for D is the degree per meter or the radian per meter.

Resolving Power

To distinguish lines whose wavelengths are close together, the line widths should also be as narrow as possible. Expressed otherwise, the grating should have a high **resolving power** R, defined as

$$R = \frac{\lambda_{av}}{\Delta\lambda} \quad \text{(resolving power defined).} \quad (37\text{-}31)$$

Here λ_{av} is the mean wavelength of two spectrum lines that can barely be recognized as separate, and $\Delta\lambda$ is the wavelength difference between them. The greater R is, the closer two emission lines can be and still be resolved. We shall show below that the resolving power of a grating is given

This is too close to the central maximum to be practical. A grating with $d \approx \lambda$ is desirable, but, since x-ray wavelengths are about equal to atomic diameters, such gratings cannot be constructed mechanically.

In 1912 it occurred to German physicist Max von Laue that a crystalline solid, which consists of a regular array of atoms, might form a natural three-dimensional "diffraction grating" for x rays. The idea is that in a crystal such as sodium chloride (NaCl) a basic unit of atoms (called the *unit cell*) repeats itself throughout the array. In NaCl four sodium ions and four chlorine ions are associated with each unit cell. Figure 37-27a represents a section through a crystal of NaCl and identifies this basic unit. The unit cell is a cube measuring a_0 on each side.

When an x-ray beam enters a crystal such as NaCl, x rays are *scattered,* that is, redirected, in all directions by the crystal structure. In some directions the scattered waves undergo destructive interference, resulting in intensity minima; in other directions the interference is constructive, resulting in intensity maxima. This process of scattering and interference is a form of diffraction, although it is unlike the diffraction of light traveling through a slit or past an edge as we discussed earlier.

Although the process of diffraction of x rays by a crystal is complicated, the maxima turn out to be in directions as *if* the x rays were reflected by a family of parallel *reflecting planes* (or *crystal planes*) that extend through the atoms within the crystal and that contain regular arrays of the atoms. (The x rays are not actually reflected; we use these fictional planes only to simplify the analysis of the actual diffraction process.)

Figure 37-27b shows three of the family of planes, with *interplanar spacing d*, from which the incident rays shown are said to reflect. Rays 1, 2, and 3 reflect from the first, second, and third planes, respectively. At each reflection the angle of incidence and the angle of reflection are represented with θ. Contrary to the custom in optics, these angles are defined relative to the *surface* of the reflecting plane rather than a normal to it. For the situation of Fig. 37-27b, the interplanar spacing happens to be equal to the unit cell dimension a_0.

Figure 37-27c shows an edge-on view of reflection from an adjacent pair of planes. The waves of rays 1 and 2 arrive at the crystal in phase. After they are reflected, they must again be in phase, because the reflections and the reflecting planes have been defined solely to explain the intensity maxima in the diffraction of x rays by a crystal. Unlike light rays, the x rays do not refract upon entering the crystal; moreover, we do not define an index of refraction for this situation. So the relative phase between the waves of rays 1 and 2 as they leave the crystal is set solely by their path length difference. For these rays to be in phase, the path length difference must be equal to an integer multiple of the wavelength λ of the x rays.

By drawing the dashed perpendiculars in Fig. 37-27c, we find that the path length difference is $2d \sin \theta$. In fact, this is true for any pair of adjacent planes in the family of planes represented in Fig. 37-27b. Thus we have, as the criterion for intensity maxima for x-ray diffraction,

$$2d \sin \theta = m\lambda, \quad \text{for } m = 1, 2, 3, \ldots$$
$$\text{(Bragg's law)}, \quad (37\text{-}34)$$

FIGURE 37-27 (a) The cubic structure of NaCl, showing the sodium and chlorine ions and a unit cell (shaded). (b) Incident x rays undergo diffraction by the structure of (a). The x rays are diffracted as if they were reflected by a family of parallel planes, with the angle of reflection equal to the angle of incidence, both angles measured relative to the planes (not relative to a normal as in optics). (c) The path length difference between waves effectively reflected by two adjacent planes is $2d \sin \theta$. (d) A different orientation of the x rays relative to the structure. A different family of parallel planes now effectively reflects the x rays.

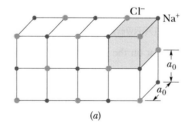

(a)

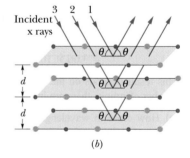

(b)

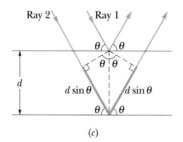

(c)

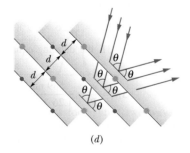

(d)

where m is the order number of an intensity maximum. Equation 37-34 is called **Bragg's law** after British physicist W. L. Bragg, who first derived it. (He and his father shared the 1915 Nobel prize for their use of x rays to study the structures of crystals.) The angle of incidence and reflection in Eq. 37-34 is called a *Bragg angle.*

Regardless of the angle at which x rays enter a crystal, there is always a family of planes from which they can be said to reflect so that we can apply Bragg's law. In Fig. 37-27d, the crystal structure has the same orientation as it does in Fig. 37-27a but the angle at which the beam enters the structure differs from that shown in Fig. 37-27b. This new angle requires a new family of planes, with a different interplanar spacing d and different Bragg angle θ, in order to explain the x-ray diffraction via Bragg's law.

Figure 37-28 shows how the interplanar spacing d can be related to the unit cell dimension a_0. For the particular family of planes shown there,

$$5d = \sqrt{5}a_0,$$

or
$$d = \frac{a_0}{\sqrt{5}}. \tag{37-35}$$

Figure 37-28 suggests how the dimensions of the unit cell can be found once the interplanar spacing has been measured by means of x-ray diffraction.

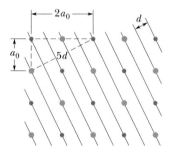

FIGURE 37-28 A family of planes through the structure of Fig. 37-27a, and a way to relate the edge length a_0 of a unit cell to the interplanar spacing d.

X-ray diffraction is a powerful tool for studying both x-ray spectra and the arrangement of atoms in crystals. To study spectra, a particular set of crystal planes, having a known spacing d, is chosen. These planes effectively reflect different wavelengths at different angles. A detector that can discriminate one angle from another can then be used to determine the wavelength of radiation reaching it. The crystal itself can be studied with a monochromatic x-ray beam, to determine not only the spacing of various crystal planes but also the structure of the unit cell.

SAMPLE PROBLEM 37-8

At what Bragg angles must x rays with $\lambda = 1.10$ Å be incident on the family of planes represented in Fig. 37-28 if effective reflections from the planes are to result in diffraction intensity maxima? Assume the material to be sodium chloride ($a_0 = 5.63$ Å).

SOLUTION: The interplanar spacing d for these planes is given by Eq. 37-35 as

$$d = \frac{a_0}{\sqrt{5}} = \frac{5.63 \text{ Å}}{\sqrt{5}} = 2.518 \text{ Å}.$$

Equation 37-34 then gives, for the Bragg angles,

$$\theta = \sin^{-1}\frac{m\lambda}{2d} = \sin^{-1}\left(\frac{(m)(1.10 \text{ Å})}{(2)(2.518 \text{ Å})}\right)$$
$$= \sin^{-1}(0.2184m).$$

Maxima are possible for $\theta = 12.6°$ ($m = 1$), $\theta = 25.9°$ ($m = 2$), $\theta = 40.9°$ ($m = 3$), and $\theta = 60.9°$ ($m = 4$). Higher order maxima cannot exist because they require that $\sin \theta$ be greater than 1.

Actually, the unit cell in cubic crystals such as NaCl has diffraction properties such that the intensity of diffracted x-ray beams corresponding to odd values of m is zero. Thus beams are expected only for $\theta = 25.9°$ and $\theta = 60.9°$.

REVIEW & SUMMARY

Diffraction

When waves encounter an edge or an obstacle or aperture with a size comparable to the wavelength of the waves, those waves spread in their direction of travel and undergo interference. This is called **diffraction.**

Single-Slit Diffraction

Waves passing through a long narrow slit of width a produce a **single-slit diffraction pattern** that includes a central maximum and other maxima, separated by minima located at angles θ to the central axis that satisfy

$$a \sin \theta = m\lambda, \quad \text{for } m = 1, 2, 3, \ldots$$
$$\text{(minima).} \tag{37-3}$$

The intensity of the diffraction pattern at any given angle θ is

$$I = I_m \left(\frac{\sin \alpha}{\alpha}\right)^2, \quad \text{where} \quad \alpha = \frac{\pi a}{\lambda} \sin \theta \tag{37-5, 37-6}$$

and I_m is the intensity at the center of the pattern.

Circular Aperture Diffraction

Diffraction by a circular aperture or a lens with diameter d produces a central maximum and concentric maxima and minima, with the first minimum at an angle θ given by

$$\sin \theta = 1.22 \frac{\lambda}{d} \qquad \text{(first minimum;}\atop\text{circular aperture).} \qquad (37\text{-}12)$$

Rayleigh's Criterion

Rayleigh's criterion suggests that two objects are on the verge of resolvability if the central diffraction maximum of one is at the first minimum of the other. Their angular separation must then be at least

$$\theta_R = 1.22 \frac{\lambda}{d} \qquad \text{(Rayleigh's criterion)}, \qquad (37\text{-}14)$$

in which d is the diameter of the aperture.

Double-Slit Diffraction

Waves passing through two slits, each of width a, whose centers are a distance d apart, display diffraction patterns whose intensity I at various diffraction angles θ is given by

$$I = I_m (\cos^2 \beta) \left(\frac{\sin \alpha}{\alpha} \right)^2 \qquad \text{(double slit)}, \qquad (37\text{-}19)$$

with $\beta = (\pi d/\lambda) \sin \theta$ and α the same as for the case of single-slit diffraction.

Multiple-Slit Diffraction

Diffraction by N (multiple) slits results in maxima (lines) at angles θ such that

$$d \sin \theta = m\lambda, \qquad \text{for } m = 0, 1, 2 \ldots$$
$$\text{(maxima)}, \qquad (37\text{-}25)$$

with the half-widths of the lines given by

$$\Delta\theta_{hw} = \frac{\lambda}{Nd \cos \theta} \qquad \text{(half-widths)}. \qquad (37\text{-}28)$$

Diffraction Gratings

A *diffraction grating* is a series of "slits" used to separate an incident wave into its component wavelengths by separating and displaying their diffraction maxima. A grating is characterized by its dispersion D and resolving power R:

$$D = \frac{\Delta\theta}{\Delta\lambda} = \frac{m}{d \cos \theta}$$

$$R = \frac{\lambda_{av}}{\Delta\lambda} = Nm. \qquad (37\text{-}29 \text{ to } 37\text{-}32)$$

X-Ray Diffraction

The regular array of atoms in a crystal is a three-dimensional diffraction grating for short-wavelength waves such as x rays. For analysis purposes, the atoms can be visualized as being arranged in planes with characteristic interplanar spacing d. Diffraction maxima (due to constructive interference) occur if the incident direction of the wave, measured from the surfaces of these planes, and the wavelength λ of the radiation satisfy **Bragg's law:**

$$2d \sin \theta = m\lambda, \qquad \text{for } m = 1, 2, 3 \ldots$$
$$\text{(Bragg's law)}. \qquad (37\text{-}34)$$

QUESTIONS

1. Light of frequency f illuminating a long narrow slit produces a diffraction pattern. (a) If we switch to light of frequency $1.3f$, does the pattern expand away from the center or contract toward it? (b) Does the pattern expand or contract if, instead, we submerge the equipment in clear corn syrup?

2. You are conducting a single-slit diffraction experiment with light of wavelength λ. What appears, on a distant viewing screen, at a point at which the top and bottom rays through the slit have a path length difference equal to (a) 5λ and (b) 4.5λ?

3. If you speak with the same intensity with and without a megaphone in front of your mouth, in which situation do you sound louder to someone directly in front of you?

4. Figure 37-29 shows four choices for the rectangular opening of a source of either sound waves or light waves. The sides have lengths of either L or $2L$, with L being 3.0 times the wavelength of the waves. Rank the openings according to the extent of (a)

left–right spreading and (b) up–down spreading of the waves due to diffraction, greatest first.

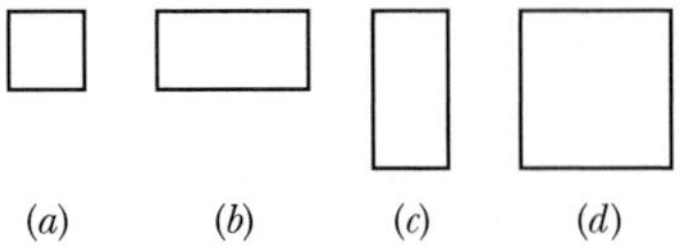

5. In a single-slit diffraction experiment, the top and bottom rays through the slit arrive at a certain point on the viewing screen with a path length difference of 4.0 wavelengths. In a phasor representation like those in Fig 37-6, how many overlapping circles does the chain of phasors make?

6. A vertical spider thread lies between you and the early morning Sun. As you move perpendicular to a line that extends through the thread and the Sun, toward that line, you begin to see the diffraction pattern of sunlight produced by the thread. As your eyes move through the first side maximum of the pattern, what color, red or blue, do you see first? (That is, in a diffraction pattern of white light, is red or blue diffracted at a greater angle?)

7. At night many people see rings (called *entoptic halos*) surrounding bright outdoor lamps in otherwise dark surroundings. The rings are the first of the side maxima in diffraction patterns produced by structures that are thought to be within the cornea (or possibly the lens) of the observer's eye. (The central maxima of such patterns overlap the lamp.) (a) Would a particular ring become smaller or larger if the lamp were switched from

FIGURE 37-29
Question 4.   *(a)*   *(b)*   *(c)*   *(d)*

blue to red light? (b) If a lamp emits white light, is blue or red on the outside edge of the ring?

8. Figure 37-30 shows the bright fringes that lie within the central diffraction envelope in two double-slit diffraction experiments using the same wavelength of light. Are (a) the slit width a (b) the slit separation d, and (c) the ratio d/a in experiment B greater than, less than, or the same as those in experiment A?

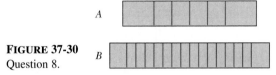

FIGURE 37-30
Question 8.

9. Figure 37-31 shows a red line and a green line of the same order in the pattern produced by a diffraction grating. If we increased the number of rulings in the grating, say, by removing tape that had covered half the rulings, would (a) the half-widths of the lines and (b) the separation of the lines increase, decrease, or remain the same? (c) Would the lines shift to the right, shift to the left, or remain in place?

FIGURE 37-31 Questions 9 and 10.

10. For the situation of Question 9 and Fig. 37-31, if instead we

increased the grating spacing, would (a) the half-widths of the lines and (b) the separation of the lines increase, decrease, or remain the same? (c) Would the lines shift to the right, shift to the left, or remain in place?

11. (a) Figure 37-32a shows the lines produced by diffraction gratings A and B using light of the same wavelength; the lines are of the same order and at the same angles θ. Which grating has the greater number of rulings? (b) Figure 37-32b shows lines of two orders produced by a single diffraction grating using light of two wavelengths, both in the red region of the spectrum. Which lines, the left pair or right pair, are in the order with greater m? Is the center of the diffraction pattern to the left or right in (c) Fig. 37-32a and (d) Fig. 37-32b?

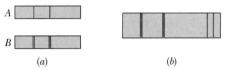

(a)   *(b)*

FIGURE 37-32 Question 11.

12. (a) For a given diffraction grating, does the least difference $\Delta\lambda$ in two wavelengths that can be resolved increase, decrease, or remain the same as the wavelength increases? (b) For a given wavelength region (say, around 500 nm), is $\Delta\lambda$ greater in the first order or in the third order?

EXERCISES & PROBLEMS

SECTION 37-2 Diffraction by a Single Slit: Locating the Minima

1E. When monochromatic light is incident on a slit 0.022 mm wide, the first diffraction minimum is observed at an angle of $1.8°$ from the direction of the incident light. What is the wavelength?

2E. Monochromatic light of wavelength 441 nm is incident on a narrow slit. On a screen 2.00 m away, the distance between the second diffraction minimum and the central maximum is 1.50 cm. (a) Calculate the angle of diffraction θ of the second minimum. (b) Find the width of the slit.

3E. Light of wavelength 633 nm is incident on a narrow slit. The angle between the first diffraction minimum on one side of the central maximum and the first minimum on the other side is $1.20°$. What is the width of the slit?

4E. A single slit is illuminated by light of wavelengths λ_a and λ_b, so chosen that the first diffraction minimum of the λ_a component coincides with the second minimum of the λ_b component. (a) What relationship exists between the two wavelengths? (b) Do any other minima in the two diffraction patterns coincide?

5E. The distance between the first and fifth minima of a single-slit diffraction pattern is 0.35 mm with the screen 40 cm away from the slit, using light of wavelength 550 nm. (a) Find the slit width. (b) Calculate the angle θ of the first diffraction minimum.

6E. What must be the ratio of the slit width to the wavelength for a single slit to have the first diffraction minimum at $\theta = 45.0°$?

7E. A plane wave of wavelength 590 nm is incident on a slit with $a = 0.40$ mm. A thin converging lens of focal length $+70$ cm is placed between the slit and a viewing screen and focuses the light on the screen. (a) How far is the screen from the lens? (b) What is the distance on the screen from the center of the diffraction pattern to the first minimum?

8P. A slit 1.00 mm wide is illuminated by light of wavelength 589 nm. We see a diffraction pattern on a screen 3.00 m away. What is the distance between the first two diffraction minima on the same side of the central diffraction maximum?

9P. Sound waves with frequency 3000 Hz and speed 343 m/s diffract through the rectangular opening of a speaker cabinet and into a large auditorium. The opening, which has a horizontal width of 30.0 cm, faces a wall 100 m away (Fig. 37-33). Where

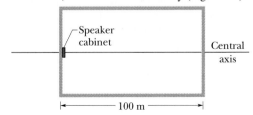

FIGURE 37-33 Problem 9.

along that wall will a listener be at the first diffraction minimum and thus have difficulty hearing the sound? (Neglect reflections from the walls.)

10P. Manufacturers of wire (and other objects of small dimensions) sometimes use a laser to continually monitor the thickness of the product. The wire intercepts the laser beam, producing a diffraction pattern like that of a single slit of the same width as the wire diameter (see Fig. 37-34). Suppose a helium–neon laser, of wavelength 632.8 nm, illuminates a wire, and the diffraction pattern appears on a screen 2.60 m away. If the desired wire diameter is 1.37 mm, what is the observed distance between the two tenth-order minima (one on each side of the central maximum)?

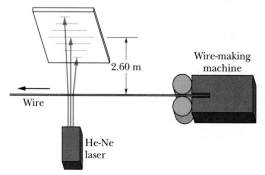

FIGURE 37-34 Problem 10.

SECTION 37-4 Intensity in Single-Slit Diffraction, Quantitatively

11E. A 0.10-mm-wide slit is illuminated by light of wavelength 589 nm. Consider rays that are diffracted at $\theta = 30°$ and calculate the phase difference at the screen of Huygens' wavelets from the top and midpoint of the slit. (*Hint:* See Eq. 37-4.)

12E. Monochromatic light with wavelength 538 nm is incident on a slit with width 0.025 mm. The distance from the slit to a screen is 3.5 m. Consider a point on the screen 1.1 cm from the central maximum. (a) Calculate θ for that point. (b) Calculate α. (c) Calculate the ratio of the intensity at this point to the intensity at the central maximum.

13P. If you double the width of a single slit, the intensity of the central maximum of the diffraction pattern increases by a factor of 4, even though the energy passing through the slit only doubles. Explain this quantitatively.

14P. *Babinet's Principle.* A monochromatic beam of parallel light is incident on a "collimating" hole of diameter $x \gg \lambda$. Point P lies in the geometrical shadow region on a *distant* screen (Fig. 37-35). Two obstacles, shown in Fig. 37-35b, are placed in turn over the collimating hole. A is an opaque circle with a hole in it and B is the "photographic negative" of A. Using superposition concepts, show that the intensity at P is identical for the two diffracting objects A and B.

15P. The full width at half-maximum (FWHM) of the central diffraction maximum is defined as the angle between the two points in the pattern where the intensity is one-half that at the center of the pattern. (See Fig. 37-7b.) (a) Show that the intensity

drops to one-half the maximum value when $\sin^2 \alpha = \alpha^2/2$. (b) Verify that $\alpha = 1.39$ radians (about 80°) is a solution to the transcendental equation of (a). (c) Show that the FWHM is $\Delta\theta = 2 \sin^{-1}(0.443\lambda/a)$. (d) Calculate the FWHM of the central maximum for slits whose widths are 1.0, 5.0, and 10 wavelengths.

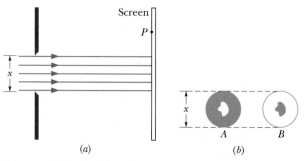

FIGURE 37-35 Problem 14.

16P. (a) Show that the values of α at which intensity maxima for single-slit diffraction occur can be found exactly by differentiating Eq. 37-5 with respect to α and equating the result to zero, obtaining the condition $\tan \alpha = \alpha$. (b) Find the values of α satisfying this relation by plotting the curve $y = \tan \alpha$ and the straight line $y = \alpha$ and finding their intersections or by using a pocket calculator to find an appropriate value of α by trial and error. (c) Find the (noninteger) values of m corresponding to successive maxima in the single-slit pattern. Note that the secondary maxima do not lie exactly halfway between minima.

17P*. Derive this expression for the intensity pattern for a three-slit "grating":

$$I = \tfrac{1}{9}I_m(1 + 4\cos\phi + 4\cos^2\phi),$$

where $\phi = (2\pi d \sin\theta)/\lambda$. Assume that $a \ll \lambda$; be guided by the derivation of the corresponding double-slit formula (Eq. 36-21).

SECTION 37-5 Diffraction by a Circular Aperture

18E. Assume that the lamp in Question 7 emits light at wavelength 550 nm. If a ring has an angular diameter of 2.5°, approximately what is the (linear) diameter of the structure in the eye that causes the ring?

19E. The two headlights of an approaching automobile are 1.4 m apart. At what (a) angular separation and (b) maximum distance will the eye resolve them? Assume that the pupil diameter is 5.0 mm, and use a wavelength of 550 nm. Also assume that diffraction effects alone limit the resolution.

20E. An astronaut in a space shuttle claims she can just barely resolve two point sources on Earth's surface, 160 km below. Calculate their (a) angular and (b) linear separation, assuming ideal conditions. Take $\lambda = 540$ nm and the pupil diameter of the astronaut's eye to be 5.0 mm.

21E. Find the separation of two points on the Moon's surface that can just be resolved by the 200 in. (= 5.1 m) telescope at Mount Palomar, assuming that this separation is determined by diffraction effects. The distance from Earth to the Moon is 3.8×10^5 km. Assume a wavelength of 550 nm.

22E. The wall of a large room is covered with acoustic tile in which small holes are drilled 5.0 mm from center to center. How far can a person be from such a tile and still distinguish the individual holes, assuming ideal conditions, the pupil diameter of the observer's eye to be 4.0 mm, and the wavelength of the room light to be 550 nm?

23E. The pupil of a person's eye has a diameter of 5.00 mm. What distance apart must two small objects be if their images are just resolved when they are 250 mm from the eye and illuminated with light of wavelength 500 nm?

24E. Under ideal conditions, estimate the linear separation of two objects on the planet Mars that can just be resolved by an observer on Earth (a) using the naked eye and (b) using the 200 in. (= 5.1 m) Mount Palomar telescope. Use the following data: distance to Mars = 8.0×10^7 km; diameter of pupil = 5.0 mm; wavelength of light = 550 nm.

25E. If Superman really had x-ray vision at 0.10 nm wavelength and a 4.0 mm pupil diameter, at what maximum altitude could he distinguish villains from heroes, assuming that he needs to resolve points separated by 5.0 cm to do this?

26E. A navy cruiser employs radar with a wavelength of 1.6 cm. The circular antenna has a diameter of 2.3 m. At a range of 6.2 km, what is the smallest distance that two speedboats can be from each other and still be resolved as two separate objects by the radar system?

27P. Nuclear-pumped x-ray lasers are seen as a possible weapon to destroy ICBM booster rockets at ranges up to 2000 km. One limitation on such a device is the spreading of the beam due to diffraction, with resulting dilution of beam intensity. Consider such a laser operating at a wavelength of 1.40 nm. The element that emits light is the end of a wire with diameter 0.200 mm. (a) Calculate the diameter of the central beam at a target 2000 km away from the beam source. (b) By what factor is the beam intensity reduced in transit to the target? (The laser is fired from space, so that atmospheric absorption can be ignored.)

28P. (a) How far from grains of red sand must you be to position yourself just at the limit of resolving the grains if your pupil diameter is 1.5 mm, the grains are spherical with radius 50 μm, and the light from the grains has wavelength 650 nm? (b) If the grains were blue and the light from them had wavelength 400 nm, would the answer to (a) be larger or smaller?

29P. The wings of tiger beetles (Fig. 37-36) are colored by interference due to thin cuticle-like layers. In addition, these layers are arranged in patches that are 60 μm across and produce different colors. The color you see is a pointillistic mixture of thin-film interference colors that varies with perspective. Approximately what viewing distance from a wing puts you at the limit of resolving the different colored patches according to Rayleigh's criterion? Use 550 nm as the wavelength of light and 3.00 mm as the diameter of your pupil.

30P. (a) A circular diaphragm 60 cm in diameter oscillates at a frequency of 25 kHz as an underwater source of sound used for submarine detection. Far from the source the sound intensity is distributed as the diffraction pattern of a circular hole whose diameter equals that of the diaphragm. Take the speed of sound in water to be 1450 m/s and find the angle between the normal to the diaphragm and the direction of the first minimum. (b) Repeat for a source having an (audible) frequency of 1.0 kHz.

FIGURE 37-36 Problem 29. Tiger beetles are colored by pointillistic mixtures of thin-film interference colors.

31P. In June 1985 a laser beam was fired from the Air Force Optical Station on Maui, Hawaii, and reflected back from the shuttle *Discovery* as it sped by, 220 mi overhead. The diameter of the central maximum of the beam at the shuttle position was said to be 30 ft, and the beam wavelength was 500 nm. What is the effective diameter of the laser aperture at the Maui ground station? (*Hint:* A laser beam spreads only because of diffraction; assume a circular exit aperture.)

32P. A spy satellite orbiting at 160 km above Earth's surface has a lens with a focal length of 3.6 m and can resolve objects on the ground as small as 30 cm; it can easily measure the size of an aircraft's air intake. What is the effective lens diameter, determined by diffraction consideration alone? Assume $\lambda = 550$ nm.

33P. Millimeter-wave radar generates a narrower beam than conventional microwave radar, making it less vulnerable to antiradar missiles. (a) Calculate the angular width of the central maximum, from first minimum to first minimum, produced by a 220 GHz radar beam emitted by a 55.0-cm-diameter circular antenna. (The frequency is chosen to coincide with a low-absorption atmospheric "window.") (b) Calculate the same quantity for the ship's radar described in Exercise 26.

34P. (a) How small is the angular separation of two stars if their images are barely resolved by the Thaw refracting telescope at the Allegheny Observatory in Pittsburgh? The lens diameter is 76 cm and its focal length is 14 m. Assume $\lambda = 550$ nm. (b) Find the distance between these barely resolved stars if each of them is 10 light-years distant from Earth. (c) For the image of a single star in this telescope, find the diameter of the first dark ring in the diffraction pattern, as measured on a photographic plate placed at the focal plane of the telescope lens. Assume that the structure of the image is associated entirely with diffraction at the lens aperture and not with lens ''errors.''

35P. A circular obstacle produces the same diffraction pattern as a circular hole of the same diameter (except very near $\theta = 0$). Airborne water drops are examples of such obstacles. When you see the Moon through suspended water drops, such as in a fog, you intercept the diffraction pattern from many drops; the composite is a bright circular pattern surrounding the Moon (Fig. 37-37). Next to the Moon, the pattern is white. (a) What color, red or blue, outlines that white pattern? (b) Suppose the outlining ring has an angular diameter that is 1.5 times the angular diameter of the Moon, which is 0.50°. Suppose also that the drops all have about the same diameter; approximately what is that diameter?

FIGURE 37-37 Problem 35. The corona around the Moon is a composite of the diffraction patterns of airborne water drops.

36P. In a joint Soviet–French experiment to monitor the Moon's surface with a light beam, pulsed radiation from a ruby laser ($\lambda = 0.69\ \mu$m) was directed to the Moon through a reflecting telescope with a mirror radius of 1.3 m. A reflector on the Moon behaved like a circular plane mirror with radius 10 cm, reflecting the light directly back toward the telescope on Earth. The reflected light was then detected after being brought to a focus by this telescope. What fraction of the original light energy was picked up by the detector? Assume that for each direction of travel all the energy is in the central diffraction peak.

SECTION 37-6 Diffraction by a Double Slit

37E. Suppose that the central diffraction envelope of a double-slit diffraction pattern contains 11 bright fringes and the first diffraction minima eliminate (are coincident with) bright fringes. How many bright fringes lie between the first and second minima of the diffraction envelope?

38E. For $d = 2a$ in Fig. 37-38, how many bright interference fringes lie in the central diffraction envelope?

FIGURE 37-38 Exercise 38 and Problem 39.

39P. If we put $d = a$ in Fig. 37-38, the two slits coalesce into a single slit of width $2a$. Show that Eq. 37-19 reduces to the diffraction pattern for such a slit.

40P. (a) In a double-slit system, what ratio of d to a causes diffraction to eliminate the fourth bright side fringe? (b) What other bright fringes are also eliminated?

41P. Two slits of width a and separation d are illuminated by a coherent beam of light of wavelength λ. What is the linear separation of the bright interference fringes observed on a screen that is at a distance D away?

42P. (a) How many fringes appear between the first diffraction-envelope minima to either side of the central maximum for a double-slit pattern if $\lambda = 550$ nm, $d = 0.150$ mm, and $a = 30.0\ \mu$m? (b) What is the ratio of the intensity of the third bright fringe to the intensity of the central fringe?

43P. Light of wavelength 440 nm passes through a double slit, yielding a diffraction pattern whose graph of intensity I versus deflection angle θ is shown in Fig. 37-39. Calculate (a) the slit width and (b) the slit separation. (c) Verify the displayed intensities of the $m = 1$ and $m = 2$ interference fringes.

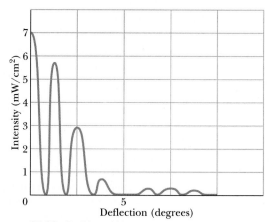

FIGURE 37-39 Problem 43.

44P. An acoustic double-slit system (of slit separation d and slit width a) is driven by two loudspeakers as shown in Fig. 37-40. By use of a variable delay line, the phase of one of the speakers may be varied. Describe in detail what changes occur in the double-slit diffraction pattern at large distances as the phase difference between the speakers is varied from zero to 2π. Take both interference and diffraction effects into account.

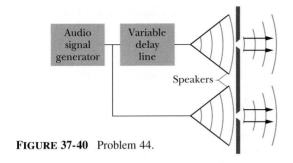

FIGURE 37-40   Problem 44.

SECTION 37-7   Diffraction Gratings

45E. A diffraction grating 20.0 mm wide has 6000 rulings. (a) Calculate the distance d between adjacent rulings. (b) At what angles will intensity maxima occur if the incident radiation has a wavelength of 589 nm?

46E. A diffraction grating has 200 rulings/mm, and it produces an intensity maximum at $\theta = 30.0°$. (a) What are the possible wavelengths of the incident visible light? (b) To what colors do they correspond?

47E. A grating has 315 rulings/mm. For what wavelengths in the visible spectrum can fifth-order diffraction be observed?

48E. Given a grating with 400 lines/mm, how many orders of the entire visible spectrum (400–700 nm) can it produce in addition to the $m = 0$ order?

49E. A diffraction grating 3.00 cm wide produces the second order at 33.0° with light of wavelength 600 nm. What is the total number of lines on the grating?

50E. Some tropical gyrinid beetles (whirligig beetles) are colored by optical interference that is due to scales whose alignment forms a diffraction grating (which uses scattered instead of transmitted light). If the incident light is perpendicular on the grating, the angle between the first-order maxima (on opposite sides of the zeroth-order maximum) is about 26°. What is the grating spacing of the beetle? Use 550 nm as the wavelength of light.

51E. A diffraction grating 1.0 cm wide has 10,000 parallel slits. Monochromatic light that is incident normally is deviated through 30° in the first order. What is the wavelength of the light?

52P. Light of wavelength 600 nm is incident normally on a diffraction grating. Two adjacent maxima occur at angles given by $\sin \theta = 0.2$ and $\sin \theta = 0.3$, respectively. The fourth-order maxima are missing. (a) What is the separation between adjacent slits? (b) What is the smallest possible individual slit width? (c) Which orders of intensity maxima are produced by the grating, assuming the values derived in (a) and (b)?

53P. A diffraction grating is made up of slits of width 300 nm with separation 900 nm. The grating is illuminated by monochromatic plane waves of wavelength $\lambda = 600$ nm at normal incidence. (a) How many diffraction maxima are there in the full pattern? (b) What is the width of a spectral line observed in the first order if the grating has 1000 slits?

54P. Assume that the limits of the visible spectrum are arbitrarily chosen as 430 and 680 nm. Calculate the number of rulings per millimeter of a grating that will spread the first-order spectrum through an angle of 20°.

55P. With light from a gaseous discharge tube incident normally on a grating with slit separation 1.73 μm, sharp maxima of green light are produced at angles $\theta = \pm 17.6°$, 37.3°, $-37.1°$, 65.2°, and $-65.0°$. Compute the wavelength of the green light that best fits these data.

56P. Light is incident on a grating at an angle ψ as shown in Fig. 37-41. Show that bright fringes occur at angles θ that satisfy the equation

$$d(\sin \psi + \sin \theta) = m\lambda, \qquad \text{for } m = 0, 1, 2, \ldots .$$

(Compare this equation with Eq. 37-25.) Only the special case $\psi = 0$ has been treated in this chapter.

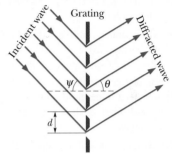

FIGURE 37-41   Problem 56.

57P. A grating with $d = 1.50$ μm is illuminated at various angles of incidence by light of wavelength 600 nm. Plot, as a function of the angle of incidence (0 to 90°), the angular deviation of the first-order maximum from the incident direction. (See Problem 56.)

58P. Two emission lines have wavelengths λ and $\lambda + \Delta\lambda$, respectively, where $\Delta\lambda \ll \lambda$. Show that their angular separation $\Delta\theta$ in a grating spectrometer is given approximately by

$$\Delta\theta = \frac{\Delta\lambda}{\sqrt{(d/m)^2 - \lambda^2}},$$

where d is the slit separation and m is the order at which the lines are observed. Note that the angular separation is greater in the higher orders than in lower orders.

59P. White light (consisting of wavelengths from 400 nm to 700 nm) is normally incident on a grating. Show that, no matter what the value of the grating spacing d, the second order and third order overlap.

60P. Show that a grating made up of alternately transparent and opaque strips of equal width eliminates all the even orders of maxima (except $m = 0$).

61P. A grating has 350 rulings per millimeter and is illuminated at normal incidence by white light. A spectrum is formed on a screen 30 cm from the grating. If a hole 10 mm square is cut in the screen, its inner edge being 50 mm from the central maximum and parallel to it, what range of wavelengths passes through the hole?

62P. Derive Eq. 37-28, the expression for the line widths.

SECTION 37-8 Gratings: Dispersion and Resolving Power

63E. The D line in the spectrum of sodium is a doublet with wavelengths 589.0 and 589.6 nm. Calculate the minimum number of lines needed in a grating that will resolve this doublet in the second-order spectrum. See Sample Problem 37-7.

64E. A grating has 600 rulings/mm and is 5.0 mm wide. (a) What is the smallest wavelength interval it can resolve in the third order at $\lambda = 500$ nm? (b) How many higher orders of maxima can be seen?

65E. A source containing a mixture of hydrogen and deuterium atoms emits red light at two wavelengths whose mean is 656.3 nm and whose separation is 0.180 nm. Find the minimum number of lines needed in a diffraction grating that can resolve these lines in the first order.

66E. (a) How many rulings must a 4.00-cm-wide diffraction grating have to resolve the wavelengths 415.496 and 415.487 nm in the second order? (b) At what angle are the maxima found?

67E. With a particular grating the sodium doublet (see Sample Problem 37-7) is viewed in the third order at 10° to the normal and is barely resolved. Find (a) the grating spacing and (b) the total width of the rulings.

68E. Show that the dispersion of a grating is $D = (\tan \theta)/\lambda$.

69E. A grating has 40,000 rulings spread over 76 mm. (a) What is its expected dispersion D for sodium light ($\lambda = 589$ nm) in the first three orders? (b) What is the grating's resolving power in these orders?

70P. Light containing a mixture of two wavelengths, 500 and 600 nm, is incident normally on a diffraction grating. It is desired (1) that the first and second maxima for each wavelength appear at $\theta \leq 30°$, (2) that the dispersion be as high as possible, and (3) that the third order for 600 nm be a missing order. (a) What should be the slit separation? (b) What is the smallest possible individual slit width? (c) For the 600 nm wavelength, which orders of intensity maxima are produced by the grating, assuming the values derived in (a) and (b)?

71P. (a) In terms of the angle θ locating a line producd by a grating, find the product of that line's half-width and the resolving power of the grating. (b) Evaluate that product for the grating of Problem 53, for first order.

72P. A diffraction grating has resolving power $R = \lambda_{av}/\Delta\lambda = Nm$. (a) Show that the corresponding frequency range Δf that can just be resolved is given by $\Delta f = c/Nm\lambda$. (b) From Fig. 37-17, show that the times required for light to travel along the two extreme rays differ by an amount $\Delta t = (Nd/c) \sin \theta$. (c) Show that $(\Delta f)(\Delta t) = 1$, this relation being independent of the various grating parameters. Assume $N \gg 1$.

SECTION 37-9 X-Ray Diffraction

73E. X rays of wavelength 0.12 nm are found to undergo second-order reflection at a Bragg angle of 28° from a lithium fluoride crystal. What is the interplanar spacing of the reflecting planes in the crystal?

74E. What is the smallest Bragg angle for x rays of wavelength 30 pm to undergo reflection from reflecting planes of spacing 0.30 nm in a calcite crystal?

75E. If first-order reflection occurs in a crystal at Bragg angle 3.4°, at what Bragg angle does second-order reflection occur from the same family of reflecting planes?

76E. Figure 37-42 is a graph of intensity versus diffraction angle for the diffraction of an x-ray beam by a crystal. The beam consists of two wavelengths, and the spacing between the reflecting planes is 0.94 nm. What are the two wavelengths?

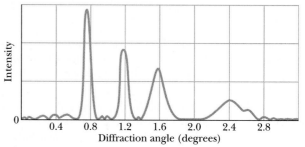

FIGURE 37-42 Exercise 76.

77E. An x-ray beam of wavelength A undergoes a first-order reflection from a crystal when its angle of incidence to a crystal face is 23°, and an x-ray beam of wavelength 97 pm undergoes third-order reflection when its angle of incidence to that face is 60°. Assuming that the two beams reflect from the same family of reflecting planes, find (a) the interplanar spacing and (b) the wavelength A.

78E. An x-ray beam of a certain wavelength is incident on a NaCl crystal, at 30.0° to a certain family of reflecting planes of spacing 39.8 pm. If the reflection from those planes is of the first order, what is the wavelength of the x rays?

79P. Prove that it is not possible to determine both wavelength of radiation and spacing of reflecting planes in a crystal by measuring the Bragg angles in several orders.

80P. In Fig. 37-43, an x-ray beam of wavelengths from 95.0 pm to 140 pm is incident on a family of reflecting planes with spacing $d = 275$ pm. At which wavelengths will these planes produce intensity maxima in their reflections?

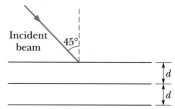

FIGURE 37-43 Problems 80 and 83.

81P. In Fig. 37-44, first-order reflection from the reflection planes shown occurs when an x-ray beam of wavelength 0.260 nm makes an angle of 63.8° with the top face of the crystal. What is the unit cell size a_0?

82P. Consider a two-dimensional square crystal structure, such as one side of the structure shown in Fig. 37-27a. One interplanar spacing of reflecting planes is the unit cell size a_0. Calculate and sketch the next five smaller interplanar spacings. (b) Show that your results in (a) obey the general formula

$$d = \frac{a_0}{\sqrt{h^2 + k^2}},$$

where h and k are relatively prime integers (they have no common factor other than unity).

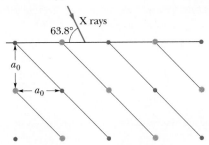

FIGURE 37-44 Problem 81.

83P. In Fig. 37-43, let a beam of x rays of wavelength 0.125 nm be incident on an NaCl crystal at an angle of 45.0° to the top face of the crystal. Let the reflecting planes have separation $d = 0.252$ nm. Through what angles must the crystal be turned about an axis that is perpendicular to the plane of the page for these reflecting planes to give intensity maxima in their reflections?

Electronic Computation

84. A computer can be used to sum the phasors corresponding to Huygens' wavelets and so find a diffraction pattern. Suppose light with a wavelength of 500 nm is incident normally on a single slit with a width of 5.00×10^{-6} m. To approximate the diffraction pattern, sum the phasors corresponding to $N = 200$ wavelets spreading from uniformly distributed sources within the slit. The horizontal and vertical components of the resultant are proportional to

$$E_h = \sum_{i=1}^{N} \cos \phi_i \quad \text{and} \quad E_v = \sum_{i=1}^{N} \sin \phi_i,$$

respectively, where ϕ_i is the phase of wavelet i. The intensity ratio is $I/I_m = (E_h^2 + E_v^2)/N^2$. The factor $1/N^2$ assures that $I/I_m = 1$ when all the wavelets have the same phase. If you consider light that is diffracted at the angle θ to the straight-ahead direction, then you may take the phase of the first wavelet to be zero and the phase of each successive wavelet to be $(2\pi/\lambda)\Delta x \sin \theta$ greater than that of the preceding wavelet. Here Δx is the distance between wavelet sources; that is, $\Delta x = a/(N - 1)$, where a is the slit width. Use this technique to search for the diffraction angles corresponding to the first three secondary maxima and find the intensity ratios for those maxima.

38
Relativity

In modern long-range navigation, the precise location and speed of moving craft are continuously monitored and updated. A system of navigation satellites called NAVSTAR permits locations and speeds anywhere on Earth to be determined to within about 16 m and 2 cm/s. However, if relativity effects were not taken into account, speeds could not be determined any closer than about 20 cm/s, which is unacceptable for modern navigation systems. How can something as abstract as Einstein's special theory of relativity be involved in something as practical as navigation ?

38-1 WHAT IS RELATIVITY ALL ABOUT?

Relativity has to do with measurements of events (things that happen): where and when they happen, and by how much any two events are separated in space and in time. In addition, relativity has to do with transforming such measurements and others between reference frames that move relative to each other. (Hence the name *relativity*.) We discussed such matters in Sections 4-8 and 4-9.

Transformations and moving reference frames were well understood and quite routine to physicists in 1905. Then Albert Einstein (Fig. 38-1) published his **special theory of relativity.** The adjective *special* means that the theory deals only with **inertial reference frames,** which are frames that move at constant velocities relative to one another. (Einstein's *general theory of relativity* treats the more challenging situation in which reference frames accelerate; in this chapter the term *relativity* implies only inertial reference frames.)

Starting with two deceivingly simple postulates, Einstein stunned the scientific world by showing that the old ideas about relativity were wrong, even though everyone was so accustomed to them that they seemed to be unquestionable common sense. This supposed common sense, however, was derived from experience only with things that move rather slowly. Einstein's relativity, which turns out to be correct for all possible speeds, predicted many effects that were, at first study, bizarre because no one had experienced them.

In particular, Einstein demonstrated that space and time are entangled; that is, the time between two events depends on how far apart they occur, and vice versa. And the entanglement is different for observers who move relative to each other. One result is that time does not pass at a fixed rate, as if it were ticked off with mechanical regularity on some master grandfather clock that controls the universe. Rather, that rate is adjustable: relative motion can change the rate at which time passes. Prior to 1905, no one but a few daydreamers would have thought that. Now, engineers and scientists take it for granted because their experience with special relativity has reshaped their common sense.

Special relativity has the reputation of being difficult. It is not difficult mathematically, at least not here. But it is difficult in that we must be very careful about *who* measures *what* about an event and just *how* that measurement is made. And it can be difficult because it can contradict experience. Before you read further, you might want to review some of the special relativity discussed earlier in this book (Table 38-1 is a guide).

38-2 THE POSTULATES

We now examine the two postulates of relativity, on which Einstein's theory is based:

---

1. The Relativity Postulate: The laws of physics are the same for observers in all inertial reference frames. No frame is preferred.

Galileo assumed that the laws of *mechanics* were the same in all inertial reference frames. (Newton's first law of motion is one important consequence.) Einstein extended that idea to include *all* the laws of physics, especially electromagnetism and optics. This postulate does *not* say that the

FIGURE 38-1 Einstein in the early 1900s, at his desk at the patent office in Bern, Switzerland, where he was employed when he published his special theory of relativity.

TABLE 38-1 EARLIER SECTIONS ON RELATIVITY

SECTION	TITLE
4-10	Relative Motion at High Speeds
7-8	Kinetic Energy at High Speeds
8-8	Mass and Energy
10-6	Reactions and Decay Processes

measured values of all physical quantities are the same for all inertial observers; most are not the same. It is the *laws of physics,* which relate these measurements to each other, that are the same.

> **2. The Speed of Light Postulate:** The speed of light in vacuum has the same value c in all directions and in all inertial reference frames.

We can also phrase this postulate to say that there is in nature an *ultimate speed c,* the same in all directions and in all inertial reference frames. Light happens to travel at this ultimate speed, as do any massless particles (neutrinos might be an example). But any entity that carries energy or information cannot exceed this limit. Moreover, any particle that does have mass cannot actually reach speed c, no matter how much or how long it is accelerated.

Both postulates have been exhaustively tested, and no exceptions have ever been found.

The Ultimate Speed

The reality of the existence of a limit to the speed of accelerated electrons was shown in a 1964 experiment of W. Bertozzi. He accelerated electrons to various measured speeds (see Fig. 38-2) and—by an independent method—also measured their kinetic energies. He found that as the force that acts on a very fast electron is increased, the electron's measured kinetic energy increases toward very large values but its speed does not increase appre-

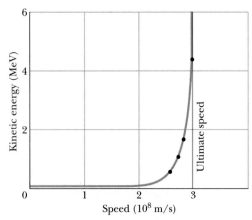

FIGURE 38-2 The dots show measured values of the kinetic energy of an electron plotted against its measured speed. No matter how much energy is given to an electron (or to any other particle having mass), its speed can never equal or exceed the ultimate limiting speed c. (The plotted curve through the dots shows the predictions of Einstein's special theory of relativity.)

ciably. Electrons have been accelerated to at least 0.999 999 999 95 times the speed of light but—close though it may be—that speed is still less than the ultimate speed c.

Testing the Speed of Light Postulate

If the speed of light is the same in all inertial reference frames, then the speed of light that is emitted by a moving source should be the same as the speed of light that is emitted by a source at rest in the laboratory. This claim has been tested directly, in an experiment of high precision. The "light source" was the *neutral pion* (symbol π^0), an unstable, short-lived particle that can be produced by collisions in a particle accelerator. It decays into two gamma rays by the process

$$\pi^0 \rightarrow \gamma + \gamma. \tag{38-1}$$

Gamma rays are part of the electromagnetic spectrum and obey the speed of light postulate, just as visible light does.

In a 1964 experiment, physicists at CERN, the European particle-physics laboratory near Geneva, generated a beam of pions moving at a speed of 0.999 75c with respect to the laboratory. The experimenters then measured the speed of the gamma rays emitted from these very rapidly moving sources. They found that the speed of the light emitted by the pions was the same as would have been measured if the pions had been at rest in the laboratory.

SAMPLE PROBLEM 38-1

An electron with a kinetic energy of 20 GeV (which is said to be a 20 GeV electron) can be shown to have a speed $v = 0.999\ 999\ 999\ 67c$. If such an electron raced a light pulse to the nearest star outside the solar system (Proxima Centauri, 4.3 light-years, or 4.0×10^{16} m, distant), by how much time would the light pulse win the race?

SOLUTION: If L is the distance to the star, the difference in travel times is

$$\Delta t = \frac{L}{v} - \frac{L}{c} = L \frac{c - v}{vc}.$$

Now v is so close to c that we can put $v = c$ in the denominator of this expression (but not in the numerator!). If we do so, we find

$$\Delta t = \frac{L}{c}\left(1 - \frac{v}{c}\right)$$
$$= \frac{(4.0 \times 10^{16}\text{ m})(1 - 0.999\ 999\ 999\ 67)}{3.00 \times 10^8\text{ m/s}}$$
$$= 0.044\text{ s} = 44\text{ ms.} \tag{Answer}$$

38-3 MEASURING AN EVENT

An **event** is something that happens, to which an observer can assign three space coordinates and one time coordinate. Among many possible events are (1) the turning on or off of a tiny lightbulb, (2) the collision of two particles, (3) the passage of a pulse of light through a specified point, (4) an explosion, and (5) the coincidence of the hand of a clock with a marker on the rim of the clock. An observer, fixed in a certain inertial reference frame, may assign to an event A the following coordinates:

RECORD OF EVENT A	
COORDINATE	VALUE
x	3.58 m
y	1.29 m
z	0 m
t	34.5 s

Because in relativity space and time are entangled with each other, we can describe these coordinates collectively as *spacetime* coordinates. The coordinate system itself is part of the reference frame of the observer.

A given event may be recorded by any number of observers, each in a different inertial reference frame. In general, different observers will assign different spacetime coordinates to the same event. Note that an event does not, in any sense, "belong" to a particular inertial reference frame. An event is just something that happens, and anyone in any reference frame may look at it and assign spacetime coordinates to it.

Making such an assignment can be complicated by a practical problem. For example, suppose a balloon bursts 1 km to your right while a firecracker pops 2 km to your left, both at 9:00 A.M. But you do not detect either event precisely at 9:00 A.M. because light from the events has not yet reached you. Because light from the pop has farther to go, it arrives at your eyes later than does light from the balloon burst and thus the pop will seem to have occurred later than the burst. To sort out the actual times and to assign 9:00 A.M. to both events, you must calculate the travel times of the light and then subtract them from the arrival times.

This procedure can be very messy in more challenging situations, and we need an easier procedure that automatically eliminates any concern about the travel time from an event to an observer. To set up such a procedure, we shall construct an imaginary array of measuring rods and clocks throughout the observer's inertial frame (the array moves rigidly with the observer). This construction may seem contrived, but it spares us much confusion and calculation and allows us to find the space coordinates, the time coordinate, and the spacetime coordinates, as follows.

1. The Space Coordinates. We imagine the observer's coordinate system fitted with a close-packed, three-dimensional array of measuring rods, one set of rods parallel to each of the three coordinate axes. These rods provide a way to determine coordinates along the axes. Thus if the event is, say, the turning on of a small lightbulb, the observer, in order to locate the position of the event, need read only the three space coordinates at the bulb's location.

2. The Time Coordinate. For the time coordinate, we imagine that every point of intersection in the array of measuring rods has a tiny clock, which the observer can read by the light generated by the event. Figure 38-3 suggests the "jungle gym" of clocks and measuring rods that we have described.

The array of clocks must be synchronized properly. It is not enough to assemble a set of identical clocks, set them all to the same time, and then move them to their assigned positions. We do not know, for example, whether moving the clocks will change their rates. (Actually, it will.) We must put the clocks in place and *then* synchronize them.

If we had a method of transmitting signals at infinite speed, synchronization would be a simple matter. However, no known signal has this property. We therefore choose light (interpreted broadly to include the entire electromagnetic spectrum) to send out our synchronizing signals because, in vacuum, light travels at the highest possible speed, the limiting speed c.

Here is one of many ways that we might synchronize an array of clocks with the help of light signals. The observer enlists the help of a large number of temporary helpers, one for each clock. The observer then stands at a point selected as the origin and sends out a pulse of light

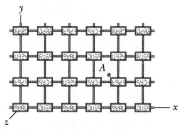

FIGURE 38-3 One section of a three-dimensional array of clocks and measuring rods by which an observer can assign spacetime coordinates to an event, such as a flash of light at point A. The space coordinates are approximately $x = 3.7$ rod lengths, $y = 1.2$ rod lengths, and $z = 0$. The time coordinate is whatever time appears on the clock closest to A at the instant of the flash.

when the origin clock reads $t = 0$. When the light pulse reaches each helper, that helper sets his or her clock to read $t = r/c$, where r is the distance between the helper and the origin. The clocks are then synchronized.

3. **The Spacetime Coordinates.** The observer can now assign spacetime coordinates to an event by simply recording the time on a clock at the event and the position as measured on the nearest measuring rods. If there are two events, the observer computes their separation in time as the difference of the times on clocks near each, and their separation in space from the differences of coordinates on rods near each. We thus avoid the practical problem of waiting for signals to reach the observer from events and then calculating the travel times of those signals.

38-4 THE RELATIVITY OF SIMULTANEITY

Suppose that one observer (Sam) notes that two independent events (event Red and event Blue) occur at the same time. Suppose also that another observer (Sally), who is moving at a constant velocity **v** with respect to Sam, also records these same two events. Will Sally also find that they occur at the same time?

The answer is that in general she will not:

> If two observers are in relative motion, they will not, in general, agree as to whether two events are simultaneous. If one observer finds them to be simultaneous, the other generally will not, and conversely.

We cannot say that one observer is right and the other wrong. Their observations are equally valid, and there is no reason to favor one over the other.

The realization that two contradictory statements about the same natural event can be correct is a seemingly strange outcome of Einstein's theory. However, in Chapter 18 we discussed another way in which motion can affect measurement, without balking at the contradictory results: in the Doppler effect, the frequency an observer measures for a sound wave depends on the relative motion of the observer and the source. So two observers moving relative to one another can measure different frequencies for the same wave. And both measurements are correct.

We conclude the following:

> Simultaneity is not an absolute concept but a relative one, depending on the motion of the observer.

If the relative speed of the observers is very much less than the speed of light, then measured departures from simultaneity are so small that they are not noticeable. Such is the case for all our experiences of daily living; this is why the relativity of simultaneity is unfamiliar.

A Closer Look at Simultaneity

Let us clarify the relativity of simultaneity with an example based on the postulates of relativity, no clocks or measuring rods being directly involved. Figure 38-4 shows two long spaceships (the SS *Sally* and the SS *Sam*), which can serve as inertial reference frames for observers Sally and Sam. The two observers are stationed at the midpoints of their ships. The ships are separating along a common x axis, the relative velocity of *Sally* with respect to *Sam* being **v**. Figure 38-4a shows the ships with the two observer stations momentarily aligned opposite each other.

Two large meteorites strike the ships, one setting off a red flare (event Red) and the other a blue flare (event Blue), not necessarily simultaneously. Each event leaves a permanent mark on each ship, at positions R, R' and B, B'.

Let us suppose that the expanding wavefronts from the two events happen to reach Sam at the same time, as Fig. 38-4c shows. Let us further suppose that, after the episode, Sam finds, by measurement, that he was indeed stationed exactly halfway between the markers B and R on his ship when the two events occurred. He will say:

SAM: Light from event Red and light from event Blue reached me at the same time. From the marks on my spaceship, I find that I was standing halfway between the two sources when the light from them reached me. Therefore event Red and event Blue are simultaneous events.

As study of Fig. 38-4 shows, however, the expanding wavefront from event Red will reach Sally *before* the expanding wavefront from event Blue does. She will say:

SALLY: Light from event Red reached me before light from event Blue did. From the marks on my spaceship, I found that I too was standing halfway between the two sources. Therefore the events were *not* simultaneous; event Red occurred first, followed by event Blue.

These reports do not agree. Nevertheless, *both* observers are correct.

Note carefully that there is only one wavefront expanding from the site of each event and that *this wavefront travels with the same speed c in both reference frames*, exactly as the speed of light postulate requires.

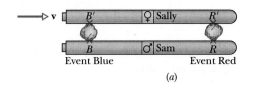

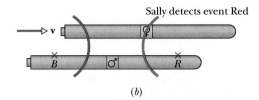

Sally detects event Red

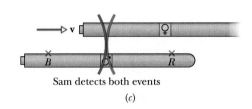

Sam detects both events

(c)

Sally detects event Blue

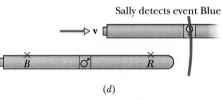

(d)

FIGURE 38-4 The spaceships of Sally and Sam and the occurrences of events from Sam's view. Sally's ship moves rightward with velocity **v**. (a) Event Red occurs at positions R, R' and event Blue occurs at positions B, B'; each event sends out a wave of light. (b) Sally detects the wave from event Red. (c) Sam simultaneously detects the waves from event Red and event Blue. (d) Sally detects the wave from event Blue.

It *might* have happened that the meteorites struck the ships in such a way that the two hits appeared to Sally to be simultaneous. If that had been the case, then Sam would have declared them not to be simultaneous. The experiences of the two observers are exactly symmetrical.

38-5 THE RELATIVITY OF TIME

If observers who move relative to each other measure the time interval (or *temporal separation*) between two events, they generally will find different results. Why? Because the spatial separation of the events can affect the time intervals measured by the observers.

> The time interval between two events depends on how far apart they are; that is, their spatial and temporal separations are entangled.

In this section we discuss an example of this entanglement by means of an example; however, the example is restricted in a crucial way: *to one of two observers, the two events occur at the same location.* We shall not get to more general examples until Section 38-7.

Figure 38-5a shows the basics of an experiment Sally conducts while she and her equipment ride in a train moving with constant velocity **v** relative to a station. A pulse of light leaves a light source B (event 1), travels vertically upward, is reflected vertically downward by a mirror, and

FIGURE 38-5 (a) Sally, on the train, measures the time interval Δt_0 between events 1 and 2 using a single clock C on the train. That clock is shown twice: first for event 1 and then for event 2. (b) Sam, watching from the station as the events occur, requires two synchronized clocks, C_1 at event 1 and C_2 at event 2, to measure the time interval between the two events; his measured time interval is Δt.

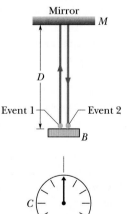

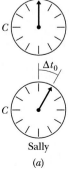

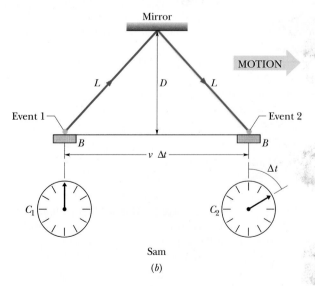

then is detected back at the source (event 2). Sally measures a certain time interval Δt_0 between the two events, related to the distance D from source to mirror by

$$\Delta t_0 = \frac{2D}{c} \qquad \text{(Sally)}. \qquad (38\text{-}2)$$

The two events occur at the same location in Sally's reference frame, and she needs only one clock C at that location to measure the time interval. Clock C is shown twice in Fig. 38-5a, at the beginning and end of the interval.

Consider now how these same two events are measured by Sam, who is standing on the station platform as the train passes. Because the equipment moves with the train during the travel time of the light, Sam sees the path of the light as shown in Fig. 38-5b. For him, the two events occur at different places in his reference frame. So to measure the time interval between events, Sam must use *two* synchronized clocks, C_1 and C_2, one at each event. According to Einstein's speed of light postulate, the light travels at the same speed c for Sam as for Sally. But now, the light travels distance $2L$ between events 1 and 2. The time interval measured by Sam between the two events is

$$\Delta t = \frac{2L}{c} \qquad \text{(Sam)}, \qquad (38\text{-}3)$$

in which

$$L = \sqrt{(\tfrac{1}{2}v\,\Delta t)^2 + D^2}. \qquad (38\text{-}4)$$

From Eq. 38-2, we can write this as

$$L = \sqrt{(\tfrac{1}{2}v\,\Delta t)^2 + (\tfrac{1}{2}c\,\Delta t_0)^2}. \qquad (38\text{-}5)$$

If we eliminate L between Eqs. 38-3 and 38-5 and solve for Δt, we find

$$\Delta t = \frac{\Delta t_0}{\sqrt{1 - (v/c)^2}}. \qquad (38\text{-}6)$$

Equation 38-6 tells us how Sam's measured interval Δt compares with Sally's interval Δt_0. Because v must be less than c, the denominator in Eq. 38-6 must be less than unity. Thus Δt must be greater than Δt_0: Sam measures a *greater* time interval between the two events than does Sally. Sam and Sally have measured the time interval between the *same* two events, but the relative motion between Sam and Sally made their measurements *different*. We conclude that relative motion can change the *rate* at which time passes between two events; the key to this effect is the fact that the speed of light is the same for both observers.

We distinguish between the measurements of Sam and Sally with the following terminology:

> When two events occur at the same location in an inertial reference frame, the time interval between them, measured in that frame, is called the **proper time interval** or the **proper time.** Measurements of the same time interval from any other inertial reference frame are always greater.

So Sally measures a proper time interval, and Sam measures a greater time interval. (The term *proper* is unfortunate in that it implies that any other measurement is improper or nonreal. That is just not so.) The increase in the time interval between two events from the proper time interval is called **time dilation.** (To dilate is to expand or stretch; here the time interval is expanded or stretched.)

Often the dimensionless ratio v/c in Eq. 38-6 is replaced with β, called the **speed parameter.** And the dimensionless inverse square root in Eq. 38-6 is often replaced with γ, called the **Lorentz factor:**

$$\gamma = \frac{1}{\sqrt{1 - \beta^2}}. \qquad (38\text{-}7)$$

With these replacements, we can rewrite Eq. 38-6 as

$$\Delta t = \gamma\,\Delta t_0 \qquad \text{(time dilation)}. \qquad (38\text{-}8)$$

The speed parameter β is always less than unity and, provided v is not zero, γ is always greater than unity. However, the difference between γ and 1 is not significant unless $v > 0.1c$. Thus, in general, "old relativity" works well enough for $v < 0.1c$, but we must use special relativity for greater values of v. As shown in Fig. 38-6, γ increases rapidly in magnitude as β approaches 1 (as v approaches c). So the greater the relative speed between Sally and Sam, the greater will be the time interval measured by Sam, until at a great enough speed, the interval takes effectively forever.

You might wonder what Sally says about Sam's having measured a greater time interval than she did. His measurement comes as no surprise to her, because to her, he failed to synchronize his clocks C_1 and C_2 in spite of his insistence that he did. Recall that observers in relative motion generally do not agree about simultaneity. Here, Sam insists that his two clocks simultaneously read the same time when event 1 occurred. To Sally, however, Sam's clock C_2 was erroneously set ahead. So when Sam read the time of event 2 on it, to Sally he was reading off a time that

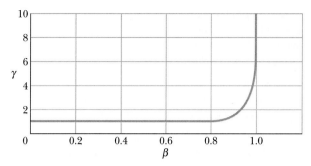

FIGURE 38-6 A plot of the Lorentz factor γ as a function of the speed parameter β ($= v/c$).

was too large, and that is why the time interval he measured between the two events was greater than the interval she measured.

Two Tests of Time Dilation

1. **Microscopic Clocks.** Subatomic particles called *muons* are unstable; that is, when a muon is produced, it lasts for only a short time before it *decays* (transforms into particles of other types). The *lifetime* of a muon is the time interval between its production (event 1) and its decay (event 2). When muons are stationary and their lifetimes are measured with stationary clocks (say, in a laboratory), their average lifetime is 2.200 μs. This is a proper time interval because, for each muon, events 1 and 2 occur at the same location in the reference frame of the muon. We can represent this proper time interval with Δt_0; moreover, we can call the reference frame in which it is measured the *rest frame* of the muon.

If, instead, the muons are moving, say, through a laboratory, then measurements of their lifetimes made with the laboratory clocks should yield a greater average lifetime (a dilated average lifetime). To check this conclusion, measurements were made of the average lifetime of muons moving with a speed of 0.9994c relative to laboratory clocks. From Eq. 38-7, with $\beta = 0.9994$, the Lorentz factor for this speed is

$$\gamma = \frac{1}{\sqrt{1 - \beta^2}} = \frac{1}{\sqrt{1 - (0.9994)^2}} = 28.87,$$

which is substantially greater than unity. Equation 38-8 then yields, for the average dilated lifetime,

$$\Delta t = \gamma \, \Delta t_0 = (28.87)(2.200 \ \mu s) = 63.5 \ \mu s.$$

The actual measured value matched this result within experimental error.

2. **Macroscopic Clocks.** In October 1977, Joseph Hafele and Richard Keating carried out what must have been a grueling experiment. They flew four portable atomic clocks twice around the world on commercial airlines, once in each direction. Their purpose was "to test Einstein's theory of relativity with macroscopic clocks." As we have just seen, the time dilation predictions of Einstein's theory have been confirmed on a microscopic scale, but there is great comfort in seeing a confirmation made with an actual clock. Such macroscopic measurements became possible only because of the very high precision of modern atomic clocks. Hafele and Keating verified the predictions of the theory to within 10%. (Einstein's *general* theory of relativity, which predicts that the rate of a clock is influenced by gravitation, also plays a role in this experiment.)

A few years later, physicists at the University of Maryland carried out a similar experiment with improved precision. They flew an atomic clock round and round over Chesapeake Bay for flights lasting 15 h and succeeded in checking the time dilation prediction to better than 1%. Today, when atomic clocks are transported from one place to another for calibration or other purposes, the time dilation caused by their motion is always taken into account.

CHECKPOINT 1: Standing beside railroad tracks, we are suddenly startled by a relativistic boxcar traveling past us as shown in the figure. Inside, a well-equipped hobo fires a laser pulse from the front of the boxcar to its rear. (a) Is our measurement of the speed of the pulse greater than, less than, or the same as that measured by the hobo? (b) Is his measurement of the flight time of the pulse a proper time? (c) Are his measurement and our measurement of the flight time related by Eq. 38-8?

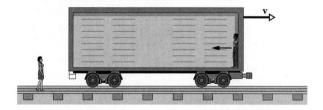

SAMPLE PROBLEM 38-2

Your starship passes Earth with a relative speed of 0.9990c. After traveling 10.0 y (your time), you stop at lookout post LP13, turn, and then travel back to Earth with the same relative speed. The trip back takes another 10.0 y (your time). How long does the round trip take according to measurements made on Earth? (Neglect any effects due to the accelerations involved with stopping and turning.)

SOLUTION: On the journey out, the start and end of the journey occur at the same location in your reference frame, namely on your ship. Hence, you measure proper time Δt_0 for the trip, which is the given 10.0 y. Equation 38-6 gives us the corresponding time Δt as measured in the Earth reference frame:

$$\Delta t = \frac{\Delta t_0}{\sqrt{1 - (v/c)^2}}$$

$$= \frac{10.0 \text{ y}}{\sqrt{1 - (0.9990c/c)^2}} = (22.37)(10.0 \text{ y}) = 224 \text{ y}.$$

On the journey back, we have the same situation and the same data. Thus the round trip requires 20 y of your time but

$$\Delta t_{\text{total}} = (2)(224 \text{ y}) = 448 \text{ y} \qquad \text{(Answer)}$$

of Earth time. In other words, you have aged 20 y while the Earth has aged 448 y. Although you cannot travel into the past (as far as we know), you can travel into the future of, say, Earth, by using high-speed relative motion to adjust the rate at which time passes.

SAMPLE PROBLEM 38-3

The elementary particle known as the *positive kaon* (K⁺) has, on average, a lifetime of 0.1237 μs when stationary, that is, when the lifetime is measured in the rest frame of the kaon. If positive kaons with a speed of 0.990c relative to a laboratory reference frame are produced, how far can they travel in that frame during their lifetime?

SOLUTION: In the laboratory frame, the distance d traveled by a kaon is related to its speed v $(= 0.990c)$ and its travel time Δt_k by $d = v\,\Delta t_k$. (This statement does not involve relativity because all quantities are measured in the same reference frame.) If special relativity did not apply, the travel time would be just the 0.1237 μs lifetime of the particle, and thus the travel distance would be just

$$d = v\,\Delta t_k = (0.990)(3.00 \times 10^8 \text{ m/s})(1.237 \times 10^{-7} \text{ s})$$

$$= 36.7 \text{ m.} \qquad \text{(Wrong Answer)}$$

However, special relativity does apply and the travel time of the kaon in the laboratory frame is its dilated lifetime Δt. With Eq. 38-6, we can find Δt from the kaon's proper lifetime Δt_0 $(= 0.1237 \mu s)$, as measured in its rest frame:

$$\Delta t = \frac{\Delta t_0}{\sqrt{1 - (v/c)^2}}$$

$$= \frac{0.1237 \times 10^{-6} \text{ s}}{\sqrt{1 - (0.990c/c)^2}} = 8.769 \times 10^{-7} \text{ s.}$$

This is about seven times longer than the kaon's proper lifetime. (This calculation does involve relativity because we must transform data from the particle's rest frame to the laboratory frame.) We can now find the travel distance in the laboratory frame as

$$d = v\,\Delta t_k = v\,\Delta t$$

$$= (0.990)(3.00 \times 10^8 \text{ m/s})(8.769 \times 10^{-7} \text{ s})$$

$$= 260 \text{ m.} \qquad \text{(Answer)}$$

This is about seven times our first (wrong) answer. Experiments like the one outlined here, which verify special relativity, became routine in physics laboratories decades ago.

38-6 THE RELATIVITY OF LENGTH

If you want to measure the length of a rod that is at rest with respect to you, you can—at your leisure—note the positions of its end points on a long stationary scale and subtract the two readings. If the rod is moving, however, you must note the positions of the end points *simultaneously* (in your reference frame) or your measurement cannot be called a length. Figure 38-7 suggests the difficulty of trying to measure the length of a moving penguin by locating its front and back at different times. Because simultaneity is relative and it enters into length measurements, length should also be a relative quantity.

Let L_0 be the length of a rod that you measure when the rod is stationary (meaning you and it are in the same

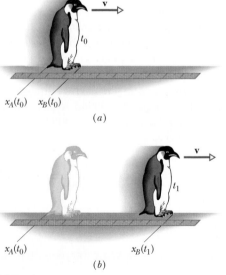

(a)

(b)

FIGURE 38-7 If you want to measure the front-to-back length of a penguin while it is moving, you must mark the positions of its front and back simultaneously (in your reference frame), as in (a), rather than at different times, as in (b).

reference frame, the rod's rest frame). If, instead, there is relative motion at speed v between you and the rod *along the length of the rod,* you measure a length L given by

$$L = L_0 \sqrt{1 - \beta^2} = \frac{L_0}{\gamma} \qquad \text{(length contraction).} \qquad (38\text{-}9)$$

Because the Lorentz factor γ is always greater than unity if there is relative motion, L is less than L_0. The relative motion causes a *length contraction,* and L is called a *contracted length.* Because γ increases with speed v, the length contraction also increases with the relative speed.

> The length L_0 of an object measured in the rest frame of the object is its **proper length** or **rest length**. Measurements of the length from any reference frame that is in relative motion parallel to that length are always less than the proper length.

Be careful: length contraction occurs only along the direction of relative motion. Also, the length that is measured does not have to be that of an object like a rod or a circle. Instead, it can be the length (or distance) between two objects in the same rest frame—for example, the Sun and a nearby star (which are, at least approximately, at rest relative to each other).

Does the object *really* shrink? Reality is based on observations and measurements; if the results are always consistent and if no error can be determined, then what is observed and measured is real. In that sense, the object really does shrink. However, a more precise statement is that the object *is really measured* to shrink—motion affects that measurement and thus reality.

Can a high-speed photograph show a contracted object? No, because what it records is not limited to the light emitted by an object at one specific instant (simultaneously). Instead it records all light from the object that happens to arrive at the camera at the instant of exposure, regardless of when the light was emitted.

When you measure a contracted length for, say, a rod, what does an observer moving with the rod say of your measurement? To that observer, you did not locate the two ends of the rod simultaneously. (Recall that observers in motion relative to each other do not agree about simultaneity.) To the observer, you first located the rod's front end and then, slightly later, its rear end, and that is why you measured a length that is less than the proper length.

Proof of Eq. 38-9

Length contraction is a direct consequence of time dilation. Consider once more our two observers. Both Sally, seated

on a train moving through a station, and Sam, again on the station platform, want to measure the length of the platform. Sam, using a tape measure, finds the length to be L_0, a proper length because the platform is at rest with respect to him. Sam also notes that Sally, on the train, moves through this length in a time $\Delta t = L_0/v$, where v is the speed of the train. That is,

$$L_0 = v \, \Delta t \qquad \text{(Sam).} \qquad (38\text{-}10)$$

This time interval Δt is not a proper time interval because the two events that define it (Sally passes back of platform and Sally passes front of platform) occur at two different places and Sam must use two synchronized clocks to measure the time interval Δt.

For Sally, however, the platform is moving. She finds that the two events measured by Sam occur *at the same place* in her reference frame. She can time them with a single stationary clock, so the interval Δt_0 that she measures is a proper time interval. To her, the length L of the platform is given by

$$L = v \, \Delta t_0 \qquad \text{(Sally).} \qquad (38\text{-}11)$$

If we divide Eq. 38-11 by Eq. 38-10 and apply Eq. 38-8, the time dilation equation, we have

$$\frac{L}{L_0} = \frac{v \, \Delta t_0}{v \, \Delta t} = \frac{1}{\gamma},$$

or $\qquad\qquad L = \frac{L_0}{\gamma}, \qquad\qquad (38\text{-}12)$

which is Eq. 38-9, the length contraction equation.

SAMPLE PROBLEM 38-4

In Fig. 38-8, Sally (at point A) and Sam's spaceship (of proper length $L_0 = 230$ m) pass each other with constant relative speed v. Sally measures a time interval of 3.57 μs for the ship to pass her (from the passage of point B to the passage of point C). What is the speed parameter β between Sally and the ship?

SOLUTION: If the relative speed v between Sally and Sam were, say, less than $0.1c$, we might have seen this situation in Chapter 2, where we would have said that a ship of length L and speed v passes Sally in a time interval

$$\Delta t = \frac{L}{v}.$$

(No relativity is involved in this statement.)

But here we probably have a relativistic problem, with $v > 0.1c$. In that case, we know that the length L that Sally would measure is not the proper length L_0 of the ship, but a contracted length, given by Eq. 38-9:

$$L = L_0 \sqrt{(1 - \beta^2}.$$

(This statement involves relativity, because we are transforming data between Sam's frame and Sally's frame.) According to Sally, the time required for the passage is now written as

$$\Delta t = \frac{\text{contracted length } L}{v} = \frac{L_0 \sqrt{(1 - \beta^2)}}{\beta c}.$$

Solving for β and then substituting the given data, we find, after a little algebra,

$$\beta = \frac{L_0}{\sqrt{(c\,\Delta t)^2 + L_0^2}}$$

$$= \frac{230 \text{ m}}{\sqrt{(3.00 \times 10^8 \text{ m/s})^2(3.57 \times 10^{-6} \text{ s})^2 + (230 \text{ m})^2}}$$

$$= 0.210. \qquad \text{(Answer)}$$

Thus the relative speed between Sally and the ship is 21% of the speed of light. Note that only the relative motion of Sally and Sam matters here; whether either is stationary relative to, say, a space station is irrelevant. In Fig. 38-8 we took Sally to be stationary, but we could instead have taken the ship to be stationary, with Sally flying past it. Nothing would have changed in our results.

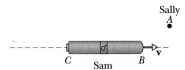

FIGURE 38-8 Sample Problem 38-4. Sally measures how long a spaceship takes to pass her at point A.

$\mathbb{C}$HECKPOINT **2**: In Sample Problem 38-4, Sally measures the passage time of the ship. If Sam does also, (a) which measurement, if either, is a proper time and (b) which measurement is smaller?

SAMPLE PROBLEM 38-5

Caught by surprise near a supernova, you race away from the explosion in your spaceship, hoping to outrun the high-speed material ejected toward you. Your Lorentz factor relative to the inertial reference frame of the local stars is 22.4.

(a) To reach a safe distance, you figure you need to cover 9.00×10^{16} m as measured in the reference frame of the local stars. How long will the flight take, as measured in that frame?

SOLUTION: The length $L_0 = 9.00 \times 10^{16}$ m is a proper length in the reference frame of the local stars, because its two ends are at rest in that frame. Figure 38-6 tells us that with such a large Lorentz factor, your speed relative to the local stars is $v \approx c$. So, with that approximation, to move through

length L_0 requires the time

$$\Delta t = \frac{L_0}{v} = \frac{L_0}{c}$$

$$= \frac{9.00 \times 10^{16} \text{ m}}{3.00 \times 10^8 \text{ m/s}} = 3.00 \times 10^8 \text{ s} = 9.49 \text{ y.} \quad \text{(Answer)}$$

(b) How long does that trip take according to you (in your reference frame)?

SOLUTION: From your reference frame, the distance you cover is a contracted length L that races past you at relative speed $v \approx c$. Equation 38-9 tells us that $L = L_0/\gamma$. So, the time you measure for the passage of that contracted length is

$$\Delta t_0 = \frac{L}{v} = \frac{L_0/\gamma}{v} = \frac{L_0}{c\gamma}$$

$$= \frac{9 \times 10^{16} \text{ m}}{(3.00 \times 10^8 \text{ m/s})(22.4)}$$

$$= 1.339 \times 10^7 \text{ s} = 0.424 \text{ y.} \quad \text{(Answer)}$$

This is a proper time, because the start and end of the passage occur at the same point in your reference frame (at your ship). You can check the validity of the two answers by substituting them into Eq. 38-8 (for time dilation) and solving for γ.

38-7 THE LORENTZ TRANSFORMATION

As Fig. 38-9 shows, inertial reference frame S' is moving with speed v relative to frame S, in the common positive direction of their horizontal axes (marked x and x'). An observer in S reports spacetime coordinates x, y, z, t for an event, and an observer in S' reports x', y', z', t' for the same event. How are these sets of numbers related?

We claim at once (although it requires proof) that the y and z coordinates, which are perpendicular to the motion, are not affected by the motion. That is, $y = y'$ and $z = z'$. Our interest then reduces to the relation between x and x' and between t and t'.

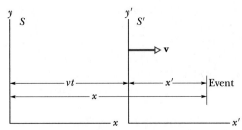

FIGURE 38-9 Two inertial reference frames: frame S' has velocity **v** relative to frame S.

The Galilean Transformation Equations

Prior to Einstein's publication of his special theory of relativity, the four coordinates of interest were assumed to be related by the *Galilean transformation equations:*

$$x' = x - vt$$
$$t' = t$$

(Galilean transformation equations; approximately valid at low speeds). (38-13)

These equations are written with the assumption that $t = t' = 0$ when the origins of S and S' coincide. You can verify the first equation with Fig. 38-9. The second equation effectively claims that time passes at the same rate for observers in both reference frames. That would have been so obviously true to a scientist prior to Einstein that it would not even have been mentioned. When speed v is small compared to c, Eqs. 38-13 generally work well.

The Lorentz Transformation Equations

We state without proof that the correct transformation equations, which remain valid for all speeds up to the speed of light, can be derived from the postulates of relativity. The results, called the **Lorentz transformation equations***, are

$$x' = \gamma(x - vt),$$
$$y' = y,$$
$$z' = z,$$
$$t' = \gamma(t - vx/c^2)$$

(Lorentz transformation equations; valid at all speeds). (38-14)

Note the spatial values x and the temporal values t are bound together in the first and last equations. This entanglement of space and time was a prime message of Einstein's theory, a message that was long rejected by many of his contemporaries.

It is a formal requirement of relativistic equations that they should reduce to familiar classical equations if we let c approach infinity. That is, if the speed of light were infinitely great, *all* finite speeds would be ''low'' and classical equations would never fail. If we let $c \to \infty$ in Eqs. 38-14, $\gamma \to 1$ and these equations reduce—as we expect—to the Galilean equations (Eqs. 38-13). You should check this.

Equations 38-14 are written in a form that is useful if we are given x and t and wish to find x' and t'. We may

*You may wonder why we do not call these the *Einstein transformation equations* (and why not the *Einstein factor* for γ). H. A. Lorentz actually derived these equations before Einstein did, but as the great Dutch physicist graciously conceded, he did not take the further bold step of interpreting these equations as describing the true nature of space and time. It is this interpretation, first made by Einstein, that is at the heart of relativity.

TABLE 38-2 THE LORENTZ TRANSFORMATION EQUATIONS FOR PAIRS OF EVENTS

1. $\Delta x = \gamma(\Delta x' + v\,\Delta t')$	1.' $\Delta x' = \gamma(\Delta x - v\,\Delta t)$
2. $\Delta t = \gamma(\Delta t' + v\,\Delta x'/c^2)$	2.' $\Delta t' = \gamma(\Delta t - v\,\Delta x/c^2)$

$$\gamma = \frac{1}{\sqrt{1 - (v/c)^2}} = \frac{1}{\sqrt{1 - \beta^2}}$$

wish to go the other way, however. In that case we simply solve Eqs. 38-14 for x and t, obtaining

$$x = \gamma(x' + vt'),$$
$$t = \gamma(t' + vx'/c^2).$$

(38-15)

Comparison shows that, starting from either Eqs. 38-14 or Eqs. 38-15, you can find the other set by interchanging primed and unprimed quantities and reversing the sign of the relative velocity v.

Equations 38-14 and 38-15 relate the coordinates of a single event as seen by two observers. Sometimes we want to know not the coordinates of a single event but the differences between coordinates for a pair of events. That is, if we label our events 1 and 2, we may want to relate

$$\Delta x = x_2 - x_1 \quad \text{and} \quad \Delta t = t_2 - t_1,$$

as measured by an observer in S, and

$$\Delta x' = x_2' - x_1' \quad \text{and} \quad \Delta t' = t_2' - t_1',$$

as measured by an observer in S'.

Table 38-2 displays the Lorentz equations in difference form, suitable for analyzing pairs of events. The equations in the table were derived by simply substituting differences (such as Δx and $\Delta x'$) for the four variables in Eqs. 38-14 and 38-15.

Be careful: when substituting values for these differences, you must be consistent and not mix the values for the first event with those for the second event. And if, say, Δx is a negative quantity, you must be certain to include the minus sign.

CHECKPOINT 3: The following figure shows three situations in which a blue reference frame and a green reference frame are in relative motion along the common direction of their x and x' axes, as indicated by the velocity vector attached to one of the frames. For each situation, if we choose the blue frame to be stationary, then is v in the equations of Table 38-2 a positive or negative quantity?

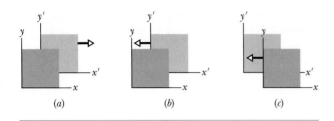

(a)   (b)   (c)

38-8 SOME CONSEQUENCES OF THE LORENTZ EQUATIONS

Here we use the transformation equations of Table 38-2 to affirm some of the conclusions that we reached earlier by arguments based directly on the postulates.

Simultaneity

Consider Eq. 2 of Table 38-2,

$$\Delta t = \gamma \left(\Delta t' + \frac{v \, \Delta x'}{c^2} \right). \qquad (38\text{-}16)$$

If two events occur at different places in reference frame S' of Fig. 38-9, then $\Delta x'$ in this equation is not zero. It follows that even if the events are simultaneous in S' (so $\Delta t' = 0$), they will not be simultaneous in frame S. The time interval between them in S will be

$$\Delta t = \gamma \frac{v \, \Delta x'}{c^2} \qquad \text{(simultaneous events in } S'\text{)}.$$

This is in accord with our conclusion in Section 38-4.

Time Dilation

Suppose now that two events occur at the same place in S' (so $\Delta x' = 0$) but at different times (so $\Delta t' \neq 0$). Equation 38-16 then reduces to

$$\Delta t = \gamma \, \Delta t' \qquad \begin{array}{c}\text{(events in same}\\ \text{place in } S'\text{).}\end{array} \qquad (38\text{-}17)$$

This confirms time dilation. Because the two events occur at the same place in S', the time interval $\Delta t'$ between them can be measured with a single clock, located at that place. Under these conditions, the measured interval is a proper time interval, and we can label it Δt_0. Thus Eq. 38-17 becomes

$$\Delta t = \gamma \, \Delta t_0 \qquad \text{(time dilation)},$$

which is exactly Eq. 38-8, the time dilation equation.

Length Contraction

Consider Eq. 1' of Table 38-2,

$$\Delta x' = \gamma(\Delta x - v \, \Delta t). \qquad (38\text{-}18)$$

If a rod lies parallel to the x, x' axes of Fig. 38-9 and is at rest in reference frame S', an observer in S' can measure its length at leisure. The value $\Delta x'$ that is obtained by subtracting the coordinates of the end points of the rod will be its proper length L_0.

Suppose the rod is moving in frame S. This means that Δx can be identified as the length L of the rod only if the coordinates of the end points are measured *simultaneously,* that is, if $\Delta t = 0$. If we put $\Delta x' = L_0$, $\Delta x = L$, and $\Delta t = 0$ in Eq. 38-18, we find

$$L = \frac{L_0}{\gamma} \qquad \text{(length contraction)}, \qquad (38\text{-}19)$$

which is exactly Eq. 38-9, the length contraction equation.

SAMPLE PROBLEM 38-6

An Earth starship has been sent to check an Earth outpost on the planet P1407, whose moon houses a battle group of the often hostile Reptulians. As the ship follows a straight-line course first past the planet and then past the moon, it detects a high-energy microwave burst at the Reptulian moon base and then, 1.10 s later, an explosion at the Earth outpost, which is 4.00×10^8 m from the Reptulian base as measured from the ship's reference frame. The Reptulians have obviously attacked the Earth outpost; so the starship begins to prepare for a confrontation with them.

(a) The speed of the ship relative to the planet and its moon is $0.980c$. What are the distance between the burst and the explosion and the time interval between them as measured in the planet–moon inertial frame (and thus according to the occupants of the stations)?

SOLUTION: The situation is shown in Fig. 38-10, where the ship's frame S is chosen to be stationary and the planet–moon frame S' is chosen to be moving with positive velocity (rightward). (This is an arbitrary choice: we could, instead, choose the planet–moon frame to be stationary. Then we would redraw **v** in Fig. 38-10 as attached to the S frame and being leftward; v would then be a negative quantity. The result would be the same.) Let subscripts e and b represent the explosion and burst, respectively. Then the given data, all in the unprimed reference frame, are

$$\Delta x = x_e - x_b = +4.00 \times 10^8 \text{ m}$$

and

$$\Delta t = t_e - t_b = +1.10 \text{ s}.$$

Here Δx is a positive quantity because in Fig. 38-10, the coordinate x_e for the explosion is greater than the coordinate x_b for the burst; Δt is also a positive quantity because the time t_e of the explosion is greater (later) than the time t_b of the burst.

We seek $\Delta x'$ and $\Delta t'$, which we shall get by transforming the given S-frame data to the planet–moon frame S'. Because we are considering a pair of events, we choose transformation equations from Table 38-2, namely Eqs. 1′ and 2′:

$$\Delta x' = \gamma(\Delta x - v\,\Delta t) \qquad (38\text{-}20)$$

and

$$\Delta t' = \gamma\left(\Delta t - \frac{v\,\Delta x}{c^2}\right). \qquad (38\text{-}21)$$

Here, $v = +0.980c$, and the Lorentz factor is

$$\gamma = \frac{1}{\sqrt{1 - (v/c)^2}} = \frac{1}{\sqrt{1 - (+0.980c/c)^2}} = 5.0252.$$

So, Eq. 38-20 becomes

$\Delta x' = (5.0252)$

$\qquad \times [4.00 \times 10^8 \text{ m} - (+0.980)(3.00 \times 10^8 \text{ m/s})(1.10 \text{ s})]$

$\qquad = 3.85 \times 10^8 \text{ m}, \qquad$ (Answer)

and Eq. 38-21 becomes

$\Delta t' = (5.0252)$

$$\times \left[(1.10 \text{ s}) - \frac{(+0.980)(3.00 \times 10^8 \text{ m/s})(4.00 \times 10^8 \text{ m})}{(3.00 \times 10^8 \text{ m/s})^2} \right]$$

$= -1.04 \text{ s}. \qquad$ (Answer)

(b) What is the meaning of the minus sign in the computed value for $\Delta t'$?

SOLUTION: Recall how we originally defined the time interval between burst and explosion: $\Delta t = t_e - t_b = +1.10$ s. To be consistent with that choice of notation, our definition of $\Delta t'$ must be $t'_e - t'_b$; thus, we have found that

$$\Delta t' = t'_e - t'_b = -1.04 \text{ s}.$$

This tells us that $t'_b > t'_e$; that is, in the planet–moon reference frame, the burst occurs 1.04 s *after* the explosion, not 1.10 s *before* the explosion as detected in the ship frame.

(c) Did the burst cause the explosion, or did the explosion cause the burst?

SOLUTION: The sequence of events measured in the planet–moon reference frame is the reverse of that measured in the ship frame. In either situation, if there is a causal relationship between the two events, information must travel from one event to cause the other. Let us check the required speed of the information. In the ship frame, this speed is

$$v_{\text{info}} = \frac{\Delta x}{\Delta t} = \frac{4.00 \times 10^8 \text{ m}}{1.10 \text{ s}} = 3.64 \times 10^8 \text{ m/s},$$

but that speed is impossible because it exceeds c. In the planet–moon frame, the speed comes out to be 3.70×10^8 m/s, also impossible. So, neither event could possibly cause the other event; that is, they are *unrelated* events. Thus the starship should not confront the Reptulians.

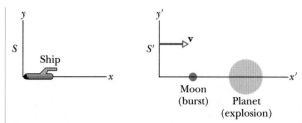

FIGURE 38-10 Sample Problem 38-6. A planet and its moon in reference frame S' move with speed v relative to a starship in reference frame S.

SAMPLE PROBLEM 38-7

Figure 38-11 shows an inertial reference frame S in which event 1 (a rock is kicked up by a truck at coordinates x_1 and t_1) causes event 2 (the rock hits you at coordinates x_2 and t_2). Is there another inertial reference frame S' from which those events can be measured to be reversed in sequence, so that the effect occurs before its cause? (Can you be injured now as a result of a future event?)

SOLUTION: To find the temporal separation $\Delta t'$ of a pair of events in frame S' when we have data for frame S, we use Eq. 2′ of Table 38-2:

$$\Delta t' = \gamma\left(\Delta t - \frac{v\,\Delta x}{c^2}\right). \qquad (38\text{-}22)$$

Recall that v is the relative velocity between S and S'. We take frame S to be stationary; frame S' then has velocity v.

Let $\Delta t = t_2 - t_1$. Then Δt is a positive quantity and, to be consistent with this notation, we must have $\Delta x = x_2 - x_1$ and $\Delta t' = t'_2 - t'_1$. As Fig. 38-11 is drawn, Δx is a positive quantity because $x_2 > x_1$.

We are interested in the possibility that $\Delta t'$ is a negative quantity, which would mean that time t'_1 of event 1 is later (and thus greater) than time t'_2 of event 2. From Eq. 38-22, we see that $\Delta t'$ can be negative only if

$$\frac{v\,\Delta x}{c^2} > \Delta t.$$

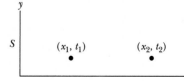

FIGURE 38-11 Sample Problem 38-7. Event 1 at spacetime coordinates (x_1, t_1) causes event 2 at spacetime coordinates (x_2, t_2). Can the sequence of cause and effect be reversed in some other reference frame?

This condition can be rearranged to produce the equivalent condition

$$\frac{\Delta x/\Delta t}{c}\frac{v}{c} > 1.$$

The ratio $\Delta x/\Delta t$ is just the speed at which information (here via a rock) travels from event 1 to produce event 2. That speed cannot exceed c. (Information could travel at c if it comes via light; rocks travel more slowly, of course.) So $(\Delta x/\Delta t)/c$ must be at most 1. And v/c cannot equal or exceed 1. Thus the left side of the last inequality must be less than 1, and the inequality cannot be satisfied.

So there is no frame S' in which event 2 occurs before its cause, event 1. More generally, although the sequence of unrelated events can sometimes be reversed in relativity (as in Sample Problem 38-6), events involving cause and effect can never be reversed.

38-9 THE RELATIVITY OF VELOCITIES

Here we wish to use the Lorentz transformation equations to compare the velocities that two observers in different inertial reference frames S and S' would measure for the same moving particle. We assume again that S' moves with velocity v relative to S.

Suppose that the particle, moving with constant velocity parallel to the x, x' axes in Fig. 38-12, sends out two signals as it moves. Each observer measures the space interval and the time interval between these two events. These four measurements are related by Eqs. 1 and 2 of Table 38-2,

$$\Delta x = \gamma(\Delta x' + v\,\Delta t')$$

and

$$\Delta t = \gamma\left(\Delta t' + \frac{v\,\Delta x'}{c^2}\right).$$

If we divide the first of these equations by the second, we find

$$\frac{\Delta x}{\Delta t} = \frac{\Delta x' + v\,\Delta t'}{\Delta t' + v\,\Delta x'/c^2}.$$

Dividing the numerator and denominator of the right side

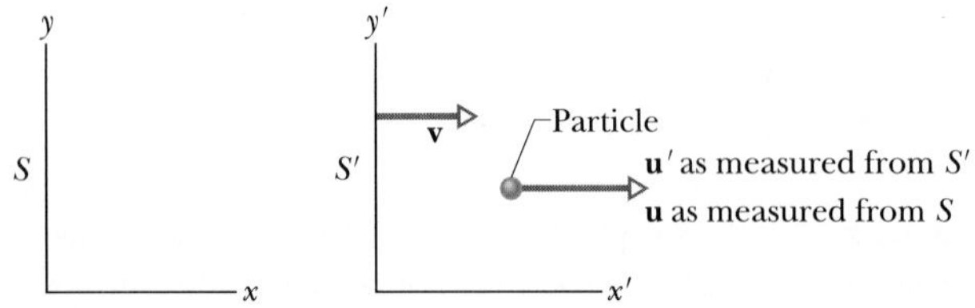

FIGURE 38-12 Reference frame S' moves with velocity $\mathbf{v}$ relative to frame S. A particle has velocity $\mathbf{u}'$ relative to reference frame S' and velocity $\mathbf{u}$ relative to reference frame S.

by $\Delta t'$, we find

$$\frac{\Delta x}{\Delta t} = \frac{\Delta x'/\Delta t' + v}{1 + v(\Delta x'/\Delta t')/c^2}.$$

But, in the differential limit, $\Delta x/\Delta t$ is u, the velocity of the particle as measured in S, and $\Delta x'/\Delta t'$ is u', the velocity of the particle as measured in S'. Then we have, finally,

$$u = \frac{u' + v}{1 + u'v/c^2} \qquad \begin{array}{l}\text{(velocity}\\ \text{transformation)}\end{array} \qquad (38\text{-}23)$$

as the relativistic velocity transformation equation. We discussed this equation, using a different notation, in Section 4-10. You may wish to reread that section and, in particular, to study Sample Problems 4-15 and 4-16. Equation 38-23 reduces to the classical, or Galilean, velocity transformation equation,

$$u = u' + v \qquad \begin{array}{l}\text{(classical velocity}\\ \text{transformation),}\end{array} \qquad (38\text{-}24)$$

when we apply the formal test of letting $c \to \infty$.

38-10 DOPPLER EFFECT FOR LIGHT

In Section 18-8 we discussed the Doppler effect (a shift in detected frequency) for sound waves traveling in air. For such waves, the Doppler effect depends on two velocities, namely, the velocities with respect to the air, which is the medium that transmits the waves, of the source and the detector.

That is not the situation with light waves, for they (and other electromagnetic waves) require no medium, being able to travel even through vacuum. The Doppler effect for light waves depends on only one velocity, the relative velocity $\mathbf{v}$ between source and detector, as measured from the reference frame of either. Let f_0 represent the **proper frequency** of the source, that is, the frequency that is measured by an observer in the rest frame of the source. Let f represent the frequency detected by an observer moving with velocity $\mathbf{v}$ relative to that rest frame. Then, when the direction of $\mathbf{v}$ is directly away from the source,

$$f = f_0 \sqrt{\frac{1 - \beta}{1 + \beta}} \qquad \begin{array}{l}\text{(source and}\\ \text{detector separating),}\end{array} \qquad (38\text{-}25)$$

where $\beta = v/c$. When the direction of $\mathbf{v}$ is directly toward the source, we must change the signs in front of both β symbols in Eq. 38-25.

According to Eq. 38-25, when the separation between source and detector is increasing, the detected frequency f is less than the proper frequency f_0. Recalling that $f = c/\lambda$, where λ is the wavelength of the light, we see that the decrease in frequency corresponds to an increase in wavelength. In Section 18-9 we called such an increase in wavelength a *red shift* (because the red portion of the visible spectrum has the longest wavelengths). Similarly, when the source–detector separation is decreasing, f is greater than f_0; this corresponds to a decrease in wavelength, that is, a *blue shift*.

For low speeds ($\beta \ll 1$), Eq. 38-25 can be expanded in a power series in β and approximated as

$$f = f_0(1 - \beta + \tfrac{1}{2}\beta^2) \qquad \text{(low speeds).} \quad (38\text{-}26)$$

The corresponding low-speed equation for sound waves (or any waves except light waves) has the same first two terms but a different coefficient in the third term. Thus the relativistic effect for low-speed light sources and detectors shows up only with the β^2 term.

As discussed briefly in Chapter 18, police radar units employ the Doppler effect with microwaves. A source in the radar unit emits a microwave signal at a certain frequency f_0 along the road. A car that is moving toward the unit intercepts a microwave signal whose frequency is shifted up to the frequency f in Eq. 38-25 (with the signs of β changed) by the Doppler effect. The car reflects that wave back toward the radar unit. Because the car is moving toward the radar unit, the detector in the unit intercepts a reflected signal that is further shifted up in frequency. The unit compares that detected frequency with f_0 and computes the speed of the car.

CHECKPOINT 4: The figure shows a source that emits light of proper frequency f_0 while moving directly toward the right with speed $c/4$ as measured from reference frame S. The figure also shows a light detector, which measures a frequency $f > f_0$ for the emitted light. (a) Is the detector moving toward the left or the right? (b) Is the speed of the detector as measured from reference frame S more than $c/4$, less than $c/4$, or equal to $c/4$?

Transverse Doppler Effect

So far, we have discussed the Doppler effect, here and in Chapter 18, only for situations in which the source and the detector move either directly toward or directly away from each other. Figure 38-13 shows a different arrangement, in

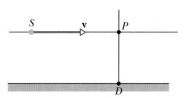

FIGURE 38-13 A light source S travels with velocity **v** past a detector at D. The special theory of relativity predicts a transverse Doppler effect as the source passes through point P, where the direction of travel is then perpendicular to the line extending through D. Classical theory predicts no such effect.

which a source S moves past a detector D. When S reaches point P, its velocity is perpendicular to the line joining S and D and, at that instant, it is moving neither toward nor away from D. If the source is emitting sound waves of frequency f, D detects that frequency (with no Doppler effect) when it intercepts the waves that were emitted at point P. However, if the source is emitting light waves, there is still a Doppler effect, called the **transverse Doppler effect.** In this situation, the detected frequency of light emitted by the source at point P is

$$f = f_0\sqrt{1 - \beta^2} \qquad \text{(transverse Doppler effect).} \quad (38\text{-}27)$$

For low speeds ($\beta \ll 1$), Eq. 38-27 can be expanded in a power series in β and approximated as

$$f = f_0(1 - \tfrac{1}{2}\beta^2) \qquad \text{(low speeds).} \quad (38\text{-}28)$$

Here the first term is what we would expect for sound waves and, again, the relativistic effect for low-speed light sources and detectors appears with the β^2 term.

In principle, a police radar unit can determine the speed of a car even when the path of the radar pulse is perpendicular (transverse) to the path of the car. However, Eq. 38-28 tells us that because β is small even for a fast car, the relativistic term $\beta^2/2$ in the transverse Doppler effect is extremely small. So $f \approx f_0$ and the radar unit computes a speed of zero. For this reason, police officers always try to direct the radar pulse along the car's path to get a Doppler shift that gives the car's actual speed. Any deviation from that alignment works in favor of the motorist, because it reduces the measured speed.

The transverse Doppler effect is really another test of time dilation. If we rewrite Eq. 38-27 in terms of the period T of oscillation of the emitted light wave instead of the frequency, we have, since $T = 1/f$,

$$T = \frac{T_0}{\sqrt{1 - \beta^2}} = \gamma T_0, \quad (38\text{-}29)$$

in which $T_0 (= 1/f_0)$ is the **proper period** of the source. As comparison with Eq. 38-8 shows, Eq. 38-29 is simply the time dilation formula, since a period is a time interval.

The NAVSTAR Navigation System

Each NAVSTAR satellite continually broadcasts radio signals giving its location, at a frequency that is set and controlled by precise atomic clocks. When the signal is sensed by the detector on, say, an aircraft, the frequency has been Doppler-shifted. By detecting the signals from several NAVSTAR satellites simultaneously, the detector can determine the direction to any one of them and the direction of the velocity of that satellite. From the Doppler shift of the signal, the detector then determines the speed of the aircraft.

Let us use some rough numbers to see how well this can be done. The speed of a NAVSTAR satellite relative to the center of Earth is about 1.0×10^4 m/s. The associated β is about 3.0×10^{-5}. Thus the term $\beta^2/2$ in Eqs. 38-26 and 38-28 (that is, the relativity term) is about 4.5×10^{-10}. In other words, relativity changes the Doppler shift of the detected signal by about 4.5 parts in 10^{10}, which seems hardly worth considering.

However, it is indeed important. The atomic clocks in the satellites are so precise that the variation in the frequency of the satellite signal is only 2 parts in 10^{12}. From Eq. 38-28, we see that β (hence v) depends on the square root of f/f_0. Thus the clock's frequency variation of 2×10^{-12} causes a variation of

$$\sqrt{2 \times 10^{-12}} = 1.4 \times 10^{-6}$$

in the measured value of the relative speed v between satellite and aircraft.

Since v is due primarily to the satellite's great speed, 1.0×10^4 m/s, this means that v (hence the aircraft's speed) can be determined to an accuracy of about

$$(1.4 \times 10^{-6})(1.0 \times 10^4 \text{ m/s}) = 1.4 \text{ cm/s}.$$

Suppose the aircraft flies for 1 h (3600 s). Knowing the speed to about 1.4 cm/s allows the location at the end of that hour to be predicted to about

$$(0.014 \text{ m/s})(3600 \text{ s}) = 50 \text{ m},$$

which is acceptable in modern navigation.

If relativity effects were not taken into account, the speed of the aircraft could not be known any closer than 21 cm/s, and its location after an hour's flight could not be predicted any better than within 760 m.

38-11 A NEW LOOK AT MOMENTUM

Suppose that a number of observers, each in a different inertial reference frame, watch an isolated collision between two particles. In classical mechanics, we have seen that—even though the observers measure different velocities for the colliding particles—they all find that the law of conservation of momentum holds. That is, they find that the momentum of the system of particles after the collision is the same as it was before the collision.

How is this situation affected by relativity? We find that if we continue to define the momentum **p** of a particle as $m\mathbf{v}$, the product of its mass and its velocity, momentum is *not* conserved for all inertial observers. We have two choices: (1) give up the law of conservation of momentum or (2) see if we can redefine the momentum of a particle in some new way so that the law of conservation of momentum still holds. We choose the second route.

Consider a particle moving with constant speed v in the x direction. Classically, its momentum has magnitude

$$p = mv = m\frac{\Delta x}{\Delta t} \qquad \text{(classical momentum)}, \qquad (38\text{-}30)$$

in which Δx is the distance covered in time Δt. To find a relativistic expression for momentum, we start with the new definition

$$p = m\frac{\Delta x}{\Delta t_0}.$$

Here, as before, Δx is the distance covered by a moving particle as viewed by an observer watching that particle. However, Δt_0 is the time required to cover that distance, measured not by the observer watching the moving particle but by an observer moving with the particle. The particle is at rest with respect to this second observer, with the result that the time this observer measures is a proper time Δt_0.

Using the time dilation formula (Eq. 38-8), we can then write

$$p = m\frac{\Delta x}{\Delta t_0} = m\frac{\Delta x}{\Delta t}\frac{\Delta t}{\Delta t_0} = m\frac{\Delta x}{\Delta t}\gamma.$$

But since $\Delta x/\Delta t$ is just the particle velocity v,

$$p = \gamma mv \qquad \text{(momentum)}. \qquad (38\text{-}31)$$

Note that this differs from the classical definition of Eq. 38-30 only by the Lorentz factor γ. However, that difference is important: unlike classical momentum, relativistic momentum approaches infinite values as v approaches c.

We can generalize the definition of Eq. 38-31 to vector form as

$$\mathbf{p} = \gamma m\mathbf{v} \qquad \text{(momentum)}. \qquad (38\text{-}32)$$

We introduced this definition without elaboration in Section 9-4 as a foretaste of things to come (see Eq. 9-24). We state without further proof that, if we adopt the definition of momentum presented in Eq. 38-32, we can continue to apply the principle of conservation of momentum up to the very highest particle speeds.

38-12 A NEW LOOK AT ENERGY

In Section 7-8 we introduced, without elaboration, a relativistic expression for the kinetic energy of a particle:

$$K = mc^2 \left(\frac{1}{\sqrt{1 - (v/c)^2}} - 1 \right).$$

We can now write this equation as

$$K = mc^2(\gamma - 1) \qquad \text{(kinetic energy).} \qquad (38\text{-}33)$$

We showed in Section 7-8 that—unlikely as it may seem—this expression reduces to the familiar classical $K = \frac{1}{2}mv^2$ at low speeds. Furthermore, Eq. 38-33 can be derived in exactly the same way as the classical kinetic energy expression: by setting the kinetic energy K equal to the work required to accelerate the particle from rest to its observed speed. Let us point out some of the consequences of Eq. 38-33.

Total Energy

We start by defining the *total energy* E of a particle as γmc^2. With the help of Eq. 38-33 we can then write

$$\begin{aligned} E &= \gamma mc^2 \\ &= mc^2 + K \end{aligned} \qquad \begin{array}{l}\text{(total energy;}\\ \text{single particle).}\end{array} \quad (38\text{-}34)$$

We interpret Eq. 38-34 as implying that the total energy E of a moving particle is made up of mc^2, which we call the **mass energy** or the **rest energy** of the particle, and K, its kinetic energy. Table 8-1 lists the rest energies of a few particles and other objects. The rest energy of an electron, for example, is 0.511 MeV, and for a proton it is 938 MeV.

The total energy of a system of n particles is

$$E = \sum_{j=1}^{n} E_j = \sum_{j=1}^{n} (\gamma_j m_j c^2) = \sum_{j=1}^{n} m_j c^2 + \sum_{j=1}^{n} K_j$$
$$\text{(total energy; system of particles).} \qquad (38\text{-}35)$$

In relativity, the principle of *conservation of energy* is stated as follows:

For an isolated system of particles, the total energy E of the system, which is defined by Eq. 38-35, remains constant, no matter what interactions may occur among the particles.

Thus in any isolated reaction or decay process involving two or more particles, the total energy of the system after the process must be equal to the total energy before the process. During the process, the total rest energy of the interacting particles may change but then the total kinetic energy must also change by an equal amount in the opposite direction to compensate.

Considerations of this sort are at the root of Einstein's well-known relation $E = mc^2$, which asserts that rest energy can be converted to other forms. All reactions—whether chemical or nuclear—in which energy is released or absorbed involve a corresponding change in the rest energy of the reactants. We discussed the relation $E = mc^2$ in detail in Section 8-8.

Momentum and Kinetic Energy

In classical mechanics, the momentum p of a particle is mv and its kinetic energy K is $\frac{1}{2}mv^2$. If we eliminate v between these two expressions, we find a direct relation between momentum and kinetic energy:

$$p^2 = 2Km \qquad \text{(classical).} \qquad (38\text{-}36)$$

We can find a similar connection in relativity by eliminating v between the relativistic definition of momentum (Eq. 38-31) and the relativistic definition of kinetic energy (Eq. 38-33). Doing so leads, after some algebra, to

$$(pc)^2 = K^2 + 2Kmc^2. \qquad (38\text{-}37)$$

With the aid of Eq. 38-34, we can transform Eq. 38-37 into a relation between the momentum p and the total energy E of a particle:

$$E^2 = (pc)^2 + (mc^2)^2. \qquad (38\text{-}38)$$

The right triangle of Fig. 38-14 helps to keep these useful relations in mind. You can also show that, in that triangle,

$$\sin \theta = \beta \quad \text{and} \quad \sin \phi = 1/\gamma. \qquad (38\text{-}39)$$

With Eq. 38-38 we can see that the product pc must have the same unit as energy E; thus we can express the unit of momentum p as an energy unit divided by c. In fact, momentum in particle physics is often reported in the units MeV/c or GeV/c.

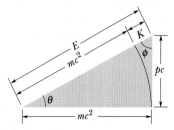

FIGURE 38-14 A useful mnemonic device for remembering the relativistic relations among the total energy E, the rest energy or mass energy mc^2, the kinetic energy K, and the momentum p.

CHECKPOINT 5: Are (a) the kinetic energy and (b) the total energy of a 1 GeV electron more than, less than, or equal to those of a 1 GeV proton?

SAMPLE PROBLEM 38-8

(a) What is the total energy E of a 2.53 MeV electron?

SOLUTION: From Eq. 38-34 we have

$$E = mc^2 + K.$$

From Table 8-1, mc^2 for an electron is 0.511 MeV; so

$$E = 0.511 \text{ MeV} + 2.53 \text{ MeV} = 3.04 \text{ MeV}. \quad \text{(Answer)}$$

(b) What is its momentum p?

SOLUTION: From Eq. 38-38,

$$E^2 = (pc)^2 + (mc^2)^2,$$

we can write

$$(3.04 \text{ MeV})^2 = (pc)^2 + (0.511 \text{ MeV})^2.$$

Then

$$pc = \sqrt{(3.04 \text{ MeV})^2 - (0.511 \text{ MeV})^2} = 3.00 \text{ MeV},$$

and, giving the momentum in units of energy divided by c, we have

$$p = 3.00 \text{ MeV}/c. \quad \text{(Answer)}$$

(c) What is the Lorentz factor γ for the electron?

SOLUTION: From Eq. 38-34, we have

$$E = \gamma mc^2.$$

With $E = 3.04$ MeV and $m = 9.11 \times 10^{-31}$ kg, we then have

$$\gamma = \frac{E}{mc^2} = \frac{(3.04 \times 10^6 \text{ eV})(1.6 \times 10^{-19} \text{ J/eV})}{(9.11 \times 10^{-31} \text{ kg})(3.0 \times 10^8 \text{ m/s})^2}$$

$$= 5.93. \quad \text{(Answer)}$$

SAMPLE PROBLEM 38-9

The most energetic proton ever detected in the cosmic rays coming in from space had an astounding kinetic energy of 3.0×10^{20} eV (enough energy to warm a teaspoon of water by a few degrees).

(a) Calculate the proton's Lorentz factor γ and speed v.

SOLUTION: Solving Eq. 38-33 for γ, we find

$$\gamma = \frac{K + mc^2}{mc^2} = \frac{K}{mc^2} + 1 = \frac{3.0 \times 10^{20} \text{ eV}}{938 \times 10^6 \text{ eV}} + 1$$

$$= 3.198 \times 10^{11} \approx 3.2 \times 10^{11}. \quad \text{(Answer)}$$

Here we used 938 MeV for a proton's rest energy.

This computed value for γ is so large that we cannot use the definition of γ (Eq. 38-7) to find v. Try it; your calculator will tell you that β is effectively equal to 1 and thus that v is effectively equal to c. Actually, v is almost c, but we want a more accurate answer, which we can obtain by first solving Eq. 38-7 for $1 - \beta$. To begin we write

$$\gamma = \frac{1}{\sqrt{1 - \beta^2}} = \frac{1}{\sqrt{(1 - \beta)(1 + \beta)}} \approx \frac{1}{\sqrt{2(1 - \beta)}},$$

where we have used the fact that β is so close to unity that $1 + \beta$ is very close to 2. Solving for $1 - \beta$ then yields

$$1 - \beta = \frac{1}{2\gamma^2} = \frac{1}{(2)(3.198 \times 10^{11})^2}$$

$$= 4.9 \times 10^{-24} \approx 5 \times 10^{-24}.$$

So

$$\beta = 1 - 5 \times 10^{-24},$$

and since $v = \beta c$,

$$v \approx 0.999\ 999\ 999\ 999\ 999\ 999\ 999\ 995c. \quad \text{(Answer)}$$

(b) Suppose that the proton travels along a diameter (9.8×10^4 ly) of the Milky Way galaxy. Approximately how long does the proton take to travel that diameter as measured from the common reference frame of Earth and the galaxy?

SOLUTION: We just saw that this *ultrarelativistic* proton is traveling at a speed barely less than c. By the definition of light-year, light takes 9.8×10^4 y to travel 9.8×10^4 ly, and this proton should take almost the same time. Thus, from our Earth–Milky Way reference frame, the trip takes

$$\Delta t = 9.8 \times 10^4 \text{ y}. \quad \text{(Answer)}$$

(c) How long does the trip take as measured in the rest frame of the proton?

SOLUTION: Because the start of the trip and the end of the trip occur at the same location in the proton's rest frame, namely, coincident with the proton itself, what we seek is the proper time of the trip. We can use the time dilation equation (Eq. 38-8) to transform Δt from the Earth–Milky Way frame to the proton rest frame:

$$\Delta t_0 = \frac{\Delta t}{\gamma} = \frac{9.8 \times 10^4 \text{ y}}{3.198 \times 10^{11}}$$

$$= 3.06 \times 10^{-7} \text{ y} = 9.7 \text{ s}. \quad \text{(Answer)}$$

In our frame, the trip takes 98,000 y. In the proton's frame, it takes 9.7 s! As promised at the start of this chapter, relative motion can alter the rate at which time passes, and we have here an extreme example.

REVIEW & SUMMARY

The Postulates

Special theory of relativity is based on two postulates:

1. The laws of physics are the same for observers in all inertial reference frames. No frame is preferred.

2. The speed of light in vacuum has the same value c in all directions and in all inertial reference frames.

The speed of light c in vacuum is an ultimate speed that cannot be exceeded by any entity carrying either energy or information.

Coordinates of an Event

Three space coordinates and one time coordinate specify an **event.** One task of special relativity is to relate these coordinates as assigned by two observers who are in uniform motion with respect to each other.

Simultaneous Events

If two observers are in relative motion, they will not, in general, agree as to whether two events are simultaneous. If one observer finds two events at different locations to be simultaneous, the other will not, and conversely. Simultaneity is *not* an absolute concept but a relative one, depending on the motion of the observer. The relativity of simultaneity is a direct consequence of the finite ultimate speed c.

Time Dilation

If two successive events occur at the same place in an inertial reference frame, the time interval Δt_0 between them, measured on a single clock where they occur, is the **proper time** between the events. *Observers in frames moving relative to that frame will measure a larger value for this interval.* For an observer moving with relative speed v, the measured time interval is

$$\Delta t = \frac{\Delta t_0}{\sqrt{1 - (v/c)^2}} = \frac{\Delta t_0}{\sqrt{1 - \beta^2}} = \gamma\, \Delta t_0$$

$$\text{(time dilation).} \quad \text{(38-6 to 38-8)}$$

Here $\beta = v/c$ is the **speed parameter** and $\gamma = 1/\sqrt{1 - \beta^2}$ is the **Lorentz factor.** An important consequence of time dilation is that moving clocks run slow as measured by an observer at rest.

Length Contraction

The length L_0 of an object measured by an observer in an inertial reference frame in which the object is at rest is called its **proper length.** *Observers in frames mvoing relative to that frame and parallel to that length will measure a shorter length.* For an observer moving with relative speed v, the measured length is

$$L = L_0\sqrt{1 - \beta^2} = \frac{L_0}{\gamma} \qquad \substack{\text{(length} \\ \text{contraction).}} \quad (38\text{-}9)$$

The Lorentz Transformation

The *Lorentz transformation* equations relate the spacetime coordinates of a single event as seen by observers in two inertial frames, S and S', where S' is moving relative to S with velocity v

in the positive x, x' direction. The four coordinates are related by

$$x' = \gamma(x - vt),$$
$$y' = y, \qquad \qquad \text{(Lorentz transformation}$$
$$\qquad \qquad \qquad \text{equations; valid} \qquad (38\text{-}14)$$
$$z' = z, \qquad \qquad \text{at all speeds).}$$
$$t' = \gamma(t - vx/c^2)$$

Relativity of Velocities

When a particle is moving with speed u' in the positive x' direction in an inertial reference frame S' that itself is moving with speed v parallel to the x direction of a second inertial frame S, the speed u of the particle as measured in S is

$$u = \frac{u' + v}{1 + u'v/c^2} \qquad \text{(relativistic velocity).} \quad (38\text{-}23)$$

Relativistic Doppler Effect

If a source emitting light waves of frequency f_0 moves directly away from a detector with relative velocity $\mathbf{v}$, the frequency f measured by the detector is

$$f = f_0 \sqrt{\frac{1 - \beta}{1 + \beta}}. \quad (38\text{-}25)$$

Transverse Doppler Effect

If the relative motion of the source is perpendicular to the source–detector line, the Doppler formula is

$$f = f_0 \sqrt{1 - \beta^2}. \quad (38\text{-}27)$$

This **transverse Doppler effect** is due to time dilation.

Momentum and Energy

The definitions of linear momentum $\mathbf{p}$, kinetic energy K, and total energy E that are valid at any possible speed are

$$\mathbf{p} = \gamma m \mathbf{v} \qquad \text{(momentum),} \quad (38\text{-}32)$$
$$K = mc^2(\gamma - 1) \qquad \text{(kinetic energy),} \quad (38\text{-}33)$$
$$E = \gamma mc^2 = mc^2 + K \qquad \substack{\text{(total energy,} \\ \text{single particle).}} \quad (38\text{-}34)$$

With these definitions, the principle of conservation of total energy for a system of particles takes the form

$$E = \sum_{j=1}^{n} (\gamma_j m_j c^2) = \sum_{j=1}^{n} m_j c^2 + \sum_{j=1}^{n} K_j$$

$$\text{(total energy, system of particles).} \quad (38\text{-}35)$$

Two additional energy relationships, derivable from Eqs. 38-22, 38-33, and 38-34, are often useful:

$$(pc)^2 = K^2 + 2Kmc^2, \quad (38\text{-}37)$$
$$E^2 = (pc)^2 + (mc^2)^2. \quad (38\text{-}38)$$

QUESTIONS

1. In Fig. 38-15, ship A sends a laser pulse to an oncoming ship B, while scout ship C races away. The indicated speeds of the ships are all measured from the same reference frame. Rank the ships according to the speed of the pulse as measured from each ship, greatest first.

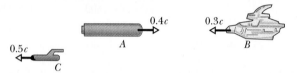

FIGURE 38-15 Questions 1 and 10.

2. Figure 38-16a shows two clocks in stationary frame S (they are synchronized in that frame) and one clock in moving frame S'. Clocks C_1 and C_1' read zero when they pass each other. When clocks C_1' and C_2 pass each other, (a) which clock has the smaller reading and (b) which clock measures a proper time?

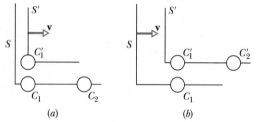

FIGURE 38-16 Questions 2 and 3.

3. Figure 38-16b shows two clocks in stationary frame S' (they are synchronized in that frame) and one clock in moving frame S. Clocks C_1 and C_1' read zero when they pass each other. When clocks C_1 and C_2' pass each other, (a) which clock has the smaller reading and (b) which clock measures a proper time?

4. Sam leaves Venus in a spaceship to Mars and passes Sally, who is on Earth, with a relative speed of $0.5c$. (a) Each measures the Venus–Mars voyage time. Who measures a proper time: Sam, Sally, or neither? (b) On the way, Sam sends a pulse of light to Mars. Each measures the travel time of the pulse. Who measures a proper time?

5. Figure 38-17 shows three situations in which a starship passes Earth (the dot) and then makes a round trip that brings it back past Earth, each at the given Lorentz factor. As measured in the rest frame of Earth, the round-trip distances are as follows: trip 1, $2D$; trip 2, $4D$; trip 3, $6D$. Without written calculation and neglecting any time needed for accelerations, rank the situations according to the travel times of the trips, greatest first, as measured from (a) the rest frame of Earth and (b) the rest frame of the starship. (*Hint:* See Sample Problem 38-5.)

FIGURE 38-17 Question 5.

6. Figure 38-18 is a map of the travel lanes allowed through a stellar region by the indigenous alien government. Each lane (between two lettered junction points) is labeled with the maximum value of γ allowed for that lane. In the rest frame of the junctions, successive junctions are separated by a distance of L or $2L$. (a) Beginning at Home Port, pick the route to Far Base that minimizes your travel time, neglecting the time needed for acceleration when γ changes. (*Hint:* See Sample Problem 38-5.) Who measures (b) less time and (c) less distance for that route, you or someone at rest relative to the junctions?

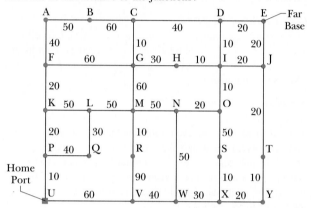

FIGURE 38-18 Question 6.

7. Figure 38-19 shows a ship (with on-board reference frame S') passing us (with reference frame S). A proton is fired at nearly the speed of light along the length of the ship, from the front to the rear. (a) Is the spatial separation $\Delta x'$ between the firing of the proton and its impact a positive or negative quantity? (b) Is the temporal separation $\Delta t'$ between those events a positive or negative quantity?

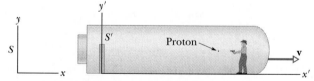

FIGURE 38-19 Question 7.

8. (a) In Fig. 38-9, suppose an observer in frame S' measures two events to be at the same location (say, at x') but not at the same time. Can an observer in frame S possibly measure them to be at the same location? (b) If two events occur simultaneously at the same place for one observer, will they be simultaneous for all other observers? (c) Will they occur at the same place for all other observers?

9. Figure 38-20 shows a starship and an asteroid. In four situations, the velocity of the starship relative to us (on a scout ship) and the velocity of the asteroid relative to the starship are, in that order, (a) $+0.4c$, $+0.4c$; (b) $+0.5c$, $+0.3c$; (c) $+0.9c$, $-0.1c$; and (d) $+0.3c$, $+0.5c$. Without written calculation, rank the situations according to the magnitude of the velocity of the asteroid relative to us, greatest first.

FIGURE 38-20 Question 9.

10. Ships A and B in Fig. 38-15 are moving directly toward each other; the velocities indicated are all measured from the same reference frame. Is the speed of ship A relative to ship B more than $0.7c$, less than $0.7c$, or equal to $0.7c$?

11. Figure 38-21 shows one of four star cruisers that are in a race. As each cruiser passes the starting line, a shuttle craft leaves the cruiser and races toward the finish line. You, judging the race, are stationary relative to the start and finish lines. The speeds v_c of the cruisers relative to you and the speeds v_s of the shuttle craft relative to their starships are, in that order, (1) $0.70c$, $0.40c$; (2) $0.40c$, $0.70c$; (3) $0.20c$, $0.90c$; (4) $0.50c$, $0.60c$. (a) Without written calculation, rank the shuttle craft according to their speeds relative to you, greatest first. (b) Still without written calculation, rank the shuttle craft according to the distances their pilots measure from the starting line to the finish line, greatest first. (c) Each starship sends a signal to its shuttle craft at a certain frequency f_0 as measured on board the starship. Again without written calculation, rank the shuttle craft according to the frequencies they detect, greatest first.

12. While on board a starship, you intercept signals from four shuttle craft that are moving either directly toward or directly away from you. The signals have the same proper frequency f_0. The speed and direction (both relative to you) of the shuttle craft are (a) $0.3c$ toward, (b) $0.6c$ toward, (c) $0.3c$ away, and (d) $0.6c$ away. Rank the shuttle craft according to the frequency you receive, greatest first.

13. Figure 38-22 shows three starships that move either left or right along the axis shown. All emit microwave signals of the same proper frequency f_0. Ship C detects the signal from ship A with a frequency $f_1 > f_0$. Ship A detects the signal from ship B with a frequency $f_2 < f_0$. Is the signal from ship B that is detected by ship C less than f_0, greater than f_1, or between f_0 and f_1?

A B C

FIGURE 38-22 Question 13.

14. The rest energy and total energy, respectively, of three particles, expressed in terms of a basic amount A are (1) A, $2A$; (2) A, $3A$; (3) $3A$, $4A$. Without written calculation, rank the particles according to (a) their mass, (b) their kinetic energy, (c) their Lorentz factor, and (d) their speed, greatest first.

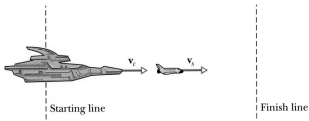

Starting line   Finish line

FIGURE 38-21 Question 11.

EXERCISES & PROBLEMS

SECTION 38-2 The Postulates

1E. What fraction of the speed of light does each of the following speeds represent; that is, what is the speed parameter β? (a) A typical rate of continental drift (1 in./y). (b) A highway speed limit of 55 mi/h. (c) A supersonic plane flying at Mach 2.5 (1200 km/h). (d) The escape speed of a projectile from the surface of Earth. (e) A typical recession speed of a distant quasar (3.0×10^4 km/s).

2E. Quite apart from effects due to Earth's rotational and orbital motions, a laboratory reference frame is not strictly an inertial frame because a particle placed at rest there will not, in general, remain at rest; it will fall. Often, however, events happen so quickly that we can ignore the gravitational acceleration and treat the frame as inertial. Consider, for example, an electron of speed $v = 0.992c$, projected horizontally into a laboratory test chamber and moving through a distance of 20 cm. (a) How long would that journey take, and (b) how far would the electron fall during this interval? What can you conclude about the suitability of the laboratory as an inertial frame in this case?

3P. Find the speed of a particle that takes 2.0 y longer than light to travel a distance of 6.0 ly.

SECTION 38-5 The Relativity of Time

4E. What must be the speed parameter β if the Lorentz factor γ is (a) 1.01, (b) 10.0, (c) 100, and (d) 1000?

5E. The mean lifetime of stationary muons is measured to be 2.2 μs. The mean lifetime of high-speed muons in a burst of cosmic rays observed from Earth is measured to be 16 μs. Find the speed of these cosmic-ray muons relative to Earth.

6P. An unstable high-energy particle enters a detector and leaves a track 1.05 mm long before it decays. Its speed relative to the detector was $0.992c$. What is its proper lifetime? That is, how long would the particle have lasted before decay had it been at rest with respect to the detector?

7P. A pion is created in the higher reaches of Earth's atmosphere when an incoming high-energy cosmic-ray particle collides with an atomic nucleus. A pion so formed descends toward Earth with a speed of $0.99c$. In a reference frame in which they are at rest, pions decay with an average life of 26 ns. As measured in a frame fixed with respect to Earth, how far (on the average) will such a pion move through the atmosphere before it decays?

8P. You wish to make a round trip from Earth in a spaceship, traveling at constant speed in a straight line for 6 months and then returning at the same constant speed. You wish further, on your return, to find Earth as it will be a thousand years in the future. (a) How fast must you travel? (b) Does it matter whether you travel in a straight line on your journey? If, for example, you traveled in a circle for 1 year, would you still find that 1000 years had elapsed by Earth clocks when you returned?

SECTION 38-6 The Relativity of Length

9E. A rod lies parallel to the x axis of reference frame S, moving along this axis at a speed of $0.630c$. Its rest length is 1.70 m. What will be its measured length in frame S?

10E. The length of a spaceship is measured to be exactly half its rest length. (a) What is the speed of the spaceship relative to the observer's frame? (b) By what factor do the spaceship's clocks run slow, compared to clocks in the observer's frame?

11E. A meter stick in frame S' makes an angle of 30° with the x' axis. If that frame moves parallel to the x axis with speed $0.90c$ relative to frame S, what is the length of the stick as measured from S?

12E. An electron of $\beta = 0.999\ 987$ moves along the axis of an evacuated tube that has a length of 3.00 m as measured by a laboratory observer S at rest relative to the tube. An observer S' at rest relative to the electron, however, would see this tube moving with speed $v\ (= \beta c)$. What length would observer S' measure for the tube?

13E. The rest radius of Earth is 6370 km, and its orbital speed about the Sun is 30 km/s. Suppose Earth moves past an observer at this speed. To the observer, by how much would Earth's diameter be contracted along the direction of motion?

14E. A spaceship of rest length 130 m races past a timing station at a speed of $0.740c$. (a) What is the length of the spaceship as measured by the timing station? (b) What time interval will the station clock record between the passage of the front and back ends of the ship?

15P. A space traveler takes off from Earth and moves at speed $0.99c$ toward the star Vega, which is 26 ly distant. How much time will have elapsed by Earth clocks (a) when the traveler reaches Vega and (b) when Earth observers receive word from the traveler that she has arrived? (c) How much older will Earth observers calculate the traveler to be (according to her) when she reaches Vega than she was when she started the trip?

16P. An airplane whose rest length is 40.0 m is moving at uniform velocity with respect to Earth, at a speed of 630 m/s. (a) By what fraction of its rest length is it shortened to an observer on Earth? (b) How long would it take, according to Earth clocks, for the airplane's clock to fall behind by 1.00 μs? (Use special relativity in your calculations.)

17P. (a) Can a person, in principle, travel from Earth to the galactic center (which is about 23,000 ly distant) in a normal lifetime? Explain, using either time-dilation or length-contraction arguments. (b) What constant speed would be needed to make the trip in 30 y (proper time)?

SECTION 38-8 Some Consequences of the Lorentz Equations

18E. Observer S assigns the spacetime coordinates

$$x = 100 \text{ km} \quad \text{and} \quad t = 200\ \mu\text{s}$$

to an event. What are the coordinates of this event in frame S', which moves in the direction of increasing x with speed $0.950c$ relative to S? Assume $x = x' = 0$ at $t = t' = 0$.

19E. Observer S reports that an event occurred on his x axis at $x = 3.00 \times 10^8$ m at time $t = 2.50$ s. (a) Observer S' is moving in the direction of increasing x at a speed of $0.400c$. Further, $x = x' = 0$ at $t = t' = 0$. What coordinates does observer S' report for the event? (b) What coordinates would S' report if she were moving in the direction of *decreasing* x at this same speed?

20E. Inertial frame S' moves at a speed of $0.60c$ with respect to frame S (Fig. 38-9). Further, $x = x' = 0$ at $t = t' = 0$. Two events are recorded. In frame S, event 1 occurs at the origin at $t = 0$ and event 2 occurs on the x axis at $x = 3.0$ km at $t = 4.0\ \mu$s. What times of occurrence does observer S' record for these same events? Explain the difference in the time order.

21E. An experimenter arranges to trigger two flashbulbs simultaneously, producing a big flash located at the origin of his reference frame and a small flash at $x = 30.0$ km. An observer, moving at a speed of $0.250c$ in the direction of increasing x, also views the flashes. (a) What time interval between them does she find? (b) Which flash does she say occurs first?

22E. In Table 38-2 the Lorentz transformation equations in the right-hand column can be derived from those in the left-hand column simply by (1) exchanging primed and unprimed quantities and (2) changing the sign of v. Verify this procedure by deriving one set of equations directly from the other by algebraic manipulation.

23P. A clock moves along the x axis at a speed of $0.600c$ and reads zero as it passes the origin. (a) Calculate the Lorentz factor. (b) What time does the clock read as it passes $x = 180$ m?

24P. An observer S sees a big flash of light 1200 m from his position and a small flash of light 720 m closer to him directly in line with the big flash. He measures the time interval between the flashes to be 5.00 μs, the big flash occurring first. (a) What is the relative velocity $\mathbf{v}$ (give both magnitude and direction) of a second observer S' who records these flashes as occurring at the same place? (b) From the point of view of S', which flash occurs first? (c) What time interval between them does S' measure?

25P. In Problem 24, observer S sees the two flashes in the same positions as before, but they now occur closer together in time. How close together in time can they be in the frame of S and still allow the possibility of finding a frame S' in which they occur at the same place?

SECTION 38-9 The Relativity of Velocities

26E. A particle moves along the x' axis of frame S' with a speed of $0.40c$. Frame S' moves with a speed of $0.60c$ with respect to frame S. What is the measured speed of the particle in frame S?

27E. Frame S' moves relative to frame S at $0.62c$ in the direction of increasing x. In frame S' a particle is measured to have a velocity of $0.47c$ in the direction of increasing x'. (a) What is the velocity of the particle with respect to frame S? (b) What would be the velocity of the particle with respect to S if the particle moved (at $0.47c$) in the direction of *decreasing* x' in the S' frame? In each case, compare your answers with the predictions of the classical velocity transformation equation.

28E. One cosmic-ray particle approaches Earth along Earth's north-south axis with a velocity of $0.80c$ toward the geographic north pole, and another approaches with a velocity $0.60c$ toward the geographic south pole (See Fig. 38-23). What is the relative speed of approach of one particle with respect to the other? (*Hint:* It is useful to consider Earth and one of the particles as the two inertial reference frames.)

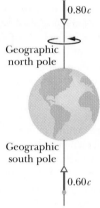

FIGURE 38-23
Exercise 28.

29E. Galaxy A is reported to be receding from us with a speed of $0.35c$. Galaxy B, located in precisely the opposite direction, is also found to be receding from us at this same speed. What recessional speed would an observer on Galaxy A find (a) for our galaxy and (b) for Galaxy B?

30E. It is concluded from measurements of the red shift of the emitted light that quasar Q_1 is moving away from us at a speed of $0.800c$. Quasar Q_2, which lies in the same direction in space but is closer to us, is moving away from us at a speed $0.400c$. What velocity for Q_2 would be measured by an observer on Q_1?

31P. A spaceship whose rest length is 350 m has a speed of $0.82c$ with respect to a certain reference frame. A micrometeorite, also with a speed of $0.82c$ in this frame, passes the spaceship on an antiparallel track. How long does it take this object to pass the spaceship as measured on the ship?

32P. To circle Earth in low orbit, a satellite must have a speed of about 17,000 mi/h. Suppose that two such satellites orbit Earth in opposite directions. (a) What is their relative speed as they pass, according to the classical Galilean velocity transformation equation? (b) What fractional error do you make in (a) by not using the (correct) relativistic transformation equation?

33P. A spaceship, at rest in a certain reference frame S, is given a speed increment of $0.50c$. Relative to its new rest frame, it is then given a further $0.50c$ increment. This process is continued until its speed with respect to its original frame S exceeds $0.999c$. How many increments does this process require?

34P. An armada of spaceships that is 1.00 ly long (in its rest system) moves with speed $0.800c$ relative to ground station S. A messenger travels from the rear of the armada to the front with a speed of $0.950c$ relative to S. How long does the trip take as measured (a) in the messenger's rest system, (b) in the armada's rest system, and (c) by an observer in system S?

SECTION 38-10 Doppler Effect for Light

35E. A spaceship, moving away from Earth at a speed of $0.900c$, reports back by transmitting on a frequency (measured in the spaceship frame) of 100 MHz. To what frequency must Earth receivers be tuned to receive the report?

36E. Some of the familiar hydrogen lines appear in the spectrum of quasar 3C9, but they are shifted so far toward the red that their wavelengths are observed to be three times longer than those observed for hydrogen atoms that are stationary in a laboratory. (a) Show that the classical Doppler equation gives a relative velocity of recession greater than c for this situation. (b) Assuming that the relative motion of 3C9 and Earth is due entirely to recession, find the recession speed that is predicted by the relativistic Doppler equation.

37E. Give the Doppler wavelength shift $\lambda - \lambda_0$, if any, for the sodium D_2 line (589.00 nm) emitted by a source moving in a circle with constant speed ($= 0.100c$) as measured by an observer fixed at the center of the circle.

38P. A spaceship is receding from Earth at a speed of $0.20c$. A light source on the rear of the ship appears blue ($\lambda = 450$ nm) to passengers on the ship. What color would the source appear to an observer on Earth monitoring the receding spaceship?

39P. A radar transmitter T is fixed to a reference frame S' that is moving to the right with speed v relative to reference frame S (see Fig. 38-24). A mechanical timer (essentially a clock) in frame S', having a period τ_0 (measured in S'), causes transmitter T to emit timed radar pulses, which travel at the speed of light and are received by R, a receiver fixed in frame S. (a) What is the period τ of the timer as detected by observer A, who is fixed in frame S? (b) Show that at the receiver R the time interval between pulses arriving from T is not τ or τ_0, but

$$\tau_R = \tau_0 \sqrt{\frac{c + v}{c - v}}.$$

(c) Explain why the receiver R and observer A, who are in the same reference frame, measure a different period for the transmitter. (*Hint:* A clock and a radar pulse are not the same thing.)

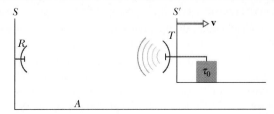

FIGURE 38-24 Problem 39.

SECTION 38-12 A New Look at Energy

40E. How much work must be done to increase the speed of an electron from rest to (a) $0.50c$, (b) $0.990c$, and (c) $0.9990c$?

41E. An electron is moving at a speed such that it could circumnavigate Earth at the equator in 1.00 s. (a) What is its speed, in terms of the speed of light? (b) What is its kinetic energy K? (c) What percent error do you make if you use the classical formula to calculate K?

42E. Find the speed parameter β and Lorentz factor γ for an electron whose kinetic energy is (a) 1.00 keV, (b) 1.00 MeV, and (c) 1.00 GeV.

43E. Find the speed parameter β and Lorentz factor γ for a particle whose kinetic energy is 10.0 MeV if the particle is (a) an electron, (b) a proton, and (c) an alpha particle.

44E. What is the speed of an electron whose kinetic energy is 100 MeV?

45E. A particle has a speed of $0.990c$ in a laboratory reference frame. What are its kinetic energy, its total energy, and its momentum if the particle is (a) a proton and (b) an electron?

46E. In 1979 the United States consumption of electrical energy was about 2.2×10^{12} kW·h. How much mass is equivalent to the consumed energy in that year? Does it make any difference to your answer if this energy is generated in oil-burning, nuclear, or hydroelectric plants?

47E. Quasars are thought to be the nuclei of active galaxies in the early stages of their formation. A typical quasar radiates energy at the rate of 10^{41} W. At what rate is the mass of this quasar being reduced to supply this energy? Express your answer in solar mass units per year, where one solar mass unit (1 smu = 2.0×10^{30} kg) is the mass of our Sun.

48P. How much work must be done to increase the speed of an electron from (a) $0.18c$ to $0.19c$ and (b) $0.98c$ to $0.99c$? Note that the speed increase is $0.01c$ in both cases.

49P. What is the speed of a particle (a) whose kinetic energy is equal to twice its rest energy and (b) whose total energy is equal to twice its rest energy?

50P. A particle with mass m has speed $c/2$ relative to inertial frame S. The particle collides with an identical particle at rest relative to frame S. What is the speed of a frame S' relative to S in which the total momentum of these particles is zero? This frame is called the *center of momentum frame*.

51P. (a) What potential difference would accelerate an electron to speed c, according to classical physics? (b) With this potential difference, what speed would the electron actually attain?

52P. A particle of mass m has a momentum equal to mc. What are (a) its Lorentz factor, (b) its speed, and (c) its kinetic energy?

53P. What must be the momentum of a particle with mass m so that the total energy of the particle is 3 times its rest energy?

54P. Consider the following, all moving in free space: a 2.0 eV photon, a 0.40 MeV electron, and a 10 MeV proton. (a) Which is moving the fastest? (b) The slowest? (c) Which has the greatest momentum? (d) The least? (*Note:* A photon, which is a particle of light, has zero mass.)

55P. A 5.00 grain aspirin tablet has a mass of 320 mg. For how many miles would the energy equivalent of this mass power an automobile? Assume 30.0 mi/gal and a heat of combustion of 1.30×10^8 J/gal for the gasoline used in the automobile.

56P. (a) If the kinetic energy K and the momentum p of a particle can be measured, it should be possible to find its mass m and thus identify the particle. Show that

$$m = \frac{(pc)^2 - K^2}{2Kc^2}.$$

(b) Show that this expression reduces to an expected result as $u/c \rightarrow 0$, in which u is the speed of the particle. (c) Find the mass of a particle whose kinetic energy is 55.0 MeV and whose momentum is 121 MeV/c. Express your answer in terms of the mass of the electron.

57P. In a high-energy collision between a cosmic-ray particle and a particle near the top of Earth's atmosphere, 120 km above sea level, a pion is created. The pion has a total energy E of 1.35×10^5 MeV and is traveling vertically downward. In the pion's rest frame, the pion decays 35.0 ns after its creation. At what altitude above sea level, as measured from Earth's reference frame, does the decay occur? The rest energy of a pion is 139.6 MeV.

58P. The average lifetime of muons at rest is 2.20 μs. A laboratory measurement on muons traveling in a beam emerging from a particle accelerator yields an average muon lifetime of 6.90 μs. What are (a) the speed of these muons in the laboratory, (b) their kinetic energy, and (c) their momentum? The mass of a muon is 207 times that of an electron.

59P. (a) How much energy is released in the explosion of a fission bomb containing 3.0 kg of fissionable material? Assume that 0.10% of the mass is converted to released energy. (b) What mass of TNT would have to explode to provide the same energy release? Assume that each mole of TNT liberates 3.4 MJ of energy on exploding. The molecular mass of TNT is 0.227 kg/mol. (c) For the same mass of explosive, how much more effective are nuclear explosions than TNT explosions? That is, compare the fractions of the mass that are converted to energy in each case.

60P. In Section 29-5 we showed that a particle of charge q and mass m moving with speed v perpendicular to a uniform magnetic field B moves in a circle of radius r given by Eq. 29-16:

$$r = \frac{mv}{qB}.$$

Also, it was demonstrated that the period T of the circular motion is independent of the speed of the particle. These results hold only if $v \ll c$. For particles moving faster, the radius of the circular path must be obtained with

$$r = \frac{p}{qB} = \frac{m(\gamma v)}{qB} = \frac{mv}{qB\sqrt{1 - \beta^2}}.$$

This equation is valid at all speeds. Compute the radius of the path of a 10.0 MeV electron moving perpendicular to a uniform 2.20 T magnetic field, using the (a) classical and (b) relativistic formulas. (c) Calculate the period $T = 2\pi r/v$ of the circular motion. Is the result independent of the speed of the electron?

61P. Ionization measurements show that a certain low-mass nuclear particle carries a double charge ($= 2e$) and is moving with a speed of $0.710c$. The radius of curvature of its path in a magnetic field of 1.00 T is 6.28 m. (The path is a circle whose plane is perpendicular to the magnetic field.) Find the mass of the particle and identify it. [*Hint:* Low-mass nuclear particles are made up of neutrons (which have no charge) and protons (charge $= +e$), in roughly equal numbers. Take the mass of each of these particles to be 1.00 u. Also, see Problem 60.]

62P. A 10 GeV proton in cosmic radiation approaches Earth with its velocity **v** perpendicular to Earth's magnetic field **B**, in a region over which Earth's average magnetic field is 55 μT. What is the radius of the proton's curved path in that region? (See Problem 60.)

63P. A 2.50 MeV electron moves perpendicular to a magnetic field in a path whose radius of curvature is 3.0 cm. What is the magnetic field B? (See Problem 60.)

64P. The proton synchrotron at Fermilab accelerates protons to a kinetic energy of 500 GeV. At such a large energy, relativistic effects are important; in particular, as the speed of the proton increases, the time the proton takes to make a trip around its circular orbit in the synchrotron also increases. In a cyclotron, where the magnetic field and the oscillator are at fixed values, this effect of time dilation will put the proton's circling out of synchronization with the oscillator. That eliminates repeated acceleration; hence, the proton will not reach an energy as great as 500 GeV. But in a synchrotron, both the magnitude of the magnetic field and the oscillation frequency are varied to allow for the increase in the time dilation.

At the energy of 500 GeV, calculate (a) the Lorentz factor, (b) the speed parameter, and (c) the magnetic field at the proton orbit, which has a radius of curvature of 750 m. (See Problem 60; use 938.3 MeV as the proton's rest energy.)

Electronic Computation

65. A space probe leaves Earth and travels at $0.97c$. Clocks on Earth and in the probe are all set to zero at launch. Every 6.0 h, mission control sends a signal (traveling at the speed of light) to the probe requesting a status report. The probe immediately sends back a reply that includes the reading on the probe clock at the time the signal was received. For each of the first five signals, calculate the time, according to Earth clocks, at which the reply is received on Earth and the time reported in the reply. Also compute the distance of the probe from Earth when each signal is received by the probe.

Additional Problems

66P. *The Car-in-the-Garage Problem.* Carman has just purchased the world's longest stretch limo, whose proper length is

$L_c = 30.5$ m. In Fig. 38-25a, it is shown parked in front of a garage, whose proper length is $L_g = 6.00$ m. The garage has a front door (shown open) and a back door (shown closed). The limo is obviously longer than the garage. Still, Garageman, who owns the garage and knows something about relativistic length contraction, makes a bet with Carman that the limo can fit in the garage with both doors closed. Carman, who dropped the physics course before reaching special relativity, says such a thing, even in principle, is impossible.

To analyze Garageman's scheme, an x_c axis is attached to the limo, with $x_c = 0$ at the rear bumper, and an x_g axis is attached to the garage, with $x_g = 0$ at the (now open) front door. Then Carman is to drive the limo directly toward the front door at a velocity of $0.9980c$ (which is, of course, both technically and financially impossible). Carman is stationary in the x_c reference frame; Garageman is stationary in the x_g reference frame.

There are two events to consider. *Event 1:* When the rear bumper clears the front door, the front door is closed. Let the time of this event be zero to both Carman and Garageman: $t_{g1} = t_{c1} = 0$. The event occurs at $x_c = x_g = 0$. Figure 38-25b shows event 1 according to the x_g reference frame. *Event 2:* When the front bumper reaches the back door, that door opens. Figure 38-25c shows event 2 according to the x_g reference frame.

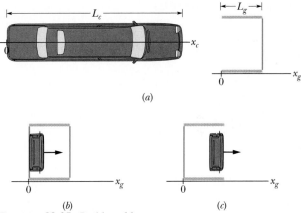

(a)

(b)   (c)

FIGURE 38-25   Problem 66.

According to Garageman, (a) what is the length of the limo and (b) what are the spacetime coordinates x_{g2} and t_{g2} of event 2? (c) For how long is the limo temporarily "trapped" inside the garage, with both doors shut?

Now consider the situation from the x_c reference frame, in which the garage comes racing past the limo at a velocity of $-0.9980c$. According to Carman, (d) what is the length of the passing garage, (e) what are the spacetime coordinates x_{c2} and t_{c2} of event 2, (f) is the limo ever in the garage with both doors shut, and (g) which event occurs first? (h) Sketch events 1 and 2 as seen by Carman. (Are the events causally related; that is, does one of them cause the other?) (i) Finally, who wins the bet?

67P. *Superluminal Jets.* Figure 38-26a shows the path taken by a knot in a jet of ionized gas that has been expelled from a galaxy.

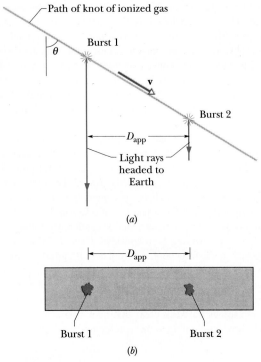

Path of knot of ionized gas

Burst 1

θ

v

Burst 2

D_{app}

Light rays headed to Earth

(a)

D_{app}

Burst 1   Burst 2

(b)

FIGURE 38-26 Problem 67.

The knot travels at a constant velocity **v** at an angle θ from the direction of Earth. The knot occasionally emits a burst of light, which is eventually detected on Earth. Two bursts are indicated in Fig. 38-26a, separated in time by t as measured in a stationary frame near the bursts. The bursts are shown in Fig. 38-26b as if they were photographed on the same film, first when the light from burst 1 arrived on Earth and then later when the light from burst 2 arrived. The apparent distance D_{app} traveled by the knot between the two bursts is the distance across an Earth-observer's view of the knot's path. The apparent time T_{app} between the bursts is the difference in the arrival times of the light from them. The apparent speed of the knot is then $V_{app} = D_{app}/T_{app}$. In terms of v, t, and θ, what are (a) D_{app} and (b) T_{app}? (c) Evaluate V_{app} for $v = 0.980c$ and $\theta = 30.0°$.

Appendix A
The International System
of Units (SI)*

1. THE SI BASE UNITS

QUANTITY	NAME	SYMBOL	DEFINITION
length	meter	m	". . . the length of the path traveled by light in vacuum in 1/299,792,458 of a second." (1983)
mass	kilogram	kg	". . . this prototype [a certain platinum–iridium cylinder] shall henceforth be considered to be the unit of mass." (1889)
time	second	s	". . . the duration of 9,192,631,770 periods of the radiation corresponding to the transition between the two hyperfine levels of the ground state of the cesium-133 atom." (1967)
electric current	ampere	A	". . . that constant current which, if maintained in two straight parallel conductors of infinite length, of negligible circular cross section, and placed 1 meter apart in vacuum, would produce between these conductors a force equal to 2×10^{-7} newton per meter of length." (1946)
thermodynamic temperature	kelvin	K	". . . the fraction 1/273.16 of the thermodynamic temperature of the triple point of water." (1967)
amount of substance	mole	mol	". . . the amount of substance of a system which contains as many elementary entities as there are atoms in 0.012 kilogram of carbon-12." (1971)
luminous intensity	candela	cd	". . . the luminous intensity, in the perpendicular direction, of a surface of 1/600,000 square meter of a blackbody at the temperature of freezing platinum under a pressure of 101.325 newtons per square meter." (1967)

*Adapted from "The International System of Units (SI)," National Bureau of Standards Special Publication 330, 1972 edition. The definitions above were adopted by the General Conference of Weights and Measures, an international body, on the dates shown. In this book we do not use the candela.

2. SOME SI DERIVED UNITS

QUANTITY	NAME OF UNIT	SYMBOL	
area	square meter	m^2	
volume	cubic meter	m^3	
frequency	hertz	Hz	s^{-1}
mass density (density)	kilogram per cubic meter	kg/m^3	
speed, velocity	meter per second	m/s	
angular velocity	radian per second	rad/s	
acceleration	meter per second per second	m/s^2	
angular acceleration	radian per second per second	rad/s^2	
force	newton	N	$kg \cdot m/s^2$
pressure	pascal	Pa	N/m^2
work, energy, quantity of heat	joule	J	$N \cdot m$
power	watt	W	J/s
quantity of electric charge	coulomb	C	$A \cdot s$
potential difference, electromotive force	volt	V	W/A
electric field strength	volt per meter (or newton per coulomb)	V/m	N/C
electric resistance	ohm	Ω	V/A
capacitance	farad	F	$A \cdot s/V$
magnetic flux	weber	Wb	$V \cdot s$
inductance	henry	H	$V \cdot s/A$
magnetic flux density	tesla	T	Wb/m^2
magnetic field strength	ampere per meter	A/m	
entropy	joule per kelvin	J/K	
specific heat	joule per kilogram kelvin	$J/(kg \cdot K)$	
thermal conductivity	watt per meter kelvin	$W/(m \cdot K)$	
radiant intensity	watt per steradian	W/sr	

3. THE SI SUPPLEMENTARY UNITS

QUANTITY	NAME OF UNIT	SYMBOL
plane angle	radian	rad
solid angle	steradian	sr

Appendix B
Some Fundamental Constants of Physics*

CONSTANT	SYMBOL	COMPUTATIONAL VALUE	BEST (1986) VALUE	
			VALUE[a]	UNCERTAINTY[b]
Speed of light in a vacuum	c	3.00×10^8 m/s	2.99792458	exact
Elementary charge	e	1.60×10^{-19} C	1.60217733	0.30
Gravitational constant	G	6.67×10^{-11} m³/s²·kg	6.67259	128
Universal gas constant	R	8.31 J/mol·K	8.314510	8.4
Avogadro constant	N_A	6.02×10^{23} mol⁻¹	6.0221367	0.59
Boltzmann constant	k	1.38×10^{-23} J/K	1.380658	8.5
Stefan-Boltzmann constant	σ	5.67×10^{-8} W/m²·K⁴	5.67051	34
Molar volume of ideal gas at STP[d]	V_m	2.24×10^{-2} m³/mol	2.241409	8.4
Permittivity constant	ϵ_0	8.85×10^{-12} F/m	8.85418781762	exact
Permeability constant	μ_0	1.26×10^{-6} H/m	1.25663706143	exact
Planck constant	h	6.63×10^{-34} J·s	6.6260755	0.60
Electron mass[c]	m_e	9.11×10^{-31} kg	9.1093897	0.59
		5.49×10^{-4} u	5.48579903	0.023
Proton mass[c]	m_p	1.67×10^{-27} kg	1.6726231	0.59
		1.0073 u	1.0072764660	0.005
Ratio of proton mass to electron mass	m_p/m_e	1840	1836.152701	0.020
Electron charge-to-mass ratio	e/m_e	1.76×10^{11} C/kg	1.75881961	0.30
Neutron mass[c]	m_n	1.68×10^{-27} kg	1.6749286	0.59
		1.0087 u	1.0086649235	0.0023
Hydrogen atom mass[c]	m_{1_H}	1.0078 u	1.0078250316	0.0005
Deuterium atom mass[c]	m_{2_H}	2.0141 u	2.0141017779	0.0005
Helium atom mass[c]	$m_{4_{He}}$	4.0026 u	4.0026032	0.067
Muon mass	m_μ	1.88×10^{-28} kg	1.8835326	0.61
Electron magnetic moment	μ_e	9.28×10^{-24} J/T	9.2847701	0.34
Proton magnetic moment	μ_p	1.41×10^{-26} J/T	1.41060761	0.34
Bohr magneton	μ_B	9.27×10^{-24} J/T	9.2740154	0.34
Nuclear magneton	μ_N	5.05×10^{-27} J/T	5.0507866	0.34
Bohr radius	r_B	5.29×10^{-11} m	5.29177249	0.045
Rydberg constant	R	1.10×10^7 m⁻¹	1.0973731534	0.0012
Electron Compton wavelength	λ_C	2.43×10^{-12} m	2.42631058	0.089

[a]Values given in this column should be given the same unit and power of 10 as the computational value.   [b]Parts per million.   [c]Masses given in u are in unified atomic mass units, where 1 u = $1.6605402 \times 10^{-27}$ kg.   [d]STP means standard temperature and pressure: 0°C and 1.0 atm (0.1 MPa).

*The values in this table were largely selected from a longer list in *Symbols, Units and Nomenclature in Physics* (IUPAP), prepared by E. Richard Cohen and Pierre Giacomo, 1986.

Appendix C
Some Astronomical Data

SOME DISTANCES FROM THE EARTH

To the moon*	3.82×10^8 m
To the sun*	1.50×10^{11} m
To the nearest star (Proxima Centauri)	4.04×10^{16} m
To the center of our galaxy	2.2×10^{20} m
To the Andromeda Galaxy	2.1×10^{22} m
To the edge of the observable universe	$\sim 10^{26}$ m

*Mean distance.

THE SUN, THE EARTH, AND THE MOON

PROPERTY	UNIT	SUN	EARTH	MOON
Mass	kg	1.99×10^{30}	5.98×10^{24}	7.36×10^{22}
Mean radius	m	6.96×10^8	6.37×10^6	1.74×10^6
Mean density	kg/m³	1410	5520	3340
Free-fall acceleration at the surface	m/s²	274	9.81	1.67
Escape velocity	km/s	618	11.2	2.38
Period of rotation[a]	—	37 d at poles[b]	23 h 56 min	27.3 d
		26 d at equator[b]		
Radiation power[c]	W	3.90×10^{26}		

[a]Measured with respect to the distant stars.

[b]The sun, a ball of gas, does not rotate as a rigid body.

[c]Just outside the Earth's atmosphere solar energy is received, assuming normal incidence, at the rate of 1340 W/m².

SOME PROPERTIES OF THE PLANETS

	MERCURY	VENUS	EARTH	MARS	JUPITER	SATURN	URANUS	NEPTUNE	PLUTO
Mean distance from sun, 10^6 km	57.9	108	150	228	778	1430	2870	4500	5900
Period of revolution, y	0.241	0.615	1.00	1.88	11.9	29.5	84.0	165	248
Period of rotation,[a] d	58.7	−243[b]	0.997	1.03	0.409	0.426	−0.451[b]	0.658	6.39
Orbital speed, km/s	47.9	35.0	29.8	24.1	13.1	9.64	6.81	5.43	4.74
Inclination of axis to orbit	<28°	≈3°	23.4°	25.0°	3.08°	26.7°	97.9°	29.6°	57.5°
Inclination of orbit to Earth's orbit	7.00°	3.39°		1.85°	1.30°	2.49°	0.77°	1.77°	17.2°
Eccentricity of orbit	0.206	0.0068	0.0167	0.0934	0.0485	0.0556	0.0472	0.0086	0.250
Equatorial diameter, km	4880	12,100	12,800	6790	143,000	120,000	51,800	49,500	2300
Mass (Earth = 1)	0.0558	0.815	1.000	0.107	318	95.1	14.5	17.2	0.002
Density (water = 1)	5.60	5.20	5.52	3.95	1.31	0.704	1.21	1.67	2.03
Surface value of g,[c] m/s²	3.78	8.60	9.78	3.72	22.9	9.05	7.77	11.0	0.5
Escape velocity,[c] km/s	4.3	10.3	11.2	5.0	59.5	35.6	21.2	23.6	1.1
Known satellites	0	0	1	2	16 + ring	18 + rings	15 + rings	8 + rings	1

[a]Measured with respect to the distant stars.

[b]Venus and Uranus rotate opposite their orbital motion.

[c]Gravitational acceleration measured at the planet's equator.

Appendix **D**
Conversion Factors

Conversion factors may be read directly from these tables. For example, 1 degree = 2.778×10^{-3} revolutions, so $16.7° = 16.7 \times 2.778 \times 10^{-3}$ rev. The SI quantities are fully capitalized.

Adapted in part from G. Shortley and D. Williams, *Elements of Physics,* Prentice-Hall, Englewood Cliffs, NJ, 1971.

PLANE ANGLE

	°	′	″	RADIAN	rev
1 degree = 1	60	3600	1.745×10^{-2}	2.778×10^{-3}
1 minute = 1.667×10^{-2}	1	60	2.909×10^{-4}	4.630×10^{-5}
1 second = 2.778×10^{-4}	1.667×10^{-2}	1	4.848×10^{-6}	7.716×10^{-7}
1 RADIAN = 57.30	3438	2.063×10^{5}	1	0.1592
1 revolution = 360	2.16×10^{4}	1.296×10^{6}	6.283	1

SOLID ANGLE

1 sphere = 4π steradians = 12.57 steradians

LENGTH

	cm	METER	km	in.	ft	mi
1 centimeter = 1	10^{-2}	10^{-5}	0.3937	3.281×10^{-2}	6.214×10^{-6}
1 METER = 100	1	10^{-3}	39.37	3.281	6.214×10^{-4}
1 kilometer = 10^{5}	1000	1	3.937×10^{4}	3281	0.6214
1 inch = 2.540	2.540×10^{-2}	2.540×10^{-5}	1	8.333×10^{-2}	1.578×10^{-5}
1 foot = 30.48	0.3048	3.048×10^{-4}	12	1	1.894×10^{-4}
1 mile = 1.609×10^{5}	1609	1.609	6.336×10^{4}	5280	1

1 angström = 10^{-10} m

1 nautical mile = 1852 m
   = 1.151 miles = 6076 ft

1 fermi = 10^{-15} m

1 light-year = 9.460×10^{12} km

1 parsec = 3.084×10^{13} km

1 fathom = 6 ft

1 Bohr radius = 5.292×10^{-11} m

1 yard = 3 ft

1 rod = 16.5 ft

1 mil = 10^{-3} in.

1 nm = 10^{-9} m

AREA

	METER^2	cm^2	ft^2	in.^{2}
1 SQUARE METER = 1	10^{4}	10.76	1550
1 square centimeter = 10^{-4}	1	1.076×10^{-3}	0.1550
1 square foot = 9.290×10^{-2}	929.0	1	144
1 square inch = 6.452×10^{-4}	6.452	6.944×10^{-3}	1

1 square mile = 2.788×10^{7} ft^2
   = 640 acres

1 barn = 10^{-28} m^2

1 acre = 43,560 ft^2

1 hectare = 10^{4} m^2 = 2.471 acres

VOLUME

	METER^3	cm^3	L	ft^3	in.^{3}
1 CUBIC METER = 1		10^6	1000	35.31	6.102×10^4
1 cubic centimeter = 10^{-6}		1	1.000×10^{-3}	3.531×10^{-5}	6.102×10^{-2}
1 liter = 1.000×10^{-3}		1000	1	3.531×10^{-2}	61.02
1 cubic foot = 2.832×10^{-2}		2.832×10^4	28.32	1	1728
1 cubic inch = 1.639×10^{-5}		16.39	1.639×10^{-2}	5.787×10^{-4}	1

1 U.S. fluid gallon = 4 U.S. fluid quarts = 8 U.S. pints = 128 U.S. fluid ounces = 231 in.^{3}

1 British imperial gallon = 277.4 in.^{3} = 1.201 U.S. fluid gallons

MASS

Quantities in the colored areas are not mass units but are often used as such. When we write, for example, 1 kg "=" 2.205 lb, this means that a kilogram is a *mass* that *weighs* 2.205 pounds at a location where g has the standard value of 9.80665 m/s^2.

	g	KILOGRAM	slug	u	oz	lb	ton
1 gram = 1		0.001	6.852×10^{-5}	6.022×10^{23}	3.527×10^{-2}	2.205×10^{-3}	1.102×10^{-6}
1 KILOGRAM = 1000		1	6.852×10^{-2}	6.022×10^{26}	35.27	2.205	1.102×10^{-3}
1 slug = 1.459×10^4		14.59	1	8.786×10^{27}	514.8	32.17	1.609×10^{-2}
1 atomic mass unit = 1.661×10^{-24}		1.661×10^{-27}	1.138×10^{-28}	1	5.857×10^{-26}	3.662×10^{-27}	1.830×10^{-30}
1 ounce = 28.35		2.835×10^{-2}	1.943×10^{-3}	1.718×10^{25}	1	6.250×10^{-2}	3.125×10^{-5}
1 pound = 453.6		0.4536	3.108×10^{-2}	2.732×10^{26}	16	1	0.0005
1 ton = 9.072×10^5		907.2	62.16	5.463×10^{29}	3.2×10^4	2000	1

1 metric ton = 1000 kg

DENSITY

Quantities in the colored areas are weight densities and, as such, are dimensionally different from mass densities. See note for mass table.

	slug/ft^3	KILOGRAM/ METER^3	g/cm^3	lb/ft^3	lb/in.^{3}
1 slug per foot^3 = 1		515.4	0.5154	32.17	1.862×10^{-2}
1 KILOGRAM per METER^3 = 1.940×10^{-3}		1	0.001	6.243×10^{-2}	3.613×10^{-5}
1 gram per centimeter^3 = 1.940		1000	1	62.43	3.613×10^{-2}
1 pound per foot^3 = 3.108×10^{-2}		16.02	1.602×10^{-2}	1	5.787×10^{-4}
1 pound per inch^3 = 53.71		2.768×10^4	27.68	1728	1

TIME

	y	d	h	min	SECOND
1 year = 1		365.25	8.766×10^3	5.259×10^5	3.156×10^7
1 day = 2.738×10^{-3}		1	24	1440	8.640×10^4
1 hour = 1.141×10^{-4}		4.167×10^{-2}	1	60	3600
1 minute = 1.901×10^{-6}		6.944×10^{-4}	1.667×10^{-2}	1	60
1 SECOND = 3.169×10^{-8}		1.157×10^{-5}	2.778×10^{-4}	1.667×10^{-2}	1

SPEED

	ft/s	km/h	METER/SECOND	mi/h	cm/s
1 foot per second = 1	1.097	0.3048	0.6818	30.48
1 kilometer per hour = 0.9113	1	0.2778	0.6214	27.78
1 METER per SECOND = 3.281	3.6	1	2.237	100
1 mile per hour = 1.467	1.609	0.4470	1	44.70
1 centimeter per second = 3.281×10^{-2}	3.6×10^{-2}	0.01	2.237×10^{-2}	1

1 knot = 1 nautical mi/h = 1.688 ft/s   1 mi/min = 88.00 ft/s = 60.00 mi/h

FORCE

Force units in the colored areas are now little used. To clarify: 1 gram-force (= 1 gf) is the force of gravity that would act on an object whose mass is 1 gram at a location where g has the standard value of 9.80665 m/s^2.

	dyne	NEWTON	lb	pdl	gf	kgf
1 dyne = 1	10^{-5}	2.248×10^{-6}	7.233×10^{-5}	1.020×10^{-3}	1.020×10^{-6}
1 NEWTON = 10^5	1	0.2248	7.233	102.0	0.1020
1 pound = 4.448×10^5	4.448	1	32.17	453.6	0.4536
1 poundal = 1.383×10^4	0.1383	3.108×10^{-2}	1	14.10	1.410×10^{-2}
1 gram-force = 980.7	9.807×10^{-3}	2.205×10^{-3}	7.093×10^{-2}	1	0.001
1 kilogram-force = 9.807×10^5	9.807	2.205	70.93	1000	1

PRESSURE

	atm	dyne/cm^2	inch of water	cm Hg	PASCAL	lb/in.^{2}	lb/ft^2
1 atmosphere = 1	1.013×10^6	406.8	76	1.013×10^5	14.70	2116
1 dyne per centimeter^2 = 9.869×10^{-7}	1	4.015×10^{-4}	7.501×10^{-5}	0.1	1.405×10^{-5}	2.089×10^{-3}
1 inch of water^a at 4°C = 2.458×10^{-3}	2491	1	0.1868	249.1	3.613×10^{-2}	5.202
1 centimeter of mercury^a at 0°C = 1.316×10^{-2}	1.333×10^4	5.353	1	1333	0.1934	27.85
1 PASCAL = 9.869×10^{-6}	10	4.015×10^{-3}	7.501×10^{-4}	1	1.450×10^{-4}	2.089×10^{-2}
1 pound per inch^2 = 6.805×10^{-2}	6.895×10^4	27.68	5.171	6.895×10^3	1	144
1 pound per foot^2 = 4.725×10^{-4}	478.8	0.1922	3.591×10^{-2}	47.88	6.944×10^{-3}	1

^{a} Where the acceleration of gravity has the standard value of 9.80665 m/s^2.

1 bar = 10^6 dyne/cm^2 = 0.1 MPa   1 millibar = 10^3 dyne/cm^2 = 10^2 Pa   1 torr = 1 mm Hg

ENERGY, WORK, HEAT

Quantities in the colored areas are not properly energy units but are included for convenience. They arise from the relativistic mass–energy equivalence formula $E = mc^2$ and represent the energy released if a kilogram or unified atomic mass unit (u) is completely converted to energy (bottom two rows) or the mass that would be completely converted to one unit of energy (rightmost two columns).

	Btu	erg	ft·lb	hp·h	JOULE	cal	kW·h	eV	MeV	kg	u
1 British thermal unit =	1	1.055×10^{10}	777.9	3.929×10^{-4}	1055	252.0	2.930×10^{-4}	6.585×10^{21}	6.585×10^{15}	1.174×10^{-14}	7.070×10^{12}
1 erg =	9.481×10^{-11}	1	7.376×10^{-8}	3.725×10^{-14}	10^{-7}	2.389×10^{-8}	2.778×10^{-14}	6.242×10^{11}	6.242×10^{5}	1.113×10^{-24}	670.2
1 foot-pound =	1.285×10^{-3}	1.356×10^{7}	1	5.051×10^{-7}	1.356	0.3238	3.766×10^{-7}	8.464×10^{18}	8.464×10^{12}	1.509×10^{-17}	9.037×10^{9}
1 horsepower-hour =	2545	2.685×10^{13}	1.980×10^{6}	1	2.685×10^{6}	6.413×10^{5}	0.7457	1.676×10^{25}	1.676×10^{19}	2.988×10^{-11}	1.799×10^{16}
1 JOULE =	9.481×10^{-4}	10^{7}	0.7376	3.725×10^{-7}	1	0.2389	2.778×10^{-7}	6.242×10^{18}	6.242×10^{12}	1.113×10^{-17}	6.702×10^{9}
1 calorie =	3.969×10^{-3}	4.186×10^{7}	3.088	1.560×10^{-6}	4.186	1	1.163×10^{-6}	2.613×10^{19}	2.613×10^{13}	4.660×10^{-17}	2.806×10^{10}
1 kilowatt-hour =	3413	3.600×10^{13}	2.655×10^{6}	1.341	3.600×10^{6}	8.600×10^{5}	1	2.247×10^{25}	2.247×10^{19}	4.007×10^{-11}	2.413×10^{16}
1 electron-volt =	1.519×10^{-22}	1.602×10^{-12}	1.182×10^{-19}	5.967×10^{-26}	1.602×10^{-19}	3.827×10^{-20}	4.450×10^{-26}	1	10^{-6}	1.783×10^{-36}	1.074×10^{-9}
1 million electron-volts =	1.519×10^{-16}	1.602×10^{-6}	1.182×10^{-13}	5.967×10^{-20}	1.602×10^{-13}	3.827×10^{-14}	4.450×10^{-20}	10^{-6}	1	1.783×10^{-30}	1.074×10^{-3}
1 kilogram =	8.521×10^{13}	8.987×10^{23}	6.629×10^{16}	3.348×10^{10}	8.987×10^{16}	2.146×10^{16}	2.497×10^{10}	5.610×10^{35}	5.610×10^{29}	1	6.022×10^{26}
1 unified atomic mass unit =	1.415×10^{-13}	1.492×10^{-3}	1.101×10^{-10}	5.559×10^{-17}	1.492×10^{-10}	3.564×10^{-11}	4.146×10^{-17}	9.320×10^{8}	932.0	1.661×10^{-27}	1

POWER

	Btu/h	ft·lb/s	hp	cal/s	kW	WATT
1 British thermal unit per hour =	1	0.2161	3.929×10^{-4}	6.998×10^{-2}	2.930×10^{-4}	0.2930
1 foot-pound per second =	4.628	1	1.818×10^{-3}	0.3239	1.356×10^{-3}	1.356
1 horsepower =	2545	550	1	178.1	0.7457	745.7
1 calorie per second =	14.29	3.088	5.615×10^{-3}	1	4.186×10^{-3}	4.186
1 kilowatt =	3413	737.6	1.341	238.9	1	1000
1 WATT =	3.413	0.7376	1.341×10^{-3}	0.2389	0.001	1

MAGNETIC FIELD

	gauss	TESLA	milligauss
1 gauss =	1	10^{-4}	1000
1 TESLA =	10^{4}	1	10^{7}
1 milligauss =	0.001	10^{-7}	1

1 tesla = 1 weber/meter^2

MAGNETIC FLUX

	maxwell	WEBER
1 maxwell =	1	10^{-8}
1 WEBER =	10^{8}	1

GEOMETRY

Circle of radius r: circumference $= 2\pi r$; area $= \pi r^2$.

Sphere of radius r: area $= 4\pi r^2$; volume $= \frac{4}{3}\pi r^3$.

Right circular cylinder of radius r and height h:
area $= 2\pi r^2 + 2\pi rh$; volume $= \pi r^2 h$.

Triangle of base a and altitude h: area $= \frac{1}{2}ah$.

QUADRATIC FORMULA

If $ax^2 + bx + c = 0$, then $x = \dfrac{-b \pm \sqrt{b^2 - 4ac}}{2a}$.

TRIGONOMETRIC FUNCTIONS OF ANGLE θ

$$\sin\theta = \frac{y}{r} \quad \cos\theta = \frac{x}{r}$$

$$\tan\theta = \frac{y}{x} \quad \cot\theta = \frac{x}{y}$$

$$\sec\theta = \frac{r}{x} \quad \csc\theta = \frac{r}{y}$$

PYTHAGOREAN THEOREM

In this right triangle,
$$a^2 + b^2 = c^2$$

TRIANGLES

Angles are A, B, C

Opposite sides are a, b, c

Angles $A + B + C = 180°$

$$\frac{\sin A}{a} = \frac{\sin B}{b} = \frac{\sin C}{c}$$

$$c^2 = a^2 + b^2 - 2ab\cos C$$

Exterior angle $D = A + C$

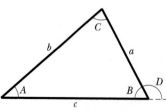

MATHEMATICAL SIGNS AND SYMBOLS

$=$	equals
$\approx$	equals approximately
$\sim$	is the order of magnitude of
$\neq$	is not equal to
$\equiv$	is identical to, is defined as
$>$	is greater than ($\gg$ is much greater than)
$<$	is less than ($\ll$ is much less than)
$\geq$	is greater than or equal to (or, is no less than)
$\leq$	is less than or equal to (or, is no more than)
$\pm$	plus or minus
$\propto$	is proportional to
Σ	the sum of
$\bar{x}$	the average value of x

TRIGONOMETRIC IDENTITIES

$$\sin(90° - \theta) = \cos\theta$$

$$\cos(90° - \theta) = \sin\theta$$

$$\sin\theta/\cos\theta = \tan\theta$$

$$\sin^2\theta + \cos^2\theta = 1$$

$$\sec^2\theta - \tan^2\theta = 1$$

$$\csc^2\theta - \cot^2\theta = 1$$

$$\sin 2\theta = 2\sin\theta\cos\theta$$

$$\cos 2\theta = \cos^2\theta - \sin^2\theta = 2\cos^2\theta - 1 = 1 - 2\sin^2\theta$$

$$\sin(\alpha \pm \beta) = \sin\alpha\cos\beta \pm \cos\alpha\sin\beta$$

$$\cos(\alpha \pm \beta) = \cos\alpha\cos\beta \mp \sin\alpha\sin\beta$$

$$\tan(\alpha \pm \beta) = \frac{\tan\alpha \pm \tan\beta}{1 \mp \tan\alpha\tan\beta}$$

$$\sin\alpha \pm \sin\beta = 2\sin\tfrac{1}{2}(\alpha \pm \beta)\cos\tfrac{1}{2}(\alpha \mp \beta)$$

$$\cos\alpha + \cos\beta = 2\cos\tfrac{1}{2}(\alpha + \beta)\cos\tfrac{1}{2}(\alpha - \beta)$$

$$\cos\alpha - \cos\beta = -2\sin\tfrac{1}{2}(\alpha + \beta)\sin\tfrac{1}{2}(\alpha - \beta)$$

BINOMIAL THEOREM

$$(1 + x)^n = 1 + \frac{nx}{1!} + \frac{n(n-1)x^2}{2!} + \cdots \qquad (x^2 < 1)$$

EXPONENTIAL EXPANSION

$$e^x = 1 + x + \frac{x^2}{2!} + \frac{x^3}{3!} + \cdots$$

LOGARITHMIC EXPANSION

$$\ln(1 + x) = x - \tfrac{1}{2}x^2 + \tfrac{1}{3}x^3 - \cdots \qquad (|x| < 1)$$

TRIGONOMETRIC EXPANSIONS
(θ in radians)

$$\sin \theta = \theta - \frac{\theta^3}{3!} + \frac{\theta^5}{5!} - \cdots$$

$$\cos \theta = 1 - \frac{\theta^2}{2!} + \frac{\theta^4}{4!} - \cdots$$

$$\tan \theta = \theta + \frac{\theta^3}{3} + \frac{2\theta^5}{15} + \cdots$$

CRAMER'S RULE

Two simultaneous equations in unknowns x and y,

$$a_1x + b_1y = c_1 \qquad \text{and} \qquad a_2x + b_2y = c_2,$$

have the solutions

$$x = \frac{\begin{vmatrix} c_1 & b_1 \\ c_2 & b_2 \end{vmatrix}}{\begin{vmatrix} a_1 & b_1 \\ a_2 & b_2 \end{vmatrix}} = \frac{c_1b_2 - c_2b_1}{a_1b_2 - a_2b_1}$$

and

$$y = \frac{\begin{vmatrix} a_1 & c_1 \\ a_2 & c_2 \end{vmatrix}}{\begin{vmatrix} a_1 & b_1 \\ a_2 & b_2 \end{vmatrix}} = \frac{a_1c_2 - a_2c_1}{a_1b_2 - a_2b_1}.$$

PRODUCTS OF VECTORS

Let $\mathbf{i}$, $\mathbf{j}$, and $\mathbf{k}$ be unit vectors in the x, y, and z directions. Then

$$\mathbf{i} \cdot \mathbf{i} = \mathbf{j} \cdot \mathbf{j} = \mathbf{k} \cdot \mathbf{k} = 1, \qquad \mathbf{i} \cdot \mathbf{j} = \mathbf{j} \cdot \mathbf{k} = \mathbf{k} \cdot \mathbf{i} = 0,$$

$$\mathbf{i} \times \mathbf{i} = \mathbf{j} \times \mathbf{j} = \mathbf{k} \times \mathbf{k} = 0,$$

$$\mathbf{i} \times \mathbf{j} = \mathbf{k}, \qquad \mathbf{j} \times \mathbf{k} = \mathbf{i}, \qquad \mathbf{k} \times \mathbf{i} = \mathbf{j},$$

Any vector $\mathbf{a}$ with components a_x, a_y, and a_z along the x, y, and z axes can be written

$$\mathbf{a} = a_x\mathbf{i} + a_y\mathbf{j} + a_z\mathbf{k}.$$

Let $\mathbf{a}$, $\mathbf{b}$, and $\mathbf{c}$ be arbitrary vectors with magnitudes a, b, and c. Then

$$\mathbf{a} \times (\mathbf{b} + \mathbf{c}) = (\mathbf{a} \times \mathbf{b}) + (\mathbf{a} \times \mathbf{c})$$

$$(s\mathbf{a}) \times \mathbf{b} = \mathbf{a} \times (s\mathbf{b}) = s(\mathbf{a} \times \mathbf{b}) \qquad (s = \text{a scalar}).$$

Let θ be the smaller of the two angles between $\mathbf{a}$ and $\mathbf{b}$. Then

$$\mathbf{a} \cdot \mathbf{b} = \mathbf{b} \cdot \mathbf{a} = a_xb_x + a_yb_y + a_zb_z = ab \cos \theta$$

$$\mathbf{a} \times \mathbf{b} = -\mathbf{b} \times \mathbf{a} = \begin{vmatrix} \mathbf{i} & \mathbf{j} & \mathbf{k} \\ a_x & a_y & a_z \\ b_x & b_y & b_z \end{vmatrix}$$

$$= \mathbf{i} \begin{vmatrix} a_y & a_z \\ b_y & b_z \end{vmatrix} - \mathbf{j} \begin{vmatrix} a_x & a_z \\ b_x & b_z \end{vmatrix} + \mathbf{k} \begin{vmatrix} a_x & a_y \\ b_x & b_y \end{vmatrix}$$

$$= (a_yb_z - b_ya_z)\mathbf{i}$$

$$+ (a_zb_x - b_za_x)\mathbf{j} + (a_xb_y - b_xa_y)\mathbf{k}$$

$$|\mathbf{a} \times \mathbf{b}| = ab \sin \theta$$

$$\mathbf{a} \cdot (\mathbf{b} \times \mathbf{c}) = \mathbf{b} \cdot (\mathbf{c} \times \mathbf{a}) = \mathbf{c} \cdot (\mathbf{a} \times \mathbf{b})$$

$$\mathbf{a} \times (\mathbf{b} \times \mathbf{c}) = (\mathbf{a} \cdot \mathbf{c})\mathbf{b} - (\mathbf{a} \cdot \mathbf{b})\mathbf{c}$$

DERIVATIVES AND INTEGRALS

In what follows, the letters u and v stand for any functions of x, and a and m are constants. To each of the indefinite integrals should be added an arbitrary constant of integration. The *Handbook of Chemistry and Physics* (CRC Press Inc.) gives a more extensive tabulation.

1. $\dfrac{dx}{dx} = 1$

2. $\dfrac{d}{dx}(au) = a\dfrac{du}{dx}$

3. $\dfrac{d}{dx}(u + v) = \dfrac{du}{dx} + \dfrac{dv}{dx}$

4. $\dfrac{d}{dx}x^m = mx^{m-1}$

5. $\dfrac{d}{dx}\ln x = \dfrac{1}{x}$

6. $\dfrac{d}{dx}(uv) = u\dfrac{dv}{dx} + v\dfrac{du}{dx}$

7. $\dfrac{d}{dx}e^x = e^x$

8. $\dfrac{d}{dx}\sin x = \cos x$

9. $\dfrac{d}{dx}\cos x = -\sin x$

10. $\dfrac{d}{dx}\tan x = \sec^2 x$

11. $\dfrac{d}{dx}\cot x = -\csc^2 x$

12. $\dfrac{d}{dx}\sec x = \tan x \sec x$

13. $\dfrac{d}{dx}\csc x = -\cot x \csc x$

14. $\dfrac{d}{dx}e^u = e^u\dfrac{du}{dx}$

15. $\dfrac{d}{dx}\sin u = \cos u\dfrac{du}{dx}$

16. $\dfrac{d}{dx}\cos u = -\sin u\dfrac{du}{dx}$

1. $\displaystyle\int dx = x$

2. $\displaystyle\int au\, dx = a\int u\, dx$

3. $\displaystyle\int (u + v)\, dx = \int u\, dx + \int v\, dx$

4. $\displaystyle\int x^m\, dx = \dfrac{x^{m+1}}{m+1} \quad (m \neq -1)$

5. $\displaystyle\int \dfrac{dx}{x} = \ln |x|$

6. $\displaystyle\int u\dfrac{dv}{dx}\, dx = uv - \int v\dfrac{du}{dx}\, dx$

7. $\displaystyle\int e^x\, dx = e^x$

8. $\displaystyle\int \sin x\, dx = -\cos x$

9. $\displaystyle\int \cos x\, dx = \sin x$

10. $\displaystyle\int \tan x\, dx = \ln |\sec x|$

11. $\displaystyle\int \sin^2 x\, dx = \tfrac{1}{2}x - \tfrac{1}{4}\sin 2x$

12. $\displaystyle\int e^{-ax}\, dx = -\dfrac{1}{a}e^{-ax}$

13. $\displaystyle\int xe^{-ax}\, dx = -\dfrac{1}{a^2}(ax + 1)e^{-ax}$

14. $\displaystyle\int x^2 e^{-ax}\, dx = -\dfrac{1}{a^3}(a^2x^2 + 2ax + 2)e^{-ax}$

15. $\displaystyle\int_0^\infty x^n e^{-ax}\, dx = \dfrac{n!}{a^{n+1}}$

16. $\displaystyle\int_0^\infty x^{2n} e^{-ax^2}\, dx = \dfrac{1 \cdot 3 \cdot 5 \cdots (2n-1)}{2^{n+1}a^n}\sqrt{\dfrac{\pi}{a}}$

17. $\displaystyle\int \dfrac{dx}{\sqrt{x^2 + a^2}} = \ln(x + \sqrt{x^2 + a^2})$

18. $\displaystyle\int \dfrac{x\, dx}{(x^2 + a^2)^{3/2}} = -\dfrac{1}{(x^2 + a^2)^{1/2}}$

19. $\displaystyle\int \dfrac{dx}{(x^2 + a^2)^{3/2}} = \dfrac{x}{a^2(x^2 + a^2)^{1/2}}$

Appendix F
Properties of the Elements

All physical properties are for a pressure of 1 atm unless otherwise specified.

ELEMENT	SYMBOL	ATOMIC NUMBER, Z	MOLAR MASS, g/mol	DENSITY, g/cm³ AT 20°C	MELTING POINT, °C	BOILING POINT, °C	SPECIFIC HEAT, J/(g·°C) AT 25°C
Actinium	Ac	89	(227)	10.06	1323	(3473)	0.092
Aluminum	Al	13	26.9815	2.699	660	2450	0.900
Americium	Am	95	(243)	13.67	1541	—	—
Antimony	Sb	51	121.75	6.691	630.5	1380	0.205
Argon	Ar	18	39.948	1.6626×10^{-3}	−189.4	−185.8	0.523
Arsenic	As	33	74.9216	5.78	817 (28 atm)	613	0.331
Astatine	At	85	(210)	—	(302)	—	—
Barium	Ba	56	137.34	3.594	729	1640	0.205
Berkelium	Bk	97	(247)	14.79	—	—	—
Beryllium	Be	4	9.0122	1.848	1287	2770	1.83
Bismuth	Bi	83	208.980	9.747	271.37	1560	0.122
Boron	B	5	10.811	2.34	2030	—	1.11
Bromine	Br	35	79.909	3.12 (liquid)	−7.2	58	0.293
Cadmium	Cd	48	112.40	8.65	321.03	765	0.226
Calcium	Ca	20	40.08	1.55	838	1440	0.624
Californium	Cf	98	(251)	—	—	—	—
Carbon	C	6	12.01115	2.26	3727	4830	0.691
Cerium	Ce	58	140.12	6.768	804	3470	0.188
Cesium	Cs	55	132.905	1.873	28.40	690	0.243
Chlorine	Cl	17	35.453	3.214×10^{-3} (0°C)	−101	−34.7	0.486
Chromium	Cr	24	51.996	7.19	1857	2665	0.448
Cobalt	Co	27	58.9332	8.85	1495	2900	0.423
Copper	Cu	29	63.54	8.96	1083.40	2595	0.385
Curium	Cm	96	(247)	13.3	—	—	—
Dysprosium	Dy	66	162.50	8.55	1409	2330	0.172
Einsteinium	Es	99	(254)	—	—	—	—
Erbium	Er	68	167.26	9.15	1522	2630	0.167
Europium	Eu	63	151.96	5.243	817	1490	0.163
Fermium	Fm	100	(237)	—	—	—	—
Fluorine	F	9	18.9984	1.696×10^{-3} (0°C)	−219.6	−188.2	0.753
Francium	Fr	87	(223)	—	(27)	—	—
Gadolinium	Gd	64	157.25	7.90	1312	2730	0.234
Gallium	Ga	31	69.72	5.907	29.75	2237	0.377
Germanium	Ge	32	72.59	5.323	937.25	2830	0.322
Gold	Au	79	196.967	19.32	1064.43	2970	0.131
Hafnium	Hf	72	178.49	13.31	2227	5400	0.144
Hahnium	Ha	105	—	—	—	—	—
Hassium	Hs	108	—	—	—	—	—

continued on next page

ELEMENT	SYMBOL	ATOMIC NUMBER, Z	MOLAR MASS, g/mol	DENSITY, g/cm³ AT 20°C	MELTING POINT, °C	BOILING POINT, °C	SPECIFIC HEAT, J/(g·°C) AT 25°C
Helium	He	2	4.0026	0.1664×10^{-3}	-269.7	-268.9	5.23
Holmium	Ho	67	164.930	8.79	1470	2330	0.165
Hydrogen	H	1	1.00797	0.08375×10^{-3}	-259.19	-252.7	14.4
Indium	In	49	114.82	7.31	156.634	2000	0.233
Iodine	I	53	126.9044	4.93	113.7	183	0.218
Iridium	Ir	77	192.2	22.5	2447	(5300)	0.130
Iron	Fe	26	55.847	7.874	1536.5	3000	0.447
Krypton	Kr	36	83.80	3.488×10^{-3}	-157.37	-152	0.247
Lanthanum	La	57	138.91	6.189	920	3470	0.195
Lawrencium	Lr	103	(257)		—	—	
Lead	Pb	82	207.19	11.35	327.45	1725	0.129
Lithium	Li	3	6.939	0.534	180.55	1300	3.58
Lutetium	Lu	71	174.97	9.849	1663	1930	0.155
Magnesium	Mg	12	24.312	1.738	650	1107	1.03
Manganese	Mn	25	54.9380	7.44	1244	2150	0.481
Meitnerium	Mt	109	—	—	—	—	—
Mendelevium	Md	101	(256)	—	—	—	—
Mercury	Hg	80	200.59	13.55	-38.87	357	0.138
Molybdenum	Mo	42	95.94	10.22	2617	5560	0.251
Neodymium	Nd	60	144.24	7.007	1016	3180	0.188
Neon	Ne	10	20.183	0.8387×10^{-3}	-248.597	-246.0	1.03
Neptunium	Np	93	(237)	20.25	637	—	1.26
Nickel	Ni	28	58.71	8.902	1453	2730	0.444
Nielsbohrium	Ns	107	—	—	—	—	—
Niobium	Nb	41	92.906	8.57	2468	4927	0.264
Nitrogen	N	7	14.0067	1.1649×10^{-3}	-210	-195.8	1.03
Nobelium	No	102	(255)	—	—	—	—
Osmium	Os	76	190.2	22.59	3027	5500	0.130
Oxygen	O	8	15.9994	1.3318×10^{-3}	-218.80	-183.0	0.913
Palladium	Pd	46	106.4	12.02	1552	3980	0.243
Phosphorus	P	15	30.9738	1.83	44.25	280	0.741
Platinum	Pt	78	195.09	21.45	1769	4530	0.134
Plutonium	Pu	94	(244)	19.8	640	3235	0.130
Polonium	Po	84	(210)	9.32	254	—	—
Potassium	K	19	39.102	0.862	63.20	760	0.758
Praseodymium	Pr	59	140.907	6.773	931	3020	0.197
Promethium	Pm	61	(145)	7.22	(1027)	—	—
Protactinium	Pa	91	(231)	15.37 (estimated)	(1230)	—	—
Radium	Ra	88	(226)	5.0	700	—	—
Radon	Rn	86	(222)	9.96×10^{-3} (0°C)	(-71)	-61.8	0.092
Rhenium	Re	75	186.2	21.02	3180	5900	0.134
Rhodium	Rh	45	102.905	12.41	1963	4500	0.243
Rubidium	Rb	37	85.47	1.532	39.49	688	0.364
Ruthenium	Ru	44	101.107	12.37	2250	4900	0.239
Rutherfordium	Rf	104	—	—	—	—	—
Samarium	Sm	62	150.35	7.52	1072	1630	0.197

continued on next page

ELEMENT	SYMBOL	ATOMIC NUMBER, Z	MOLAR MASS, g/mol	DENSITY, g/cm³ AT 20°C	MELTING POINT, °C	BOILING POINT, °C	SPECIFIC HEAT, J/(g·°C) AT 25°C
Scandium	Sc	21	44.956	2.99	1539	2730	0.569
Seaborgium	Sg	106	—	—	—	—	—
Selenium	Se	34	78.96	4.79	221	685	0.318
Silicon	Si	14	28.086	2.33	1412	2680	0.712
Silver	Ag	47	107.870	10.49	960.8	2210	0.234
Sodium	Na	11	22.9898	0.9712	97.85	892	1.23
Strontium	Sr	38	87.62	2.54	768	1380	0.737
Sulfur	S	16	32.064	2.07	119.0	444.6	0.707
Tantalum	Ta	73	180.948	16.6	3014	5425	0.138
Technetium	Tc	43	(99)	11.46	2200	—	0.209
Tellurium	Te	52	127.60	6.24	449.5	990	0.201
Terbium	Tb	65	158.924	8.229	1357	2530	0.180
Thallium	Tl	81	204.37	11.85	304	1457	0.130
Thorium	Th	90	(232)	11.72	1755	(3850)	0.117
Thulium	Tm	69	168.934	9.32	1545	1720	0.159
Tin	Sn	50	118.69	7.2984	231.868	2270	0.226
Titanium	Ti	22	47.90	4.54	1670	3260	0.523
Tungsten	W	74	183.85	19.3	3380	5930	0.134
Uranium	U	92	(238)	18.95	1132	3818	0.117
Vanadium	V	23	50.942	6.11	1902	3400	0.490
Xenon	Xe	54	131.30	5.495×10^{-3}	−111.79	−108	0.159
Ytterbium	Yb	70	173.04	6.965	824	1530	0.155
Yttrium	Y	39	88.905	4.469	1526	3030	0.297
Zinc	Zn	30	65.37	7.133	419.58	906	0.389
Zirconium	Zr	40	91.22	6.506	1852	3580	0.276

The values in parentheses in the column of molar masses are the mass numbers of the longest-lived isotopes of those elements that are radioactive. Melting points and boiling points in parentheses are uncertain.

The data for gases are valid only when these are in their usual molecular state, such as H_2, He, O_2, Ne, etc. The specific heats of the gases are the values at constant pressure.

Source: Adapted from Wehr, Richards, Adair, *Physics of the Atom,* 4th ed., Addison-Wesley, Reading, MA, 1984, and from J. Emsley, *The Elements,* 2nd ed., Clarendon Press, Oxford, 1991.

Appendix G
Periodic Table of the Elements

Metals

Metalloids

Nonmetals

Transition metals

Alkali metals — IA

Noble gases — 0

THE HORIZONTAL PERIODS

IA	IIA											IIIA	IVA	VA	VIA	VIIA	0
1 H																	2 He
3 Li	4 Be											5 B	6 C	7 N	8 O	9 F	10 Ne
11 Na	12 Mg	IIIB	IVB	VB	VIB	VIIB	VIIIB			IB	IIB	13 Al	14 Si	15 P	16 S	17 Cl	18 Ar
19 K	20 Ca	21 Sc	22 Ti	23 V	24 Cr	25 Mn	26 Fe	27 Co	28 Ni	29 Cu	30 Zn	31 Ga	32 Ge	33 As	34 Se	35 Br	36 Kr
37 Rb	38 Sr	39 Y	40 Zr	41 Nb	42 Mo	43 Tc	44 Ru	45 Rh	46 Pd	47 Ag	48 Cd	49 In	50 Sn	51 Sb	52 Te	53 I	54 Xe
55 Cs	56 Ba	57-71 *	72 Hf	73 Ta	74 W	75 Re	76 Os	77 Ir	78 Pt	79 Au	80 Hg	81 Tl	82 Pb	83 Bi	84 Po	85 At	86 Rn
87 Fr	88 Ra	89-103 †	104 Rf	105 Ha	106 Sg	107 Ns	108 Hs	109 Mt	110	111	112						

Inner transition metals

Lanthanide series *

57 La	58 Ce	59 Pr	60 Nd	61 Pm	62 Sm	63 Eu	64 Gd	65 Tb	66 Dy	67 Ho	68 Er	69 Tm	70 Yb	71 Lu

Actinide series †

89 Ac	90 Th	91 Pa	92 U	93 Np	94 Pu	95 Am	96 Cm	97 Bk	98 Cf	99 Es	100 Fm	101 Md	102 No	103 Lr

The names for elements 104–109 (Rutherfordium, Hahnium, Seaborgium, Nielsbohrium, Hassium, and Meitnerium, respectively) are those recommended by the American Chemical Society Nomenclature Committee. As of 1996, the names and symbols for elements 104–108 have not yet been approved by the appropriate international body. Elements 110, 111 and 112 have been discovered but, as of 1996, have not been provisionally named.

1901 Wilhelm Konrad Röntgen *(1845–1923)* for the discovery of x rays

1902 Hendrik Antoon Lorentz *(1853–1928)* and Pieter Zeeman *(1865–1943)* for their researches into the influence of magnetism upon radiation phenomena

1903 Antoine Henri Becquerel *(1852–1908)* for his discovery of spontaneous radioactivity

Pierre Curie *(1859–1906)* and Marie Sklowdowska-Curie *(1867–1934)* for their joint researches on the radiation phenomena discovered by Becquerel

1904 Lord Rayleigh (John William Strutt) *(1842–1919)* for his investigations of the densities of the most important gases and for his discovery of argon

1905 Philipp Eduard Anton von Lenard *(1862–1947)* for his work on cathode rays

1906 Joseph John Thomson *(1856–1940)* for his theoretical and experimental investigations on the conduction of electricity by gases

1907 Albert Abraham Michelson *(1852–1931)* for his optical precision instruments and metrological investigations carried out with their aid

1908 Gabriel Lippmann *(1845–1921)* for his method of reproducing colors photographically based on the phenomena of interference

1909 Guglielmo Marconi *(1874–1937)* and Carl Ferdinand Braun *(1850–1918)* for their contributions to the development of wireless telegraphy

1910 Johannes Diderik van der Waals *(1837–1932)* for his work on the equation of state for gases and liquids

1911 Wilhelm Wien *(1864–1928)* for his discoveries regarding the laws governing the radiation of heat

1912 Nils Gustaf Dalén *(1869–1937)* for his invention of automatic regulators for use in conjunction with gas accumulators for illuminating lighthouses and buoys

1913 Heike Kamerlingh Onnes *(1853–1926)* for his investigations of the properties of matter at low temperatures which led, among other things, to the production of liquid helium

1914 Max von Laue *(1879–1960)* for his discovery of the diffraction of Röntgen rays by crystals

1915 William Henry Bragg *(1862–1942)* and William Lawrence Bragg *(1890–1971)* for their services in the analysis of crystal structure by means of x rays

1917 Charles Glover Barkla *(1877–1944)* for his discovery of the characteristic x rays of the elements

1918 Max Planck *(1858–1947)* for his discovery of energy quanta

1919 Johannes Stark *(1874–1957)* for his discovery of the Doppler effect in canal rays and the splitting of spectral lines in electric fields

1920 Charles-Édouard Guillaume *(1861–1938)* for the service he rendered to precision measurements in physics by his discovery of anomalies in nickel steel alloys

1921 Albert Einstein *(1879–1955)* for his services to theoretical physics, and especially for his discovery of the law of the photoelectric effect

1922 Niels Bohr *(1885–1962)* for the investigation of the structure of atoms, and of the radiation emanating from them

1923 Robert Andrews Millikan *(1868–1953)* for his work on the elementary charge of electricity and on the photoelectric effect

1924 Karl Manne Georg Siegbahn *(1888–1979)* for his discoveries and research in the field of x-ray spectroscopy

1925 James Franck *(1882–1964)* and Gustav Hertz *(1887–1975)* for their discovery of the laws governing the impact of an electron upon an atom

1926 Jean Baptiste Perrin *(1870–1942)* for his work on the discontinuous structure of matter, and especially for his discovery of sedimentation equilibrium

1927 Arthur Holly Compton *(1892–1962)* for his discovery of the effect named after him

Charles Thomson Rees Wilson (1869–1959) for his method of making the paths of electrically charged particles visible by condensation of vapor

1928 Owen Willans Richardson *(1879–1959)* for his work on the thermionic phenomenon and especially for the discovery of the law named after him

1929 Prince Louis Victor de Broglie *(1892–1987)* for his discovery of the wave nature of electrons

*See *Nobel Lectures, Physics,* 1901–1970, Elsevier Publishing Company, for biographies of the awardees and for lectures given by them on receiving the prize.

A17

1930 Sir Chandrasekhara Venkata Raman *(1888–1970)* for his work on the scattering of light and for the discovery of the effect named after him

1932 Werner Heisenberg *(1901–1976)* for the creation of quantum mechanics, the application of which has, among other things, led to the discovery of the allotropic forms of hydrogen

1933 Erwin Schrödinger *(1887–1961)* and Paul Adrien Maurice Dirac *(1902–1984)* for the discovery of new productive forms of atomic theory

1935 James Chadwick *(1891–1974)* for his discovery of the neutron

1936 Victor Franz Hess *(1883–1964)* for the discovery of cosmic radiation

Carl David Anderson *(1905–1991)* for his discovery of the positron

1937 Clinton Joseph Davisson *(1881–1958)* and George Paget Thomson *(1892–1975)* for their experimental discovery of the diffraction of electrons by crystals

1938 Enrico Fermi *(1901–1954)* for his demonstrations of the existence of new radioactive elements produced by neutron irradiation, and for his related discovery of nuclear reactions brought about by slow neutrons

1939 Ernest Orlando Lawrence *(1901–1958)* for the invention and development of the cyclotron and for results obtained with it, especially for artificial radioactive elements

1943 Otto Stern *(1888–1969)* for his contribution to the development of the molecular-ray method and his discovery of the magnetic moment of the proton

1944 Isidor Isaac Rabi *(1898–1988)* for his resonance method for recording the magnetic properties of atomic nuclei

1945 Wolfgang Pauli *(1900–1958)* for the discovery of the Exclusion Principle (also called Pauli Principle)

1946 Percy Williams Bridgman *(1882–1961)* for the invention of an apparatus to produce extremely high pressures and for the discoveries he made therewith in the field of high-pressure physics

1947 Sir Edward Victor Appleton *(1892–1965)* for his investigations of the physics of the upper atmosphere, especially for the discovery of the so-called Appleton layer

1948 Patrick Maynard Stuart Blackett *(1897–1974)* for his development of the Wilson cloud-chamber method, and his discoveries therewith in nuclear physics and cosmic radiation

1949 Hideki Yukawa *(1907–1981)* for his prediction of the existence of mesons on the basis of theoretical work on nuclear forces

1950 Cecil Frank Powell *(1903–1969)* for his development of the photographic method of studying nuclear processes and his discoveries regarding mesons made with this method

1951 Sir John Douglas Cockcroft *(1897–1967)* and Ernest Thomas Sinton Walton *(1903–   )* for their pioneer work on the transmutation of atomic nuclei by artificially accelerated atomic particles

1952 Felix Bloch *(1905–1983)* and Edward Mills Purcell *(1912–   )* for their development of new nuclear-magnetic precision methods and discoveries in connection therewith

1953 Frits Zernike *(1888–1966)* for his demonstration of the phase-contrast method, especially for his invention of the phase-contrast microscope

1954 Max Born *(1882–1970)* for his fundamental research in quantum mechanics, especially for his statistical interpretation of the wave function

Walther Bothe *(1891–1957)* for the coincidence method and his discoveries made therewith

1955 Willis Eugene Lamb *(1913–   )* for his discoveries concerning the fine structure of the hydrogen spectrum

Polykarp Kusch *(1911–1993)* for his precision determination of the magnetic moment of the electron

1956 William Shockley *(1910–1989)*, John Bardeen *(1908–1991)* and Walter Houser Brattain *(1902–1987)* for their researches on semiconductors and their discovery of the transistor effect

1957 Chen Ning Yang *(1922–   )* and Tsung Dao Lee *(1926–   )* for their penetrating investigation of the parity laws which has led to important discoveries regarding the elementary particles

1958 Pavel Aleksejevič Čerenkov *(1904–   )*, Il' ja Michajlovič Frank *(1908–1990)* and Igor' Evgen' evič Tamm *(1895–1971)* for the discovery and interpretation of the Cerenkov effect

1959 Emilio Gino Segrè *(1905–1989)* and Owen Chamberlain *(1920–   )* for their discovery of the antiproton

1960 Donald Arthur Glaser *(1926–   )* for the invention of the bubble chamber

1961 Robert Hofstadter *(1915–1990)* for his pioneering studies of electron scattering in atomic nuclei and for his thereby achieved discoveries concerning the structure of the nucleons

Rudolf Ludwig Mössbauer *(1929–   )* for his researches concerning the resonance absorption of γ rays and his discovery in this connection of the effect which bears his name

1962 Lev Davidovič Landau *(1908–1968)* for his pioneering theories of condensed matter, especially liquid helium

1963 Eugene P. Wigner *(1902–1995)* for his contributions to the theory of the atomic nucleus and the elementary particles, particularly through the discovery and application of fundamental symmetry principles

Maria Goeppert Mayer *(1906–1972)* and J. Hans D. Jensen *(1907–1973)* for their discoveries concerning nuclear shell structure

1964 Charles H. Townes *(1915–   )*, Nikolai G. Basov *(1922–   )* and Alexander M. Prochorov *(1916–   )* for fundamental work in the field of quantum electronics which has led to the construction of oscillators and amplifiers based on the maser–laser principle

1965 Sin-itiro Tomonaga *(1906–1979)*, Julian Schwinger *(1918–1994)* and Richard P. Feynman *(1918–1988)* for their fundamental work in quantum electrodynamics, with deep-ploughing consequences for the physics of elementary particles

1966 Alfred Kastler *(1902–1984)* for the discovery and development of optical methods for studying Hertzian resonance in atoms

1967 Hans Albrecht Bethe *(1906–   )* for his contributions to the theory of nuclear reactions, especially his discoveries concerning the energy production in stars

1968 Luis W. Alvarez *(1911–1988)* for his decisive contribution to elementary particle physics, in particular the discovery of a large number of resonance states, made possible through his development of the techniques of using the hydrogen bubble chamber and its data analysis

1969 Murray Gell-Mann *(1929–   )* for his contributions and discoveries concerning the classification of elementary particles and their interactions

1970 Hannes Alfvén *(1908–1995)* for fundamental work and discoveries in magneto-hydrodynamics with fruitful applications in different parts of plasma physics

Louis Néel *(1904–   )* for fundamental work and discoveries concerning antiferromagnetism and ferrimagnetism which have led to important applications in solid state physics

1971 Dennis Gabor *(1900–1979)* for his discovery of the principles of holography

1972 John Bardeen *(1908–1991)*, Leon N. Cooper *(1930–   )* and J. Robert Schrieffer *(1931–   )* for their development of a theory of superconductivity

1973 Leo Esaki *(1925–   )* for his discovery of tunneling in semiconductors

Ivar Giaever *(1929–   )* for his discovery of tunneling in superconductors

Brian D. Josephson *(1940–   )* for his theoretical prediction of the properties of a supercurrent through a tunnel barrier

1974 Antony Hewish *(1924–   )* for the discovery of pulsars

Sir Martin Ryle *(1918–1984)* for his pioneering work in radioastronomy

1975 Aage Bohr *(1922–   )*, Ben Mottelson *(1926–   )* and James Rainwater *(1917–1986)* for the discovery of the connection between collective motion and particle motion and the development of the theory of the structure of the atomic nucleus based on this connection

1976 Burton Richter *(1931–   )* and Samuel Chao Chung Ting *(1936–   )* for their (independent) discovery of an important fundamental particle

1977 Philip Warren Anderson *(1923–   )*, Nevill Francis Mott *(1905–   )* and John Hasbrouck Van Vleck *(1899–1980)* for their fundamental theoretical investigations of the electronic structure of magnetic and disordered systems

1978 Peter L. Kapitza *(1894–1984)* for his basic inventions and discoveries in low-temperature physics

Arno A. Penzias *(1933–   )* and Robert Woodrow Wilson *(1936–   )* for their discovery of cosmic microwave background radiation

1979 Sheldon Lee Glashow *(1932–   )*, Abdus Salam *(1926–   )*, and Steven Weinberg *(1933–   )* for their unified model of the action of the weak and electromagnetic forces and for their prediction of the existence of neutral currents

1980 James W. Cronin *(1931–   )* and Val L. Fitch *(1923–   )* for the discovery of violations of fundamental symmetry principles in the decay of neutral K mesons

1981 Nicolaas Bloembergen *(1920–   )* and Arthur Leonard Schawlow *(1921–   )* for their contribution to the development of laser spectroscopy

Kai M. Siegbahn *(1918–   )* for his contribution to high-resolution electron spectroscopy

1982 Kenneth Geddes Wilson *(1936–   )* for his method of analyzing the critical phenomena inherent in the changes of matter under the influence of pressure and temperature

1983 Subrehmanyan Chandrasekhar *(1910–1995)* for his theoretical studies of the structure and evolution of stars

William A. Fowler *(1911–1995)* for his studies of the formation of the chemical elements in the universe

1984 Carlo Rubbia *(1934–   )* and Simon van der Meer *(1925–   )* for their decisive contributions to the Large Project, which led to the discovery of the field particles W and Z, communicators of the weak interaction

1985 Klaus von Klitzing *(1943–   )* for his discovery of the quantized Hall resistance

1986 Ernst Ruska *(1906–1988)* for his invention of the electron microscope

Gerd Binnig *(1947–   )*, Heinrich Rohrer *(1933–   )* for their invention of the scanning tunneling microscope

1987 Karl Alex Müller *(1927–   )* and J. George Bednorz *(1950–   )* for their discovery of a new class of superconductors

1988 Leon M. Lederman *(1922–   )*, Melvin Schwartz *(1932–   )* and Jack Steinberger *(1921–   )* for the first use of a neutrino beam and the discovery of the muon neutrino

1989 Norman Ramsey *(1915–   )*, Hans Dehmelt *(1922–   )* and Wolfgang Paul *(1913–1993)* for their work that led to the development of atomic clocks and precision timing

1990 Jerome I. Friedman *(1930–   )*, Henry W. Kendall *(1926–   )* and Richard E. Taylor *(1929–   )* for demonstrating that protons and neutrons consist of quarks

1991 Pierre de Gennes *(1932–   )* for studies of order phenomena, such as in liquid crystals and polymers

1992 George Charpak *(1924–   )* for his invention of fast electronic detectors for high-energy particles

1993   Joseph H. Taylor *(1941–   )* and Russell A. Hulse *(1950–   )* for the discovery and interpretation of the first binary pulsar.

1994   Bertram N. Brockhouse *(1918–   )* and Clifford G. Shull *(1915–   )* for the development of neutron scattering techniques

1995   Martin L. Perl *(1927–   )* for the discovery of the tau lepton

Frederick Reines *(1918–   )* for the detection of the neutrino

Answers to Checkpoints, Odd-Numbered Questions, Exercises, and Problems

Chapter 1

EP **3.** (a) 186 mi; (b) 3.0×10^8 mm **5.** (a) 10^9; (b) 10^{-4}; (c) 9.1×10^5 **7.** 32.2 km **9.** 0.020 km³ **11.** (a) 250 ft²; (b) 23.3 m²; (c) 3060 ft³; (d) 86.6 m³ **13.** 8×10^2 km **15.** (a) 11.3 m²/L; (b) 1.13×10^4 m⁻¹; (c) 2.17×10^{-3} gal/ft² **17.** (a) $d_{Sun}/d_{Moon} = 400$; (b) $V_{Sun}/V_{Moon} = 6.4 \times 10^7$; (c) 3.5×10^3 km **19.** (a) 0.98 ft/ns; (b) 0.30 mm/ps **21.** 3.156×10^7 s **23.** 5.79×10^{12} days **25.** (a) 0.013; (b) 0.54; (c) 10.3; (d) 31 m/s **27.** 15° **29.** 3.3 ft **31.** 2 days 5 hours **33.** (a) 2.99×10^{-26} kg; (b) 4.68×10^{46} **35.** 1.3×10^9 kg **37.** (a) 10^3 kg/m³; (b) 158 kg/s **39.** (a) 1.18×10^{-29} m³; (b) 0.282 nm

Chapter 2

CP **1.** b and c **2.** zero **3.** (a) 1 and 4; (b) 2 and 3; (c) 3 **4.** (a) plus; (b) minus; (c) minus; (d) plus **5.** 1 and 4 **6.** (a) plus; (b) minus; (c) $a = -g = -9.8$ m/s² **Q** **1.** (a) yes; (b) no; (c) yes; (d) yes **3.** (a) 2, 3; (b) 1, 3; (c) 4 **5.** all tie (see Eq. 2-16) **7.** (a) $-g$; (b) 2 m/s upward **9.** same **11.** $x = t^2$ and $x = 8(t - 2) + (1.5)(t - 2)^2$ **13.** increase **EP** **1.** (a) Lewis: 10.0 m/s, Rodgers: 5.41 m/s; (b) 1 h 10 min **3.** 309 ft **5.** 2 cm/y **7.** 6.71×10^8 mi/h, 9.84×10^8 ft/s, 1.00 ly/y **9.** (a) 5.7 ft/s; (b) 7.0 ft/s **11.** (a) 45 mi/h (72 km/h); (b) 43 mi/h (69 km/h); (c) 44 mi/h (71 km/h); (d) 0 **13.** (a) 28.5 cm/s; (b) 18.0 cm/s; (c) 40.5 cm/s; (d) 28.1 cm/s; (e) 30.3 cm/s **15.** (a) mathematically, an infinite number; (b) 60 km **17.** (a) 4 s $> t > 2$ s; (b) 3 s $> t > 0$; (c) 7 s $> t > 3$ s; (d) $t = 3$ s **19.** 100 m **23.** (a) The signs of v and a are: AB: +, −; BC: 0, 0; CD: +, +; DE: +, 0; (b) no; (c) no **25.** (e) situations (a), (b), and (d) **27.** (a) 80 m/s; (b) 110 m/s; (c) 20 m/s² **29.** (a) 1.10 m/s, 6.11 mm/s²; (b) 1.47 m/s, 6.11 mm/s² **31.** (a) 2.00 s; (b) 12 cm from left edge of screen; (c) 9.00 cm/s², to the left; (d) to the right; (e) to the left; (f) 3.46 s **33.** 0.556 s **35.** each, 0.28 m/s² **37.** 2.8 m/s² **39.** 1.62×10^{15} m/s² **41.** $21g$ **43.** (a) $25g$; (b) 400 m **45.** 90 m **47.** (a) 5.0 m/s²; (b) 4.0 s; (c) 6.0 s; (d) 90 m **49.** (a) 5.00 m/s; (b) 1.67 m/s²; (c) 7.50 m **51.** (a) 0.74 s; (b) -20 ft/s² **53.** (a) 0.75 s; (b) 50 m **55.** (a) 34.7 ft; (b) 41.6 s **57.** (a) 3.26 ft/s² **61.** (a) 31 m/s; (b) 6.4 s **63.** (a) 48.5 m/s; (b) 4.95 s; (c) 34.3 m/s; (d) 3.50 s **65.** (a) 5.44 s; (b) 53.3 m/s; (d) 5.80 m **67.** (a) 3.2 s; (b) 1.3 s **69.** 4.0 m/s **71.** (a) 350 ms; (b) 82 ms (each is for ascent and descent through the 15 cm) **73.** 857 m/s², upward **75.** (a) 1.23 cm; (b) 4 times, 9 times, 16 times, 25 times **77.** (a) 8.85 m/s; (b) 1.00 m **79.** 22 cm and 89 cm below the nozzle **81.** (a) 3.41 s; (b) 57 m **83.** (a) 40.0 ft/s **85.** 1.5 s **87.** (a) 5.4 s; (b) 41 m/s **89.** 20.4 m

91. (a) $d = v_i^2/2a' + T_R v_i$; (b) 9.0 m/s²; (c) 0.66 s. **93.** (a) $v_f^2 = 2a'd_0(j - 1) + v_1^2$; (c) 7.0 m/s²; (d) 14 m.

Chapter 3

CP **1.** (a) 7 m; (b) 1 m **2.** c, d, f **3.** (a) +, +; (b) +, −; (c) +, + **4.** (a) 90°; (b) 0 (vectors are parallel); (c) 180° (vectors are antiparallel) **5.** (a) 0° or 180°; (b) 90° **Q** **1.** **A** and **B** **3.** No, but **a** and $-\mathbf{b}$ are commutative: $\mathbf{a} + (-\mathbf{b}) = (-\mathbf{b}) + \mathbf{a}$. **5.** (a) **a** and **b** are parallel; (b) $\mathbf{b} = 0$; (c) **a** and **b** are perpendicular **7.** (a)–(c) yes (example: 5**i** and $-2\mathbf{i}$) **9.** all but e **11.** (a) minus, minus; (b) minus, minus **13.** (a) **B** and **C**, **D** and **E**; (b) **D** and **E** **15.** no (their orientations can differ) **17.** (a) 0 (vectors are parallel); (b) 0 (vectors are antiparallel) **EP** **1.** The displacements should be (a) parallel, (b) antiparallel, (c) perpendicular **3.** (b) 3.2 km, 41° south of west **5.** $\mathbf{a} + \mathbf{b}$: 4.2, 40° east of north; $\mathbf{b} - \mathbf{a}$: 8.0, 24° north of west **7.** (a) 38 units at 320°; (b) 130 units at 1.2°; (c) 62 units at 130° **9.** $a_x = -2.5$, $a_y = -6.9$ **11.** $r_x = 13$ m, $r_y = 7.5$ m **13.** (a) 14 cm, 45° left of straight down; (b) 20 cm, vertically up; (c) zero **15.** 4.74 km **17.** 168 cm, 32.5° above the floor **19.** $r_x = 12$, $r_y = -5.8$, $r_z = -2.8$ **21.** (a) 8**i** + 2**j**, 8.2, 14°; (b) 2**i** − 6**j**, 6.3, $-72°$ relative to **i** **23.** (a) 5.0, $-37°$; (b) 10, 53°; (c) 11, 27°; (d) 11, 80°; (e) 11, 260°; the angles are relative to $+x$, the last two vectors are in opposite directions **25.** 4.1 **27.** (a) $r_x = 1.59$, $r_y = 12.1$; (b) 12.2; (c) 82.5° **29.** 3390 ft, horizontally **31.** (a) -2.83 m, -2.83 m, $+5.00$ m, 0 m, 3.00 m, 5.20 m; (b) 5.17 m, 2.37 m; (c) 5.69 m, 24.6° north of east; (d) 5.69 m, 24.6° south of west **35.** (a) $a_x = 9.51$ m, $a_y = 14.1$ m; (b) $a_x' = 13.4$ m, $a_y' = 10.5$ m **37.** (a) $+y$; (b) $-y$; (c) 0; (d) 0; (e) $+z$; (f) $-z$; (g) ab, both; (h) ab/d, $+z$ **39.** yes **41.** (a) up, unit magnitude; (b) zero; (c) south, unit magnitude; (d) 1.00; (e) 0 **43.** (a) -18.8; (b) 26.9, $+z$ direction **45.** (a) 12, out of page; (b) 12, into page; (c) 12, out of page **47.** (a) 11**i** + 5**j** − 7**k**; (b) 120° **51.** (a) 57°; (b) $c_x = \pm 2.2$, $c_y = \mp 4.5$ **53.** (a) -21; (b) -9; (c) 5**i** − 11**j** − 9**k**

Chapter 4

CP **1.** (a) (8**i** − 6**j**) m; (b) yes, the xy plane **2.** (a) first; (b) third **3.** (1) and (3) a_x and a_y are both constant and thus **a** is constant; (2) and (4) a_y is constant but a_x is not, thus **a** is not **4.** 4 m/s³, -2 m/s, 3 m **5.** (a) v_x constant; (b) v_y initially positive, decreases to zero, and then becomes progressively more negative; (c) $a_x = 0$ throughout; (d) $a_y = -g$ throughout **6.** (a) $-(4$ m/s$)\mathbf{i}$; (b) $-(8$ m/s²$)\mathbf{j}$ **7.** (1) 0, distance not changing; (2) $+70$ km/h, distance increasing; (3) $+80$ km/h, distance decreasing **Q** **1.** (1) and (3) a_y is constant but a_x is not and thus **a** is not; (2) a_x is constant but a_y

is not and thus **a** is not; (4) a_x and a_y are both constant and thus **a** is constant; -2 m/s², 3 m/s **3.** (a) highest point; (b) lowest point **5.** (a) all tie; (b) 1 and 2 tie (the rocket is shot upward), then 3 and 4 tie (it is shot into the ground!) **7.** $(2\mathbf{i} - 4\mathbf{j})$ m/s **9.** (a) all tie; (b) all tie; (c) c, b, a; (d) c, b, a **11.** (a) no; (b) same **13.** (a) in your hands; (b) behind you; (c) in front of you **15.** (a) straight down; (b) curved; (c) more curved **17.** (a) 3; (b) 4. **EP** **1.** (a) $(-5.0\mathbf{i} + 8.0\mathbf{j})$ m; (b) 9.4 m, 122° from $+x$; (d) $(8\mathbf{i} - 8\mathbf{j})$ m; (e) 11 m, $-45°$ from $+x$ **3.** (a) $(-7.0\mathbf{i} + 12\mathbf{j})$ m; (b) xy plane **5.** (a) 671 mi, 63.4° south of east; (b) 298 mi/h, 63.4° south of east; (c) 400 mi/h **7.** (a) 6.79 km/h; (b) 6.96° **9.** (a) $(3\mathbf{i} - 8t\mathbf{j})$ m/s; (b) $(3\mathbf{i} - 16\mathbf{j})$ m/s; (c) 16 m/s, $-79°$ to $+x$ **11.** (a) $(8t\mathbf{j} + \mathbf{k})$ m/s; (b) $8\mathbf{j}$ m/s² **13.** $(-2.10\mathbf{i} + 2.81\mathbf{j})$ m/s² **15.** (a) $-1.5\mathbf{j}$ m/s; (b) $(4.5\mathbf{i} - 2.25\mathbf{j})$ m **17.** 60.0° **19.** (a) 63 ms; (b) 1.6×10^3 ft/s **21.** (a) 2.0 ns; (b) 2.0 mm; (c) $(1.0 \times 10^9\mathbf{i} - 2.0 \times 10^8\mathbf{j})$ cm/s **23.** (a) 3.03 s; (b) 758 m; (c) 29.7 m/s **25.** (a) 16 m/s, 23° above the horizontal; (b) 27 m/s, 57° below the horizontal **27.** (a) 32.4 m; (b) -37.7 m **29.** (b) 76° **31.** (a) 51.8 m; (b) 27.4 m/s; (c) 67.5 m **33.** (a) 194 m/s; (b) 38° **35.** 1.9 in. **37.** (a) 11 m; (b) 23 m; (c) 17 m/s, 63° below horizontal **41.** (a) 73 ft; (b) 7.6°; (c) 1.0 s **43.** 23 ft/s **45.** (a) 11 m; (b) 45 m/s **47.** 30 m above the release point **49.** 19 ft/s **51.** (a) 202 m/s; (b) 806 m; (c) 161 m/s, -171 m/s **53.** (a) 20 cm; (b) no, the ball hits the net only 4.4 cm above the ground **55.** yes; its center passes about 4.1 ft above the fence **57.** (a) 9.00×10^{22} m/s², toward the center; (b) 1.52×10^{-16} s **59.** (a) 6.7×10^6 m/s; (b) 1.4×10^{-7} s **61.** (a) 7.49 km/s; (b) 8.00 m/s² **63.** (a) 0.94 m; (b) 19 m/s; (c) 2400 m/s², toward center; (d) 0.05 s **65.** (a) 1.3×10^5 m/s; (b) 7.9×10^5 m/s² or $(8.0 \times 10^4)g$, toward the center; (c) both answers increase **67.** (a) 0.034 m/s²; (b) 84 min **69.** 2.58 cm/s² **71.** 160 m/s² **73.** 36 s, no **75.** 0.018 mi/s² from either frame **77.** 130° **79.** 60° **81.** (a) 5.8 m/s; (b) 16.7 m; (c) 67° **83.** 185 km/h, 22° south of west **85.** (a) from 75° east of south; (b) 30° east of north; substitute west for east to get second solution **87.** (a) 30° upstream; (b) 69 min; (c) 80 min; (d) 80 min; (e) perpendicular to the current, the shortest possible time is 60 min **89.** $0.83c$ **91.** (a) $0.35c$; (b) $0.62c$ **93.** For launch angles from 5° to 70°, it always moves away from the launch site. For a 75° launch angle, it moves toward the site from 11.5 s to 18.5 s after launch. For an 80° launch angle, it moves toward the site from 10.5 s to 20.5 s after launch. For an 85° launch angle, it moves toward the site from 10.5 s to 20.5 s after launch. For a 90° launch angle, it moves toward the site from 10 s to 20.5 s after launch. **95.** (a) 1.6 s; (b) no; (c) 14 m/s; (d) yes **97.** (a) $\Delta\mathbf{D} = (1.0\text{ m})\mathbf{i} - (2.0\text{ m})\mathbf{j} + (1.0\text{ m})\mathbf{k}$; (b) 2.4 m; (c) $\bar{\mathbf{v}} = (0.025\text{ m/s})\mathbf{i} - (0.050\text{ m/s})\mathbf{j} + (0.025\text{ m/s})\mathbf{k}$ (d) cannot be determined without additional information

Chapter 5

CP **1.** c, d, and e **2.** (a) and (b) 2 N, leftward (acceleration is zero in each situation) **3.** (a) and (b) 1, 4, 3, 2 **4.** (a) equal; (b) greater (acceleration is upward, thus net force on body must be upward) **5.** (a) equal; (b) greater; (c) less **6.** (a) increase; (b) yes; (c) same; (d) yes **7.** (a) $F \sin \theta$; (b) increase **8.** 0 **Q** **1.** (a) yes; (b) yes; (c) yes; (d) yes **3.** (a) 2 and 4; (b) 2 and 4 **5.** (a) 50 N, upward; (b) 150 N, upward **7.** (a) less; (b) greater **9.** (a) no; (b) no; (c) no **11.** (a) increases; (b) increases; (c) decreases; (d) decreases **13.** (a) 20 kg; (b) 18 kg; (c) 10 kg; (d) all tie; (e) 3, 2, 1 **15.** d, c, a, b **EP** **1.** (a) $F_x = 1.88$ N, $F_y = 0.684$ N; (b) $(1.88\mathbf{i} + 0.684\mathbf{j})$ N **3.** (a) $(-6.26\mathbf{i} - 3.23\mathbf{j})$ N; (b) 7.0 N, 207° relative to $+x$ **5.** $(-2\mathbf{i} + 6\mathbf{j})$ N **7.** (a) 0; (b) $+20$ N; (c) -20 N; (d) -40 N; (e) -60 N **9.** (a) $(1\mathbf{i} - 1.3\mathbf{j})$ m/s²; (b) 1.6 m/s² at $-50°$ from $+x$ **11.** (a) $\mathbf{F}_2$ and $\mathbf{F}_3$ are in the $-x$ direction, $\mathbf{a} = 0$; (b) $\mathbf{F}_2$ and $\mathbf{F}_3$ are in the $-x$ direction, $\mathbf{a}$ is on the x axis, $a = 0.83$ m/s²; (c) $\mathbf{F}_2$ and $\mathbf{F}_3$ are at 34° from $-x$ direction; $\mathbf{a} = 0$ **13.** (a) 22 N, 2.3 kg; (b) 1100 N, 110 kg; (c) 1.6×10^4 N, 1.6×10^3 kg **15.** (a) 11 N, 2.2 kg; (b) 0, 2.2 kg **17.** (a) 44 N; (b) 78 N; (c) 54 N; (d) 152 N **19.** 1.18×10^4 N **21.** 1.2×10^5 N **23.** 16 N **25.** (a) 13 ft/s²; (b) 190 lb **27.** (a) 42 N; (b) 72 N; (c) 4.9 m/s² **29.** (a) 0.02 m/s²; (b) 8×10^4 km; (c) 2×10^3 m/s **31.** (a) 1.1×10^{-15} N; (b) 8.9×10^{-30} N **33.** (a) 5500 N; (b) 2.7 s; (c) 4 times as far; (d) twice the time **35.** (a) 4.9×10^5 N; (b) 1.5×10^6 N **37.** (a) 110 lb, up; (b) 110 lb, down **39.** (a) 0.74 m/s²; (b) 7.3 m/s² **41.** (a) $\cos \theta$; (b) $\sqrt{\cos \theta}$ **43.** 1.8×10^4 N **45.** (a) 4.6×10^3 N; (b) 5.8×10^3 N **47.** (a) 250 m/s²; (b) 2.0×10^4 N **49.** 23 kg **51.** (a) 620 N; (b) 580 N **53.** 1.9×10^5 lb **55.** (a) rope breaks; (b) 1.6 m/s² **57.** 4.6 N **59.** (a) allow a downward acceleration with magnitude ≥ 4.2 ft/s²; (b) 13 ft/s or greater **61.** 195 N, up **63.** (a) 566 N; (b) 1130 N **65.** 18,000 N **67.** (a) 1.4×10^4 N; (b) 1.1×10^4 N; (c) 2700 N, toward the counterweight **69.** 6800 N, at 21° to the line of motion of the barge **71.** (a) 4.6 m/s²; (b) 2.6 m/s² **73.** (b) $Fl/(m + M)$; (c) $MFl/(m + M)$; (d) $F(m + 2M)/2(m + M)$ **75.** $T_1 = 13$ N, $T_2 = 20$ N, $a = 3.2$ m/s²

Chapter 6

CP **1.** (a) zero (because there is no attempt at sliding); (b) 5 N; (c) no; (d) yes **2.** (a) same (10 N); (b) decreases; (c) decreases **3.** greater **4.** (a) **a** downward; **N** upward; (b) **a** and **N** upward **5.** (a) $4R_1$; (b) $4R_1$ **6.** (a) same; (b) increases; (c) increases **Q** **1.** They slide at the same angle for all orders. **3.** (a) upward; (b) horizontal, toward you; (c) no change; (d) increases; (e) increases **5.** The frictional force $\mathbf{f}_s$ is initially directed up the ramp, decreases in magnitude to zero, and then is directed down the ramp, increasing in magnitude until the magnitude reaches $f_{s,\text{max}}$; thereafter, the magnitude of the frictional force is f_k, which is a constant smaller value. **7.** (a) decreases; (b) decreases; (c) increases; (d) increases **9.** (a) zero; (b) infinite **11.** 4, 3; then 1, 2, and 5 tie **13.** (a) less; (b) greater **EP** **1.** (a) 200 N; (b) 120 N **3.** 2° **5.** 440 N

7. (a) 110 N; (b) 130 N; (c) no; (d) 46 N; (e) 17 N
9. (a) 90 N; (b) 70 N; (c) 0.89 m/s^2 **11.** (a) no; (b) $(-12\mathbf{i} + 5\mathbf{j})$ N **13.** 20° **15.** (a) 0.13 N; (b) 0.12 **17.** $\mu_s = 0.58$, $\mu_k = 0.54$ **19.** (a) 0.11 m/s^2, 0.23 m/s^2; (b) 0.041, 0.029 **21.** 36 m **23.** (a) 300 N; (b) 1.3 m/s^2 **25.** (a) 66 N; (b) 2.3 m/s^2 **27.** (a) $\mu_k mg/(\sin\theta - \mu_k \cos\theta)$; (b) $\theta_0 = \tan^{-1}\mu_s$ **29.** (b) 3.0×10^7 N **31.** 100 N **33.** 3.3 kg **35.** (a) 11 ft/s^2; (b) 0.46 lb; (c) blocks move independently **37.** (a) 27 N; (b) 3.0 m/s^2 **39.** (a) 6.1 m/s^2, leftward; (b) 0.98 m/s^2, leftward **41.** (a) 3.0×10^5 N; (b) 1.2° **43.** 9.9 s **45.** 3.75 **47.** 12 cm **49.** 68 ft **51.** (a) 3210 N; (b) yes **53.** 0.078 **55.** (a) 0.72 m/s; (b) 2.1 m/s^2; (c) 0.50 N **57.** $\sqrt{Mgr/m}$ **59.** (a) 30 cm/s; (b) 180 cm/s^2, radially inward; (c) 3.6×10^{-3} N, radially inward; (d) 0.37 **61.** (a) 275 N; (b) 877 N **63.** 874 N **65.** (a) at the bottom of the circle; (b) 31 ft/s **67.** (a) 9.5 m/s; (b) 20 m **69.** 13° **71.** (a) 0.0338 N; (b) 9.77 N

Chapter 7

CP **1.** (a) decrease; (b) same; (c) negative, zero **2.** d, c, b, a **3.** (a) same; (b) smaller **4.** (a) positive; (b) negative; (c) zero **5.** zero **Q** **1.** all tie **3.** (a) increasing; (b) same; (c) same; (d) increasing **5.** (a) positive; (b) negative; (c) negative **7.** (a) positive; (b) zero; (c) negative; (d) negative; (e) zero; (f) positive **9.** all tie **11.** c, d, a and b tie; then f, e. **13.** (a) 3 m; (b) 3 m; (c) 0 and 6 m; (d) negative direction of x **15.** (a) A; (b) B **17.** twice **EP** **1.** 1.8×10^{13} J; **3.** (a) 3610 J; (b) 1900 J; (c) 1.1×10^{10} J **5.** (a) 1×10^5 megatons TNT; (b) 1×10^7 bombs **7.** father, 2.4 m/s; son, 4.8 m/s **9.** (a) 200 N; (b) 700 m; (c) -1.4×10^5 J; (d) 400 N, 350 m, -1.4×10^5 J **11.** 5000 J **13.** 47 keV **15.** 7.9 J **17.** 530 J **19.** -37 J **21.** (a) 314 J; (b) -155 J; (c) 0; (d) 158 J **23.** (a) 98 N; (b) 4.0 cm; (c) 3.9 J; (d) -3.9 J **25.** (a) $-3Mgd/4$; (b) Mgd; (c) $Mgd/4$ (d) $\sqrt{gd/2}$ **27.** 25 J **31.** -6 J **33.** (a) 12 J; (b) 4.0 m; (c) 18 J **35.** (a) -0.043 J; (b) -0.13 J **37.** (a) 6.6 m/s; (b) 4.7 m **39.** (a) up; (b) 5.0 cm; (c) 5.0 J **41.** 270 kW **43.** 235 kW **45.** 490 W **47.** (a) 100 J; (b) 67 W; (c) 33 W **49.** 0.99 hp **51.** (a) 0; (b) -350 W **53.** (a) 79.4 keV; (b) 3.12 MeV; (c) 10.9 MeV **55.** (a) 32 J; (b) 8 W; (c) 78°

Chapter 8

CP **1.** no **2.** 3, 1, 2 **3.** (a) all tie; (b) all tie **4.** (a) CD, AB, BC (zero); (b) positive direction of x **5.** 2, 1, 3 **6.** decrease **7.** (a) seventh excited state, with energy E_7; (b) 1.3 eV **Q** **1.** -40 J **3.** (c) and (d) tie; then (a) and (b) tie **5.** (a) all tie; (b) all tie **7.** (a) 3, 2, 1; (b) 1, 2, 3 **9.** less than (smaller decrease in potential energy) **11.** (a) $E < 3$ J, $K < 2$ J; (b) $E < 5$ J, $K < 4$ J **13.** (a) increasing; (b) decreasing; (c) decreasing; (d) constant in AB and BC, decreasing in CD **EP** **1.** 15 J **3.** (a) 167 J; (b) -167 J; (c) 196 J; (d) 29 J **5.** (a) 0; (b) $mgh/2$; (c) mgh; (d) $mgh/2$; (e) mgh **7.** (a) -0.80 J; (b) -0.80 J; (c) $+1.1$ J **9.** (a) $mgL(1 - \cos\theta)$; (b) $-mgL(1 - \cos\theta)$; (c) $mgL(1 - \cos\theta)$ **11.** (a) 18 J;

(b) 0; (c) 30 J; (d) 0; (e) parts b and d **13.** (a) 2.08 m/s; (b) 2.08 m/s **15.** (a) $\sqrt{2gL}$; (b) $2\sqrt{gL}$; (c) $\sqrt{2gL}$ **17.** 830 ft **19.** (a) 6.75 J; (b) -6.75 J; (c) 6.75 J; (d) 6.75 J; (e) -6.75 J; (f) 0.459 m **21.** (a) 21.0 m/s; (b) 21.0 m/s **23.** (a) 0.98 J; (b) -0.98 J; (c) 3.1 N/cm **25.** (a) 39.2 J; (b) 39.2 J; (c) 4.00 m **27.** (a) 54 m/s; (b) 52 m/s; (c) 76 m, below **29.** (a) 39 ft/s; (b) 4.3 in. **31.** (a) 300 J; (b) 93.8 J; (c) 6.38 m **33.** (a) 4.8 m/s; (b) 2.4 m/s **35.** (a) $[v_0^2 + 2gL(1 - \cos\theta_0)]^{1/2}$; (b) $(2gL\cos\theta_0)^{1/2}$; (c) $[gL(3 + 2\cos\theta_0)]^{1/2}$ **37.** (a) $U(x) = -Gm_1m_2/x$; (b) $Gm_1m_2d/x_1(x_1 + d)$ **39.** (a) $8mg$ leftward and mg downward; (b) $2.5R$ **43.** $mgL/32$ **47.** (a) $1.12(A/B)^{1/6}$; (b) repulsive; (c) attractive **49.** (a) turning point on left, none on right; molecule breaks apart; (b) turning points on both left and right; molecule does not break apart; (c) -1.2×10^{-19} J; (d) 2.2×10^{-19} J; (e) $\approx 1 \times 10^{-9}$ on each, directed toward the other; (f) $r < 0.2$ nm; (g) $r > 0.2$ nm; (h) $r = 0.2$ nm **51.** -25 J **53.** (a) 2200 J; (b) -1500 J; (c) 700 J **55.** 17 kW **57.** (a) -0.74 J; (b) -0.53 J **59.** -12 J **61.** 54% **63.** 880 MW **65.** (a) 39 kW; (b) 39 kW **67.** (a) 1.5 MJ; (b) 0.51 MJ; (c) 1.0 MJ; (d) 63 m/s **69.** (a) 67 J; (b) 67 J; (c) 46 cm **71.** Your force on the cabbage does work. **73.** (a) -0.90 J; (b) 0.46 J; (c) 1.0 m/s **75.** (a) 18 ft/s; (b) 18 ft **77.** 4.3 m **79.** (a) 31.0 J; (b) 5.35 m/s; (c) conservative **81.** 1.2 m **85.** in the center of the flat part **87.** (a) 24 ft/s; (b) 3.0 ft; (c) 9.0 ft; (d) 49 ft **89.** (a) 216 J; (b) 1180 N; (c) 432 J; (d) motor also supplies thermal energy to crate and belt **91.** (a) 1.1×10^{17} J; (b) 1.2 kg **93.** 7.28 MeV **95.** (a) release; (b) 17.6 MeV **97.** (a) 5.3 eV; (b) 0.9 eV **99.** (a) 7.2 J; (b) -7.2 J; (c) 86 cm; (d) 26 cm

Chapter 9

CP **1.** (a) origin; (b) fourth quadrant; (c) on y axis below origin; (d) origin; (e) third quadrant; (f) origin **2.** (a) to (c) at the center of mass, still at the origin (their forces are internal to the system and cannot move the center of mass) **3.** (a) 1, 3, and then 2 and 4 tie (zero force); (b) 3 **4.** (a) 0; (b) no; (c) negative x **5.** (a) 500 km/h; (b) 2600 km/h; (c) 1600 km/h **6.** (a) yes; (b) no **Q** **1.** point 4 **3.** (a) at the center of the sled; (b) $L/4$, to the right; (c) not at all (no net external force); (d) $L/4$, to the left; (e) L; (f) $L/2$; (g) $L/2$ **5.** (a) $ac, cd,$ and bc; (b) bc; (c) bd and ad **7.** (a) 2 N, rightward; (b) 2 N, rightward; (c) greater than 2 N, rightward **9.** b, c, a **11.** (a) yes; (b) 6 kg·m/s in $-x$ direction; (c) can't tell **EP** **1.** (a) 4600 km; (b) $0.73R_e$ **3.** (a) $x_{cm} = 1.1$ m, $y_{cm} = 1.3$ m; (b) shifts toward topmost particle **5.** $x_{cm} = -0.25$ m, $y_{cm} = 0$ **7.** in the iron, at midheight and midwidth, 2.7 cm from midlength **9.** $x_{cm} = y_{cm} = 20$ cm, $z_{cm} = 16$ cm **11.** (a) $H/2$; (b) $H/2$; (c) descends to lowest point and then ascends to $H/2$; (d) $(HM/m)(\sqrt{1 + m/M} - 1)$ **13.** 72 km/h **15.** (a) center of mass does not move; (b) 0.75 m **17.** 4.8 m/s **19.** (a) 22 m; (b) 9.3 m/s **21.** 53 m **23.** 13.6 ft **25.** (a) 52.0 km/h; (b) 28.8 km/h **27.** a proton **29.** (a) 30°; (b) $-0.572\mathbf{j}$ kg·m/s **31.** (a) $(-4.0 \times 10^4\,\mathbf{i})$ kg·m/s; (b) west; (c) 0 **33.** $0.707c$ **35.** 0.57 m/s, toward center of mass **37.** it increases by

4.4 m/s   **39.** (a) rocket case: 7290 m/s, payload: 8200 m/s; (b) before: 1.271×10^{10} J, after: 1.275×10^{10} J
41. (a) -1; (b) 1830; (c) 1830; (d) same   **43.** 14 m/s, 135° from the other pieces   **45.** 190 m/s
47. (a) $0.200v_{rel}$; (b) $0.210v_{rel}$; (c) $0.209v_{rel}$ **49.** (a) 1.57×10^6 N; (b) 1.35×10^5 kg; (c) 2.08 km/s   **51.** 108 m/s
53. 2.2×10^{-3} **57.** fast barge: 46 N more; slow barge: no change   **59.** (a) 7.8 MJ; (b) 6.2   **61.** 690 W
63. 5.5×10^6 N   **65.** 24 W   **67.** 100 m   **69.** (a) 860 N; (b) 2.4 m/s   **71.** (a) 3.0×10^5 J; (b) 10 kW; (c) 20 kW
73. (a) 2.1×10^6 kg; (b) $\sqrt{100 + 1.5t}$ m/s; (c) $(1.5 \times 10^6)/\sqrt{100 + 1.5t}$ N; (d) 6.7 km   **75.** $t = (3d/2)^{2/3}(m/2P)^{1/3}$

Chapter 10

CP   **1.** (a) unchanged; (b) unchanged; (c) decreased
2. (a) zero; (b) positive; (c) positive direction of y **3.** (a) 4 kg·m/s; (b) 8 kg·m/s; (c) 3 J   **4.** (a) 0; (b) 4 kg·m/s
5. (a) 10 kg·m/s; (b) 14 kg·m/s; (c) 6 kg·m/s   **6.** (a) 2 kg·m/s; (b) 3 kg·m/s   **7.** (a) increases; (b) increases
Q   **1.** all tie   **3.** b and c   **5.** (a) one stationary; (b) 2; (c) 5; (d) equal (pool player's result)   **7.** (a) 1 and 4 tie; then 2 and 3 tie; (b) 1; 3 and 4 tie; then 2   **9.** (a) rightward; (b) rightward; (c) smaller   **11.** positive direction of x axis
EP   **1.** (a) 750 N; (b) 6.0 m/s   **3.** 6.2×10^4 N   **5.** 3000 N ($= 660$ lb)   **7.** 1.1 m   **9.** (a) 42 N·s; (b) 2100 N
11. (a) $(7.4 \times 10^3\,\mathbf{i} - 7.4 \times 10^3\,\mathbf{j})$ N·s; (b) $(-7.4 \times 10^3\,\mathbf{i})$ N·s; (c) 2.3×10^3 N; (d) 2.1×10^4 N; (e) $-45°$
13. (a) 1.0 kg·m/s; (b) 250 J; (c) 10 N; (d) 1700 N
15. 5 N   **17.** $2\mu v$ **19.** 990 N   **21.** (a) 1.8 N·s, to the left; (b) 180 N, to the right   **25.** 8 m/s   **27.** 38 km/s
29. 4.2 m/s   **31.** (a) 99 g; (b) 1.9 m/s; (c) 0.93 m/s
33. (a) 1.2 kg; (b) 2.5 m/s   **35.** 7.8 kg   **37.** (a) 1/3; (b) $4h$ **39.** 35 cm   **41.** 3.0 m/s   **43.** (a) $(10\mathbf{i} + 15\mathbf{j})$ m/s; (b) 500 J lost   **45.** (a) 2.7 m/s; (b) 1400 m/s   **47.** (a) A: 4.6 m/s, B: 3.9 m/s; (b) 7.5 m/s   **49.** 20 J for the heavy particle, 40 J for the light particle   **51.** $mv^2/6$ **53.** 13 tons
55. 25 cm   **57.** 0.975 m/s, 0.841 m/s   **59.** (a) 4.1 ft/s; (b) 1700 ft·lb; (c) $v_{24} = 5.3$ ft/s, $v_{32} = 3.3$ ft/s   **61.** (a) 30° from the incoming proton's direction; (b) 250 m/s and 430 m/s   **63.** (a) 41°; (b) 4.76 m/s; (c) no   **65.** $v = V/4$
67. (a) 117° from the final direction of B; (b) no   **69.** 120°
71. (a) 1.9 m/s, 30° to initial direction; (b) no   **73.** (a) 3.4 m/s, deflected by 17° to the right; (b) 0.95 MJ   **75.** (a) 117 MeV; (b) equal and opposite momenta; (c) π^-
77. (a) 4.94 MeV; (b) 0; (c) 4.85 MeV; (d) 0.09 MeV

Chapter 11

CP   **1.** (b) and (c)   **2.** (a) and (d)   **3.** (a) yes; (b) no; (c) yes; (d) yes   **4.** all tie   **5.** 1, 2, 4, 3   **6.** (a) 1 and 3 tie, 4; then 2 and 5 tie (zero)   **7.** (a) downward in the figure; (b) less   **Q**   **1.** (a) positive; (b) zero; (c) negative; (d) negative   **3.** (a) 2 and 3; (b) 1 and 3; (c) 4   **5.** (a) and (c)   **7.** (a) all tie; (b) 2, 3; then 1 and 4 tie   **9.** b, c, a
11. less   **13.** 90°; then 70° and 110° tie   **15.** Finite angular

displacements are not commutative.   **EP**   **1.** (a) 1.50 rad; (b) 85.9°; (c) 1.49 m   **3.** (a) 0.105 rad/s; (b) 1.75×10^{-3} rad/s; (c) 1.45×10^{-4} rad/s   **5.** (a) $\omega(2) = 4.0$ rad/s, $\omega(4) = 28$ rad/s; (b) 12 rad/s²; (c) $\alpha(2) = 6.0$ rad/s², $\alpha(4) = 18$ rad/s²   **7.** (a) $\omega_0 + at^4 - bt^3$; (b) $\theta_0 + \omega_0 t + at^5/5 - bt^4/4$ **9.** 11 rad/s   **11.** (a) 9000 rev/min²; (b) 420 rev
13. (a) 30 s; (b) 1800 rad   **15.** 200 rev/min
17. (a) 2.0 rad/s²; (b) 5.0 rad/s; (c) 10 rad/s; (d) 75 rad
19. (a) 13.5 s; (b) 27.0 rad/s   **21.** (a) 340 s; (b) -4.5×10^{-3} rad/s²; (c) 98 s   **23.** (a) 1.0 rev/s²; (b) 4.8 s; (c) 9.6 s; (d) 48 rev   **25.** 6.1 ft/s² (1.8 m/s²), toward the center
27. 0.13 rad/s   **29.** 5.6 rad/s²   **31.** (a) 5.1 h; (b) 8.1 h
33. (a) 2.50×10^{-3} rad/s; (b) 20.2 m/s²; (c) 0   **35.** (a) -1.1 rev/min²; (b) 9900 rev; (c) -0.99 mm/s²; (d) 31 m/s²
37. (a) 310 m/s; (b) 340 m/s   **39.** (a) 1.94 m/s²; (b) 75.1°, toward the center of the track   **41.** 16 s   **43.** (a) 73 cm/s²; (b) 0.075; (c) 0.11   **45.** 12.3 kg·m²   **47.** first cylinder: 1100 J; second cylinder: 9700 J   **49.** (a) 221 kg·m²; (b) 1.10×10^4 J   **51.** (a) 6490 kg·m²; (b) 4.36 MJ
53. 0.097 kg·m²   **57.** (a) 1300 g·cm²; (b) 550 g·cm²; (c) 1900 g·cm²; (d) $A + B$ **59.** (a) 49 MJ; (b) 100 min
61. 4.6 N·m   **63.** (a) $r_1 F_1 \sin \theta_1 - r_2 F_2 \sin \theta_2$; (b) -3.8 N·m   **65.** 1.28 kg·m²   **67.** 9.7 rad/s², counterclockwise
69. (a) 155 kg·m²; (b) 64.4 kg   **71.** (a) 420 rad/s²; (b) 500 rad/s   **73.** small sphere: (a) 0.689 N·m and (b) 3.05 N; large sphere: (a) 9.84 N·m and (b) 11.5 N   **75.** 1.73 m/s²; 6.92 m/s²   **77.** (a) 1.4 m/s; (b) 1.4 m/s   **79.** (a) 19.8 kJ; (b) 1.32 kW   **81.** (a) 8.2×10^{28} N·m; (b) 2.6×10^{29} J; (c) 3.0×10^{21} kW   **83.** $\sqrt{9g/4\ell}$ **85.** (a) 4.8×10^5 N; (b) 1.1×10^4 N·m; (c) 1.3×10^6 J   **87.** (a) $3g(1 - \cos \theta)$; (b) $\frac{3}{2}g \sin \theta$; (c) 41.8°   **89.** (a) 5.6 rad/s²; (b) 3.1 rad/s
91. (a) 42.1 km/h; (b) 3.09 rad/s²; (c) 7.57 kW   **93.** (a) 3.4×10^5 g·cm²; (b) 2.9×10^5 g·cm²; (c) 6.3×10^5 g·cm²; (d) $(1.2$ cm$)\,\mathbf{i} + (5.9$ cm$)\,\mathbf{j}$

Chapter 12

CP   **1.** (a) same; (b) less   **2.** less   **3.** (a) $\pm z$; (b) $+y$; (c) $-x$ **4.** (a) 1 and 3 tie, then 2 and 4 tie, then 5 (zero); (b) 2 and 3   **5.** (a) 3, 1; then 2 and 4 tie (zero); (b) 3
6. (a) all tie (same τ, same t, thus same ΔL); (b) sphere, disk, hoop (reverse order of I)   **7.** (a) decreases; (b) same; (c) increases   **Q**   **1.** (a) same; (b) block; (c) block
3. (a) greater; (b) same   **5.** (a) L; (b) $1.5L$ **7.** b, then c and d tie; then a and e tie (zero)   **9.** a, then b and c tie; then e, d (zero)   **11.** (a) same; (b) increases, because of decrease in rotational inertia   **13.** (a) 30 units clockwise; (b) 2 then 4, then the others; or 4 then 2, then the others   **15.** (a) spins in place; (b) rolls toward you; (c) rolls away from you
EP   **1.** 1.00   **3.** (a) 59.3 rad/s; (b) -9.31 rad/s²; (c) 70.7 m   **5.** (a) -4.11 m/s²; (b) -16.4 rad/s²; (c) -2.54 N·m   **7.** (a) 8.0°; (b) $0.14g$ **9.** (a) 4.0 N, to the left; (b) 0.60 kg·m²   **11.** (a) $\frac{1}{2}mR^2$; (b) a solid circular cylinder
13. (a) $mg(R - r)$; (b) 2/7; (c) $(17/7)mg$ **15.** (a) $2.7R$; (b) $(50/7)mg$ **17.** (a) 13 cm/s²; (b) 4.4 s; (c) 55 cm/s; (d) 1.8×10^{-2} J; (e) 1.4 J; (f) 27 rev/s   **21.** (a) 24 N·m, in $+y$ direction; (b) 24 N·m, $-y$; (c) 12 N·m, $+y$;

(d) 12 N·m, $-y$ **23.** (a) $(-1.5\mathbf{i} - 4.0\mathbf{j} - \mathbf{k})$ N·m;
(b) $(-1.5\mathbf{i} - 4.0\mathbf{j} - \mathbf{k})$ N·m   **25.** $-2.0\mathbf{i}$ N·m   **27.** 9.8
kg·m²/s   **29.** (a) 12 kg·m²/s, out of page; (b) 3.0 N·m,
out of page   **31.** (a) 0; (b) $(8.0\mathbf{i} + 8.0\mathbf{k})$ N·m   **33.** (a) mvd;
(b) no; (c) 0, yes   **35.** (a) 3.15×10^{43} kg·m²/s; (b) 0.616
37. 4.5 N·m, parallel to xy plane at $-63°$ from $+x$
39. (a) 0; (b) 0; (c) $30t^3$ kg·m²/s, $90t^2$ N·m, both in $-z$
direction; (d) $30t^3$ kg·m²/s, $90t^2$ N·m, both in $+z$ direction
41. (a) $\frac{1}{2}mgt^2 v_0 \cos\theta_0$; (b) $mgtv_0 \cos\theta_0$; (c) $mgtv_0 \cos\theta_0$
43. (a) -1.47 N·m; (b) 20.4 rad; (c) -29.9 J; (d) 19.9 W
45. (a) 12.2 kg·m²; (b) 308 kg·m²/s, down   **47.** (a) 1/3;
(b) 1/9   **49.** $\omega_0 R_1 R_2 I_1/(I_1 R_2^2 + I_2 R_1^2)$ **51.** (a) 3.6 rev/s;
(b) 3.0; (c) work done by man in moving weights inward
53. (a) 267 rev/min; (b) 2/3   **55.** 3.0 min   **57.** 2.6 rad/s
59. (a) they revolve in a circle of 1.5 m radius at 0.93 rad/s;
(b) 8.4 rad/s; (c) $K_a = 98$ J, $K_b = 880$ J; (d) from the work
done in pulling inward   **61.** $m/(M + m)(v/R)$
63. (a) $mvR/(I + MR^2)$; (b) $mvR^2/(I + MR^2)$ **65.** 1300 m/s
67. (a) 18 rad/s; (b) 0.92
69. $\theta = \cos^{-1}\left[1 - \dfrac{6m^2 h}{\ell(2m + M)(3m + M)}\right]$
71. 5.28×10^{-35} J·s   **73.** Any three are spin up; the other is
spin down.   **75.** (a) The magnitude of the angular momentum
increases in proportion to t^2 and the magnitude of the torque
increases in proportion to t, in agreement with the second law
for rotation. (b) The magnitudes of the angular momentum and
torque again increase with time. But the change in the
magnitude of the angular momentum in any interval is less than
is predicted by proportionality to t^2 law and the change in the
torque is less than is predicted by proportionality to t. At any
position of the projectile the torque is less when drag is present
than when it is not.

Chapter 13

CP   **1.** c, e, f **2.** (a) no; (b) at site of $\mathbf{F}_1$, perpendicular to
plane of figure; (c) 45 N   **3.** (a) at C (to eliminate forces there
from a torque equation); (b) plus; (c) minus; (d) equal   **4.** d
5. (a) equal; (b) B; (c) B **Q**   **1.** (a) yes; (b) yes; (c) yes;
(d) no   **3.** b **5.** (a) yes; (b) no; (c) no (it could balance the
torques but the forces would then be unbalanced)   **7.** (a) a,
then b and c tie, then d **9.** (a) 20 N (the key is the pulley
with the 20 N weight); (b) 25 N   **11.** (a) $\sin\theta$; (b) same;
(c) larger   **13.** tie of A and B, then C **EP**   **1.** (a) two;
(b) seven   **3.** (a) 2.5 m; (b) 7.3°   **5.** 120°   **7.** 7920 N
9. (a) 840 N; (b) 530 N   **11.** 0.536 m   **13.** (a) 2770 N;
(b) 3890 N   **15.** (a) 1160 N, down; (b) 1740 N, up; (c) left,
stretched; (d) right, compressed   **17.** (a) 280 N; (b) 880 N,
71° above the horizontal   **19.** bars BC, CD, and DA are under
tension due to forces T, diagonals AC and BD are compressed
due to forces $\sqrt{2}T$ **21.** (a) 1800 lb; (b) 822 lb; (c) 1270 lb
23. (a) 49 N; (b) 28 N; (c) 57 N; (d) 29°   **25.** (a) 1900 N, up;
(b) 2100 N, down   **27.** (a) 340 N; (b) 0.88 m; (c) increases,
decreases   **29.** $W\sqrt{2rh - h^2}/(r - h)$ **31.** (a) $L/2$; (b) $L/4$;
(c) $L/6$; (d) $L/8$; (e) $25L/24$ **33.** (a) 6630 N; (b) $F_h = 5740$ N;
(c) $F_v = 5960$ N   **35.** 2.20 m   **37.** (a) 1.50 m; (b) 433 N;

(c) 250 N   **39.** (a) $a_1 = L/2$, $a_2 = 5L/8$, $h = 9L/8$;
(b) $b_1 = 2L/3$, $b_2 = L/2$, $h = 7L/6$ **41.** (a) 47 lb; (b) 120 lb;
(c) 72 lb   **43.** (a) 445 N; (b) 0.50; (c) 315 N
45. (a) 3.9 m/s²; (b) 2000 N on each rear wheel, 3500 N on
each front wheel; (c) 790 N on each rear wheel, 1410 N on
each front wheel   **47.** (a) 1.9×10^{-3}; (b) 1.3×10^7 N/m²;
(c) 6.9×10^9 N/m²   **49.** 3.1 cm   **51.** 2.4×10^9 N/m²
53. (a) 1.8×10^7 N; (b) 1.4×10^7 N; (c) 16   **55.** (a) 867 N;
(b) 143 N; (c) 0.165

Chapter 14

CP   **1.** all tie   **2.** (a) 1, tie of 2 and 4, then 3; (b) line d
3. negative y direction   **4.** (a) increase; (b) negative
5. (a) 2; (b) 1   **6.** (a) path 1 (decreased E (more negative)
gives decreased a); (b) less than (decreased a gives decreased T)
Q   **1.** (a) between, closer to less massive particle; (b) no;
(c) no (other than infinity)   **3.** $3GM^2/d^2$, leftward   **5.** b, tie
of a and c, then d **7.** b, a, c **9.** (a) negative; (b) negative;
(c) postive; (d) all tie   **11.** (a) all tie; (b) all tie
13. (a) same; (b) greater   **EP**   **1.** 19 m   **3.** 2.16
5. 1/2   **7.** 3.4×10^5 km   **9.** (a) 3.7×10^{-5} N, increasing y **11.** $M = m$ **13.** 3.2×10^{-7} N   **15.** $(GmM/d^2) \times$
$\left[1 - \dfrac{1}{8(1 - R/2d)^2}\right]$ **17.** 2.6×10^6 m   **19.** (a) $1.3 \times$
10^{12} m/s²; (b) 1.6×10^6 m/s   **21.** (a) 17 N; (b) 2.5
23. (b) 1.9 h   **27.** (a) $a_g = (3.03 \times 10^{43}$ kg·m/s²$)/M_h$;
(b) decrease; (c) 9.82 m/s²; (d) 7.30×10^{-15} m/s²; (e) no
29. 7.91 km/s   **31.** (a) $(3.0 \times 10^{-7}m)$ N; (b) $(3.3 \times 10^{-7}m)$
N; (c) $(6.7 \times 10^{-7}mr)$ N   **33.** (a) 9.83 m/s²; (b) 9.84 m/s²;
(c) 9.79 m/s²   **35.** (a) -1.4×10^{-4} J; (b) less; (c) positive;
(d) negative   **37.** (a) 0.74; (b) 3.7 m/s²; (c) 5.0 km/s
39. (a) 0.0451; (b) 28.5   **41.** $-Gm(M_E/R + M_M/r)$
43. (a) 5.0×10^{-11} J; (b) -5.0×10^{-11} J   **45.** (a) 1700 m/s;
(b) 250 km; (c) 1400 m/s   **47.** (a) 2.2×10^{-7} rad/s;
(b) 90 km/s   **51.** (a) -1.67×10^{-8} J; (b) 0.56×10^{-8} J
55. 6.5×10^{23} kg   **57.** 5×10^{10} **59.** (a) 7.82 km/s;
(b) 87.5 min   **61.** (a) 6640 km; (b) 0.0136   **63.** (a) 39.5
AU³/M_S·y²; (b) $T^2 = r^3/M$ **65.** (a) 1.9×10^{13} m;
(b) $3.5R_P$ **67.** south, at 35.4° above the horizon
71. $2\pi r^{3/2}/\sqrt{G(M + m/4)}$ **73.** $\sqrt{GM/L}$ **75.** (a) 2.8 y;
(b) 1.0×10^{-4} **77.** (a) 1/2; (b) 1/2; (c) B, by 1.1×10^8 J
79. (a) 54 km/s; (b) 960 m/s; (c) $R_p/R_a = v_a/v_p$ **81.** (a) $4.6 \times$
10^5 J; (b) 260   **83.** (a) 7.5 km/s; (b) 97 min; (c) 410 km;
(d) 7.7 km/s; (e) 92 min; (f) 3.2×10^{-3} N; (g) if the satellite–
Earth system is considered isolated, its $\mathbf{L}$ is conserved
85. (a) 5540 s; (b) 7.68 km/s; (c) 7.60 km/s; (d) 5.78×10^{10} J;
(e) -11.8×10^{10} J; (f) -6.02×10^{10} J; (g) $6.63 \times$
10^6 m; (h) 170 s, new orbit   **87.** (a) $(-7.0$ mm$)\mathbf{i} +$
$(3.0$ cm$)\mathbf{j}$; (b) $(-0.19$ m/s$)\mathbf{i} + (0.40$ m/s$)\mathbf{j}$ **89.** (a) $1.98 \times$
10^{30} kg; (b) 1.96×10^{30} kg

Chapter 15

CP   **1.** all tie   **2.** (a) all tie; (b) $0.95\rho_0$, ρ_0, $1.1\rho_0$
3. 13 cm³/s, outward   **4.** (a) all tie; (b) 1, then 2 and 3 tie, 4;

(c) 4, 3, 2, 1   **Q** **1.** e, then b and d tie, then a and c tie
3. (a) 1, 3, 2; (b) all tie; (c) no (you must consider the weight exerted on the scale via the walls)   **5.** 3, 4, 1, 2
7. (a) downward; (b) downward; (c) same   **9.** (a) same;
(b) same; (c) lower; (d) higher   **11.** (a) block 1, counterclockwise; block 2, clockwise; (b) block 1, tip more; block 2, right itself   **EP** **1.** 1000 kg/m³   **3.** 1.1×10^5 Pa or 1.1 atm   **5.** 2.9×10^4 N   **7.** 6.0 lb/in.²   **9.** 1.90×10^4 Pa   **11.** 5.4×10^4 Pa   **13.** 0.52 m   **15.** (a) 6.06×10^9 N;
(b) 20 atm   **17.** 0.412 cm   **19.** $\frac{1}{4}\rho g A(h_2 - h_1)^2$
21. 44 km   **23.** (a) $\rho g W D^2/2$; (b) $\rho g W D^3/6$; (c) $D/3$
25. (a) 2.2; (b) 2.4   **27.** -3.9×10^{-3} atm   **29.** (a) fA/a;
(b) 20 lb   **31.** 1070 g   **33.** 1.5 g/cm³   **35.** 600 kg/m³
37. (a) 670 kg/m³; (b) 740 kg/m³   **39.** 390 kg
41. (a) 1.2 kg; (b) 1300 kg/m³   **43.** 0.126 m³   **45.** five
47. (a) 1.80 m³; (b) 4.75 m³   **49.** 2.79 g/cm³
51. (a) 9.4 N; (b) 1.6 N   **53.** 4.0 m   **55.** 28 ft/s   **57.** 43
cm/s   **59.** (a) 2.40 m/s; (b) 245 Pa   **61.** (a) 12 ft/s; (b) 13
lb/in.²   **63.** 0.72 ft·lb/ft³   **65.** (a) 2; (b) $R_1/R_2 = \frac{1}{2}$; (c) drain
it until $h_2 = h_1/4$ **67.** 116 m/s   **69.** (a) 6.4 m³; (b) 5.4 m/s;
(c) 9.8×10^4 Pa   **71.** (a) 560 Pa; (b) 5.0×10^4 N
73. 40 m/s   **75.** (b) $H - h$; (c) $H/2$ **77.** (b) 5.4 ft³/s
79. (b) 63.3 m/s

Chapter 16

CP **1.** (a) $-x_m$; (b) $+x_m$; (c) 0   **2.** a   **3.** (a) 5 J; (b) 2 J;
(c) 5 J   **4.** all tie (in Eq. 16-32, m is included in I)
5. 1, 2, 3 (the ratio m/b matters; k does not)   **Q** **1.** c
3. (a) 0; (b) between 0 and $+x_m$; (c) between $-x_m$ and 0;
(d) between $-x_m$ and 0   **5.** (a) toward $-x_m$; (b) toward $+x_m$;
(c) between $-x_m$ and 0; (d) between $-x_m$ and 0; (e) decreasing;
(f) increasing   **7.** (a) 3, 2, 1; (b) all tie   **9.** 3, 2, 1
11. system with spring A **13.** b (infinite period; does not
oscillate), c, a **15.** (a) same; (b) same; (c) same; (d) smaller;
(e) smaller; (f) and (g) larger ($T = \infty$)   **EP** **1.** (a) 0.50 s;
(b) 2.0 Hz; (c) 18 cm   **3.** (a) 245 N/m; (b) 0.284 s
5. 708 N/m   **7.** $f > 500$ Hz   **9.** (a) 100 N/m; (b) 0.45 s
11. (a) 6.28×10^5 rad/s; (b) 1.59 mm   **13.** (a) 1.0 mm;
(b) 0.75 m/s; (c) 570 m/s²   **15.** (a) 1.29×10^5 N/m;
(b) 2.68 Hz   **17.** (a) 4.0 s; (b) $\pi/2$ rad/s; (c) 0.37 cm;
(d) (0.37 cm) cos $\frac{\pi}{2}t$; (e) (-0.58 cm/s) sin $\frac{\pi}{2}t$; (f) 0.58 cm/s;
(g) 0.91 cm/s²; (h) 0; (i) 0.58 cm/s   **19.** (b) 12.47 kg;
(c) 54.43 kg   **21.** 1.6 kg   **23.** (a) 1.6 Hz; (b) 1.0 m/s, 0;
(c) 10 m/s², ±10 cm; (d) (-10 N/m)x **25.** 22 cm
27. (a) 25 cm; (b) 2.2 Hz   **29.** (a) 0.500 m; (b) -0.251 m;
(c) 3.06 m/s   **31.** (a) 0.183A; (b) same direction
37. (a) $k_1 = (n + 1)k/n$, $k_2 = (n + 1)k$; (b) $f_1 = \sqrt{(n + 1)/n}f$,
$f_2 = \sqrt{n + 1}f$ **39.** (b) 42 min   **41.** (a) 200 N/m;
(b) 1.39 kg; (c) 1.91 Hz   **43.** (a) 130 N/m; (b) 0.62 s; (c) 1.6
Hz; (d) 5.0 cm; (e) 0.51 m/s   **45.** (a) 3/4; (b) 1/4; (c) $x_m/\sqrt{2}$
47. (a) 3.5 m; (b) 0.75 s   **49.** (a) 0.21 m; (b) 1.6 Hz;
(c) 0.10 m   **51.** (a) 0.0625 J; (b) 0.03125 J   **53.** 12 s
55. (a) 39.5 rad/s; (b) 34.2 rad/s; (c) 124 rad/s²   **57.** (a) 8.3 s;
(b) no   **59.** 9.47 m/s²   **61.** 8.77 s   **63.** 5.6 cm
65. $2\pi\sqrt{(R^2 + 2d^2)/2gd}$ **67.** (a) 0.205 kg·m²; (b) 47.7 cm;
(c) 1.50 s   **71.** (a) $2\pi\sqrt{(L^2 + 12x^2)/12gx}$; (b) 0.289 m

73. 9.78 m/s²   **75.** $2\pi\sqrt{m/3k}$ **77.** $(1/2\pi)(\sqrt{g^2 + v^4/R^2}/L)^{1/2}$
79. (b) smaller   **81.** (a) 2.0 s; (b) 18.5 N·m/rad
83. 0.29L **85.** 0.39   **87.** (a) 0.102 kg/s; (b) 0.137 J
89. $k = 490$ N/cm, $b = 1100$ kg/s   **91.** 1.9 in.
93. (a) $y_m = 8.8 \times 10^{-4}$ m, $T = 0.18$ s, $\omega = 35$ rad/s;
(b) $y_m = 5.6 \times 10^{-2}$ m, $T = 0.48$ s, $\omega = 13$ rad/s;
(c) $y_m = 3.3 \times 10^{-2}$ m, $T = 0.31$ s, $\omega = 20$ rad/s

Chapter 17

CP **1.** a, 2; b, 3; c, 1   **2.** (a) 2, 3, 1; (b) 3, then 1 and 2 tie
3. a **4.** 0.20 and 0.80 tie, then 0.60, 0.45   **5.** (a) 1; (b) 3;
(c) 2   **6.** (a) 75 Hz; (b) 525 Hz   **Q** **1.** 7d **3.** tie of A
and B, then C, D **5.** intermediate (closer to fully destructive
interference)   **7.** a and d tie, then b and c tie   **9.** (a) 8;
(b) antinode; (c) longer; (d) lower   **11.** (a) integer multiples
of 3; (b) node; (c) node   **13.** string A **15.** decrease
EP **1.** (a) 75 Hz; (b) 13 ms   **3.** (a) 7.5×10^{14} to 4.3×10^{14}
Hz; (b) 1.0 to 200 m; (c) 6.0×10^{16} to 3.0×10^{19} Hz
5. $y = 0.010$ sin $\pi(3.33x + 1100t)$, with x and y in m and t
in s   **11.** (a) $z = 3.0$ sin($60y - 10\pi t$), with z in mm, y in cm,
and t in s; (b) 9.4 cm/s   **13.** (a) $y = 2.0$ sin $2\pi(0.10x - 400t)$, with x and y in cm and t in s; (b) 50 m/s; (c) 40 m/s
15. (b) 2.0 cm/s; (c) $y = (4.0$ cm) sin($\pi x/10 - \pi t/5 + \pi)$,
where x is in cm and t is in s; (d) -2.5 cm/s   **17.** 129 m/s
19. 135 N   **23.** (a) 15 m/s; (b) 0.036 N   **25.** $y = 0.12$ sin($141x + 628t$), with y in mm, x in m, and t in s
27. (a) 5.0 cm; (b) 40 cm; (c) 12 m/s; (d) 0.033 s; (e) 9.4 m/s;
(f) 5.0 sin($16x + 190t + 0.79$), with x in m, y in cm, and t in s
29. (a) $v_1 = 28.6$ m/s, $v_2 = 22.1$ m/s; (b) $M_1 = 188$ g, $M_2 = 313$ g   **31.** (a) $\sqrt{k(\Delta l)(l + \Delta l)/m}$ **33.** (a) $P_2 = 2P_1$;
(b) $P_2 = P_1/4$ **35.** (a) 3.77 m/s; (b) 12.3 N; (c) zero; (d) 46.3
W; (e) zero; (f) zero; (g) ±0.50 cm   **37.** 82.8°, 1.45 rad,
0.23 wavelength   **39.** 5.0 cm   **41.** (a) 4.4 mm; (b) 112°
43. (a) 0.83y_1; (b) 37°   **45.** (a) $2f_3$; (b) λ_3 **47.** 10 cm
49. (a) 82.0 m/s; (b) 16.8 m; (c) 4.88 Hz   **51.** 240 cm, 120 cm,
80 cm   **53.** 7.91 Hz, 15.8 Hz, 23.7 Hz   **55.** $f_{1A} = f_{4B}$,
$f_{2A} = f_{8B}$ **57.** (a) 2.0 Hz, 200 cm, 400 cm/s; (b) $x = 50$ cm,
150 cm, 250 cm, etc.; (c) $x = 0$, 100 cm, 200 cm, etc.
63. (a) 1.3 m; (b) $y' = 0.002$ sin($9.4x$) cos($3800t$), with x and
y in m and t in s   **67.** (b) in the positive x direction;
interchange the amplitudes of the original two traveling waves;
(c) largest at $x = \lambda/4 = 6.26$ cm; smallest at $x = 0$ and
$x = \lambda/2 = 12.5$ cm; (d) the largest amplitude is 4.0 mm, which
is the sum of the amplitudes of the original traveling waves; the
smallest amplitude is 1.0 mm, which is the difference of the
amplitudes of the original traveling waves

Chapter 18

CP **1.** beginning to decrease (example: mentally move the
curves of Fig. 18-6 rightward past the point at $x = 42$ m)
2. (a) fully constructive, 0; (b) fully constructive, 4   **3.** (a) 1
and 2 tie, then 3; (b) 3, then 1 and 2 tie   **4.** second
5. loosen   **6.** a, greater; b, less; c, can't tell; d, can't tell;
e, greater; f, less   **7.** (a) 222 m/s; (b) 262 m/s   **Q** **1.** pulse

along path 2 **3.** (a) 2.0 wavelengths; (b) 1.5 wavelengths; (c) fully constructive, fully destructive **5.** (a) exactly out of phase; (b) exactly out of phase **7.** 70 dB **9.** (a) two; (b) antinode **11.** all odd harmonics **13.** 501, 503, and 508 Hz; or 505, 507, and 508 Hz **EP 1.** (a) $\approx 6\%$ **3.** the radio listener by about 0.85 s **5.** 7.9×10^{10} Pa **7.** If only the length is uncertain, it must be known to within 10^{-4} m. If only the time is imprecise, the uncertainty must be no more than one part in 10^8. **9.** 43.5 m **11.** 40.7 m **13.** 100 kHz **15.** (a) 2.29, 0.229, 22.9 kHz; (b) 1.14, 0.114, 11.4 kHz **17.** (a) 6.0 m/s; (b) $y = 0.30 \sin(\pi x/12 + 50\pi t)$, with x and y in cm and t in s **19.** 4.12 rad **21.** (a) $343 \times (1 + 2m)$ Hz, with m being an integer from 0 to 28; (b) $686m$ Hz, with m being an integer from 1 to 29 **23.** (a) eight; (b) eight **25.** 64.7 Hz, 129 Hz **27.** (a) 0.080 W/m²; (b) 0.013 W/m² **29.** 36.8 nm **31.** (a) 1000; (b) 32 **33.** (a) 39.7 μW/m²; (b) 171 nm; (c) 0.893 Pa **35.** (a) 59.7; (b) 2.81×10^{-4} **37.** $s_m \propto r^{-1/2}$ **39.** (a) 5000; (b) 71; (c) 71 **41.** 171 m **43.** 3.16 km **45.** (a) 5200 Hz; (b) amplitude$_{SAD}$/amplitude$_{SBD}$ = 2 **47.** 20 kHz **49.** by a factor of 4 **51.** water filled to a height of $\frac{7}{8}, \frac{5}{8}, \frac{3}{8}, \frac{1}{8}$ m **53.** (a) 5.0 cm from one end; (b) 1.2; (c) 1.2 **55.** (a) 1130, 1500, and 1880 Hz **57.** (a) 230 Hz; (b) higher **59.** (a) node; (c) 22 s **61.** 387 Hz **63.** 0.02 **65.** 3.8 Hz **67.** (a) 380 mi/h, away from owner; (b) 77 mi/h, away from owner **69.** 15.1 ft/s **71.** 2.6×10^8 m/s **73.** (a) 77.6 Hz; (b) 77.0 Hz **75.** 33.0 km **79.** (a) 970 Hz; (b) 1030 Hz; (c) 60 Hz, no **81.** (a) 1.02 kHz; (b) 1.04 kHz **83.** 1540 m/s **85.** 41 kHz **87.** (a) 2.0 kHz; (b) 2.0 kHz **89.** (a) 485.8 Hz; (b) 500.0 Hz; (c) 486.2 Hz; (d) 500.0 Hz **91.** 1×10^6 m/s, receding **93.** $0.13c$

Chapter 19

CP 1. (a) all tie; (b) 50°X, 50°Y, 50°W **2.** (a) 2 and 3 tie, then 1, then 4; (b) 3, 2, then 1 and 4 tie **3.** A **4.** c and e **5.** (a) zero; (b) positive **6.** b and d tie, then a, c **Q 1.** 25 S°, 25 U°, 25 R° **3.** c, then the rest tie **5.** B, then A and C tie **7.** (a) both clockwise; (b) both clockwise **9.** c, a, b **11.** upward (with liquid water on the exterior and at the bottom, $\Delta T = 0$ horizontally and downward) **13.** at the temperature of your fingers **15.** 3, 2, 1 **EP 1.** 2.71 K **3.** 0.05 kPa, nitrogen **5.** (a) 320°F; (b) -12.3°F **7.** (a) -96°F; (b) 56.7°C **9.** (a) -40°; (b) 575°; (c) Celsius and Kelvin cannot give the same reading **11.** (a) Dimensions are inverse time **13.** 4.4×10^{-3} cm **15.** 0.038 in. **17.** (a) 9.996 cm; (b) 68°C **19.** 170 km **21.** 0.32 cm² **23.** 29 cm³ **25.** 0.432 cm³ **27.** -157°C **29.** 360°C **35.** +0.68 s/h **37.** (b) use 39.3 cm of steel and 13.1 cm of brass **39.** (a) 523 J/kg·K; (b) 0.600; (c) 26.2 J/mol·K **41.** 94.6 L **43.** 109 g **45.** 1.30 MJ **47.** 1.9 times as great **49.** (a) 33.9 Btu; (b) 172 F° **51.** (a) 52 MJ; (b) 0°C **53.** (a) 411 g; (b) 3.1¢ **55.** 0.41 kJ/kg·K **57.** 3.0 min **59.** 73 kW **61.** 2.17 g **63.** 33 m³ **65.** 33 g **67.** (a) 0°C; (b) 2.5°C **69.** 2500 J/kg·K **71.** A: 120 J, B: 75 J, C: 30 J **73.** (a) -200 J; (b) -293 J;

(c) -93 J **75.** -5.0 J **77.** 33.3 kJ **79.** 766°C **81.** (a) 1.2 W/m·K, 0.70 Btu/ft·F°·h; (b) 0.030 ft²·F°·h/Btu **83.** 1660 J/s **87.** arrangement b **89.** (a) 2.0 MW; (b) 220 W **91.** (a) 17 kW/m²; (b) 18 W/m² **93.** -6.1 nW **95.** 0.40 cm/h **97.** Cu-Al, 84.3°C; Al-brass, 57.6°C

Chapter 20

CP 1. all but c **2.** (a) all tie; (b) 3, 2, 1 **3.** gas A **4.** 5 (greatest change in T), then tie of 1, 2, 3, and 4 **5.** 1, 2, 3 ($Q_3 = 0$, Q_2 goes into work W_2, but Q_1 goes into greater work W_1 and increases gas temperature) **Q 1.** increased but less than doubled **3.** a, c, b **5.** 1180 J **7.** d, tie of a and b, then c **9.** constant-volume process **11.** (a) same; (b) increases; (c) decreases; (d) increases **13.** -4 J **15.** (a) 1, polyatomic; 2, diatomic; 3, monatomic; (b) more **EP 1.** (a) 0.0127; (b) 7.65×10^{21} **3.** 6560 **5.** number of molecules in the ink $\approx 3 \times 10^{16}$; number of people $\approx 5 \times 10^{20}$; statement is wrong, by a factor of about 20,000 **7.** (a) 5.47×10^{-8} mol; (b) 3.29×10^{16} **9.** (a) 106; (b) 0.892 m³ **11.** 27.0 lb/in.² **13.** (a) 2.5×10^{25}; (b) 1.2 kg **15.** 5600 J **17.** 1/5 **19.** (a) -45 J; (b) 180 K **21.** 100 cm³ **23.** 198°F **25.** 2.0×10^5 Pa **27.** 180 m/s **29.** 9.53×10^6 m/s **31.** 313°C **33.** 1.9 kPa **35.** (a) 0.0353 eV, 0.0483 eV; (b) 3400 J, 4650 J **37.** 9.1×10^{-6} **39.** (a) 6.75×10^{-20} J; (b) 10.7 **41.** 0.32 nm **43.** 15 cm **45.** (a) 3.27×10^{10}; (b) 172 m **47.** (a) 22.5 L; (b) 2.25; (c) 8.4×10^{-5} cm; (d) same as (c) **51.** (a) 3.2 cm/s; (b) 3.4 cm/s; (c) 4.0 cm/s **53.** (a) v_P, v_{rms}, $\bar{v}$ (b) reverse ranking **55.** (a) 1.0×10^4 K, 1.6×10^5 K; (b) 440 K, 7000 K **57.** 4.7 **59.** (a) $2N/3v_0$; (b) $N/3$; (c) $1.22v_0$; (d) $1.31v_0$ **61.** $RT \ln(V_f/V_i)$ **63.** (a) 15.9 J; (b) 34.4 J/mol·K; (c) 26.1 J/mol·K **65.** $(n_1C_1 + n_2C_2 + n_3C_3)/(n_1 + n_2 + n_3)$ **67.** (a) -5.0 kJ; (b) 2.0 kJ; (c) 5.0 kJ **69.** (a) 0.375 mol; (b) 1090 J; (c) 0.714 **71.** (a) 14 atm; (b) 620 K **79.** 0.63 **81.** (a) monatomic; (b) 2.7×10^4 K; (c) 4.5×10^4 mol; (d) 3.4 kJ, 340 kJ; (e) 0.01 **83.** 5 m³ **85.** (a) in joules, in the order Q, ΔE_{int}, W: $1 \rightarrow 2$: 3740, 3740, 0; $2 \rightarrow 3$: 0, -1810, 1810; $3 \rightarrow 1$: -3220, -1930, -1290; cycle: 520, 0, 520; (b) $V_2 = 0.0246$ m³, $p_2 = 2.00$ atm, $V_3 = 0.0373$ m³, $p_3 = 1.00$ atm

Chapter 21

CP 1. a, b, c **2.** smaller **3.** c, b, a **4.** a, d, c, b **5.** b **Q 1.** increase **3.** (a) increase; (b) same **5.** equal **7.** lower the temperature of the low temperature reservoir **9.** (a) same; (b) increase; (c) decrease **11.** (a) same; (b) increase; (c) decrease **13.** more than the age of the universe **EP 1.** 1.86×10^4 J **3.** 2.75 mol **7.** (a) 5.79×10^4 J; (b) 173 J/K **9.** +3.59 J/K **11.** (a) 14.6 J/K; (b) 30.2 J/K **13.** (a) 4.45 J/K; (b) no **15.** (a) 4500 J; (b) -5000 J; (c) 9500 J **17.** (a) 57.0°C; (b) -22.1 J/K; (c) +24.9 J/K; (d) +2.8 J/K **19.** (a) -710 mJ/K; (b) +710 mJ/K; (c) +723 mJ/K;

(d) -723 mJ/K; (e) $+13$ mJ/K; (f) 0 **23.** (a) (I) constant T, $Q = pV \ln 2$; constant V, $Q = 4.5pV$; (II) constant T, $Q = -pV \ln 2$; constant p, $Q = 7.5pV$; (b) (I) constant T, $W = pV \ln 2$; constant V, $W = 0$; (II) constant T, $W = -pV \ln 2$; constant p, $W = 3pV$; (c) $4.5pV$ for either case; (d) $4R \ln 2$ for either case **25.** 0.75 J/K **27.** (a) -943 J/K; (b) $+943$ J/K; (c) yes **29.** (a) $3p_0V_0$; (b) $6RT_0$, $(3/2)R \ln 2$; (c) both are zero **33.** (a) 31%; (b) 16 kJ **35.** engine A, first; engine B, first and second; engine C, second; engine D, neither **37.** 97 K **39.** 99.99995% **41.** 7.2 J/cycle **43.** (a) 7200 J; (b) 960 J; (c) 13% **45.** (a) 2270 J; (b) 14,800 J; (c) 15.4%; (d) 75.0%, greater **49.** (a) 78%; (b) 81 kg/s **55.** (a) 49 kJ; (b) 7.4 kJ **57.** 21 J **59.** (a) 0.071 J; (b) 0.50 J; (c) 2.0 J; (d) 5.0 J **61.** 1.08 MJ **63.** $[1 - (T_2/T_1)] \div [1 - (T_4/T_3)]$ **67.** (a) 1.27×10^{30}; (b) 7.9%; (c) 7.3%; (d) 7.3%; (e) 1.1%; (f) 0.0023% **69.** (a) $W = N!/(n_1! \, n_2! \, n_3!)$; (b) $[(N/2)! \, (N/2)!]/[(N/3)! \, (N/3)! \, (N/3)!]$; (c) 4.2×10^{16}

Chapter 22

CP **1.** C and D attract; B and D attract **2.** (a) leftward; (b) leftward; (c) leftward **3.** (a) a, c, b; (b) less than **4.** $-15e$ (net charge of $-30e$ is equally shared) **Q** **1.** No, only for charged particles, charged particle-like objects, and spherical shells (including solid spheres) of uniform charge **3.** a and b **5.** two points: one to the left of the particles and one between the protons **7.** $6q^2/4\pi\epsilon_0 d^2$, leftward **9.** (a) same; (b) less than; (c) cancel; (d) add; (e) the adding components; (f) positive direction of y; (g) negative direction of y; (h) positive direction of x; (i) negative direction of x **11.** (a) $A, B,$ and D; (b) all four; (c) Connect A and D; disconnect them; then connect one of them to B. (There are two more solutions.) **13.** (a) possibly; (b) definitely **15.** same **17.** D **EP** **1.** 0.50 C **3.** 2.81 N on each **5.** (a) 4.9×10^{-7} kg; (b) 7.1×10^{-11} C **7.** $3F/8$ **9.** (a) 1.60 N; (b) 2.77 N . **11.** (a) $q_1 = 9q_2$; (b) $q_1 = -25q_2$ **13.** either -1.00 μC and $+3.00$ μC or $+1.00$ μC and -3.00 μC **15.** (a) 36 N, $-10°$ from the x axis; (b) $x = -8.3$ cm, $y = +2.7$ cm **17.** (a) 5.7×10^{13} C, no; (b) 6.0×10^5 kg **19.** (a) $Q = -2\sqrt{2}q$; (b) no **21.** 3.1 cm **23.** 2.89×10^{-9} N **25.** -1.32×10^{13} C **27.** (a) 3.2×10^{-19} C; (b) two **29.** (a) 8.99×10^{-19} N; (b) 625 **31.** 5.1 m below the electron **33.** 1.3 days **35.** (a) 0; (b) 1.9×10^{-9} N **37.** 10^{18} N **39.** (a) ^{9}B; (b) ^{13}N; (c) ^{12}C **41.** (a) $F = (Q^2/4\pi\epsilon_0 d^2)\alpha(1 - \alpha)$; (c) 0.5; (d) 0.15 and 0.85

Chapter 23

CP **1.** (a) rightward; (b) leftward; (c) leftward; (d) rightward (p and e have same charge magnitude, and p is farther) **2.** all tie **3.** (a) toward positive y; (b) toward positive x; (c) toward negative y **4.** (a) leftward; (b) leftward; (c) decrease **5.** (a) all tie; (b) 1 and 3 tie, then 2 and 4 tie **Q** **1.** (a) toward positive x; (b) downward and to the right;

(c) A **3.** two points: one to the left of the particles, the other between the protons **5.** (a) yes; (b) toward; (c) no (the field vectors are not along the same line); (d) cancel; (e) add; (f) adding components; (g) toward negative y **7.** (a) 3, then 1 and 2 tie (zero); (b) all tie; (c) 1 and 2 tie, then 3 **9.** (a) rightward; (b) $+q_1$ and $-q_3$, increase; q_2, decrease; n, same **11.** a, b, c **13.** (a) 4, 3, 1, 2; (b) 3, then 1 and 4 tie, then 2 **EP** **1.** (a) 6.4×10^{-18} N; (b) 20 N/C **3.** to the right in the figure **7.** 56 pC **9.** 3.07×10^{21} N/C, radially outward **13.** (a) $q/8\pi\epsilon_0 d^2$, to the left; $3q/\pi\epsilon_0 d^2$, to the right; $7q/16\pi\epsilon_0 d^2$, to the left **15.** 0 **17.** 9:30 **19.** $E = q/\pi\epsilon_0 a^2$, along bisector, away from triangle **21.** $7.4q/4\pi\epsilon_0 d^2$, 28° counterclockwise to $+x$ **23.** 6.88×10^{-28} C·m **25.** $(1/4\pi\epsilon_0)(p/r^3)$, antiparallel to **p** **29.** $R/\sqrt{2}$ **31.** $(1/4\pi\epsilon_0)(4q/\pi R^2)$, toward increasing y **37.** (a) 0.10 μC; (b) 1.3×10^{17}; (c) 5.0×10^{-6} **39.** 3.51×10^{15} m/s^2 **41.** 6.6×10^{-15} N **43.** 2.03×10^{-7} N/C, up **45.** (a) -0.029 C; (b) repulsive forces would explode the sphere **47.** (a) 1.92×10^{12} m/s^2; (b) 1.96×10^5 m/s **49.** (a) 8.87×10^{-15} N; (b) 120 **51.** 1.64×10^{-19} C ($\approx$3% high) **53.** (a) 0.245 N, 11.3° clockwise from the $+x$ axis; (b) $x = 108$ m, $y = -21.6$ m **55.** 27μm **57.** (a) yes; (b) upper plate, 2.73 cm **59.** (a) 0; (b) 8.5×10^{-22} N·m; (c) 0 **61.** $(1/2\pi)\sqrt{pE/I}$ **63.** (a) $E = (2q/4\pi\epsilon_0 d^2)(\alpha/(1 + \alpha^2)^{3/2})$; (c) 0.707; (d) 0.21 and 1.9

Chapter 24

CP **1.** (a) $+EA$; (b) $-EA$; (c) 0; (d) 0 **2.** (a) 2; (b) 3; (c) 1 **3.** (a) equal; (b) equal; (c) equal **4.** (a) $+50e$; (b) $-150e$ **5.** 3 and 4 tie, then 2, 1 **Q** **1.** (a) 8 N·m^2/C; (b) 0 **3.** (a) all tie (zero); (b) all tie **5.** $+13q/\epsilon_0$ **7.** all tie **9.** all tie **11.** $2\sigma, \sigma, 3\sigma$; or $3\sigma, \sigma, 2\sigma$ **13.** (a) all tie ($E = 0$); (b) all tie **15.** (a) same ($E = 0$); (b) decrease; (c) decrease (to zero); (d) same **EP** **1.** (a) 693 kg/s; (b) 693 kg/s; (c) 347 kg/s; (d) 347 kg/s; (e) 575 kg/s **3.** (a) 0; (b) -3.92 N·m^2/C; (c) 0; (d) 0 for each field **5.** (a) enclose $2q$ and $-2q$, or enclose all four charges; (b) enclose $2q$ and q; (c) not possible **7.** 2.0×10^5 N·m^2/C **9.** $q/6\epsilon_0$ **11.** (a) $-\pi R^2 E$; (b) $\pi R^2 E$ **13.** -4.2×10^{-10} C **15.** 0 through each of the three faces meeting at q, $q/24\epsilon_0$ through each of the other faces **17.** 2.0 μC/m^2 **19.** (a) 4.5×10^{-7} C/m^2; (b) 5.1×10^4 N/C **21.** (a) -3.0×10^{-6} C; (b) $+1.3 \times 10^{-5}$ C **23.** (a) 0.32 μC; (b) 0.14 μC **27.** (a) $E = q/2\pi\epsilon_0 Lr$, radially inward; (b) $-q$ on both inner and outer surfaces; (c) $E = q/2\pi\epsilon_0 Lr$, radially outward **29.** 3.6 nC **31.** (b) $\rho R^2/2\epsilon_0 r$ **33.** (a) 5.3×10^7 N/C; (b) 60 N/C **35.** 5.0 nC/m^2 **37.** 0.44 mm **39.** (a) 4.9×10^{-22} C/m^2; (b) downward **41.** (a) px/ϵ_0; (b) $pd/2\epsilon_0$ **43.** (a) -750 N·m^2/C; (b) -6.64 nC **45.** (a) 4.0×10^6 N/C; (b) 0 **47.** (a) 0; (b) $q_a/4\pi\epsilon_0 r^2$; (c) $(q_a + q_b)/4\pi\epsilon_0 r^2$ **51.** (a) $-q$; (b) $+q$; (c) $E = q/4\pi\epsilon_0 r^2$, radially outward; (d) $E = 0$; (e) $E = q/4\pi\epsilon_0 r^2$, radially outward; (f) 0; (g) $E = q/4\pi\epsilon_0 r^2$, radially outward; (h) yes, charge is induced; (i) no; (j) yes; (k) no; (l) no **53.** (a) $E = (q/4\pi\epsilon_0 a^3)r$; (b) $E = q/4\pi\epsilon_0 r^2$; (c) 0; (d) 0; (e) inner, $-q$; outer, 0 **55.** $q/2\pi a^2$ **59.** $\alpha = 0.80$

Chapter 25

CP **1.** (a) negative; (b) increase **2.** (a) positive; (b) higher **3.** (a) rightward; (b) 1, 2, 3, 5: positive; 4, negative; (c) 3, then 1, 2, and 5 tie, then 4 **4.** all tie **5.** a, c (zero), b **6.** (a) 2, then 1 and 3 tie; (b) 3; (c) accelerate leftward **7.** closer (half of 9.23 fm) **Q** **1.** (a) higher; (b) positive; (c) negative; (d) all tie **3.** (a) 1 and 2; (b) none; (c) no; (d) 1 and 2 yes, 3 and 4 no **5.** b, then a, c, and d tie **7.** (a) negative; (b) zero **9.** (a) 1, then 2 and 3 tie; (b) 3 **11.** left **13.** a, b, c **15.** (a) c, b, a; (b) zero **17.** (a) positive; (b) positive; (c) negative; (d) all tie **19.** (a) no; (b) yes **21.** no (a particle at the intersection would have two different potential energies) **23.** (a)–(b) all tie; (c) C, B, A; (d) all tie **EP** **1.** 1.2 GeV **3.** (a) 3.0×10^{10} J; (b) 7.7 km/s; (c) 9.0×10^4 kg **7.** 2.90 kV **9.** 8.8 mm **11.** (a) 136 MV/m; (b) 8.82 kV/m **13.** (b) because $V = 0$ point is chosen differently; (c) $q/(8\pi\epsilon_0 R)$; (d) potential differences are independent of the choice for the $V = 0$ point **15.** (a) -4500 V; (b) -4500 V **17.** 843 V **19.** 2.8×10^5 **21.** $x = d/4$ and $x = -d/2$ **23.** none **25.** (a) 3.3 nC; (b) 12 nC/m² **27.** 6.4×10^8 V **29.** 190 MV **31.** (a) -4.8 nm; (b) 8.1 nm; (c) no **33.** 16.3 μV **35.** (a) $\dfrac{2\lambda}{4\pi\epsilon_0} \ln\left[\dfrac{L/2 + (L^2/4 + d^2)^{1/2}}{d}\right]$; (b) 0 **37.** (a) $-5Q/4\pi\epsilon_0 R$; (b) $-5Q/4\pi\epsilon_0(z^2 + R^2)^{1/2}$ **39.** $0.113\sigma R/\epsilon_0$ **41.** $(Q/4\pi\epsilon_0 L) \ln(1 + L/d)$ **43.** 670 V/m **45.** $p/2\pi\epsilon_0 r^3$ **47.** 39 V/m, $-x$ direction **51.** (a) $\dfrac{c}{4\pi\epsilon_0}[\sqrt{L^2 + y^2} - y]$; (b) $\dfrac{c}{4\pi\epsilon_0}\left[1 - \dfrac{y}{\sqrt{L^2 + y^2}}\right]$ **53.** (a) 2.5 MV; (b) 5.1 J; (c) 6.9 J **55.** (a) 2.72×10^{-14} J; (b) 3.02×10^{-31} kg, about $\frac{1}{3}$ of accepted value **57.** (a) 0.484 MeV; (b) 0 **59.** 2.1 d **61.** 0 **63.** (a) 27.2 V; (b) -27.2 eV; (c) 13.6 eV; (d) 13.6 eV **65.** 1.8×10^{-10} J **67.** 1.48×10^7 m/s **69.** $qQ/4\pi\epsilon_0 K$ **71.** 0.32 km/s **73.** 1.6×10^{-9} m **77.** (a) $V_1 = V_2$; (b) $q_1 = q/3$, $q_2 = 2q/3$; (c) 2 **79.** (a) -0.12 V; (b) 1.8×10^{-8} N/C, radially inward **81.** (a) 12,000 N/C; (b) 1800 V; (c) 5.8 cm **83.** (c) 4.24 V

Chapter 26

CP **1.** (a) same; (b) same **2.** (a) decreases; (b) increases; (c) decreases **3.** (a) V, $q/2$; (b) $V/2$, q **4.** (a) $q_0 = q_1 + q_{34}$; (b) equal (C_3 and C_4 are in series) **5.** (a) same; (b)–(d) increase; (e) same (same potential difference across same plate separation) **6.** (a) same; (b) decrease; (c) increase **Q** **1.** a, 2; b, 1; c, 3 **3.** (a) increase; (b) same **5.** (a) parallel; (b) series **7.** (a) $C/3$; (b) $3C$; (c) parallel **9.** (a) equal; (b) less **11.** (a)–(d) less **13.** (a) 2; (b) 3; (c) 1 **15.** Increase plate separation d, but also plate area A, keeping A/d constant. **EP** **1.** 7.5 pC **3.** 3.0 mC **5.** (a) 140 pF; (b) 17 nC **7.** (a) 84.5 pF; (b) 191 cm² **9.** (a) 11 cm²; (b) 11 pF; (c) 1.2 V **13.** (b) 4.6×10^{-5}/K **15.** 7.33 μF **17.** 315 mC **19.** (a) 10.0 μF; (b) $q_2 = 0.800$ mC, $q_1 = 1.20$ mC; (c) 200 V for both **21.** (a) $d/3$; (b) $3d$ **25.** (a) five in series; (b) three

arrays as in (a) in parallel (and other possibilities) **27.** 43 pF **29.** (a) 50 V; (b) 5.0×10^{-5} C; (c) 1.5×10^{-4} C **31.** (a) $q_1 = 9.0$ μC, $q_2 = 16$ μC, $q_3 = 9.0$ μC, $q_4 = 16$ μC; (b) $q_1 = 8.4$ μC, $q_2 = 17$ μC, $q_3 = 11$ μC, $q_4 = 14$ μC **33.** 99.6 nJ **35.** 72 F **37.** 4.9% **39.** 0.27 J **41.** 0.11 J/m³ **43.** (a) 2.0 J **45.** (a) $q_1 = 0.21$ mC, $q_2 = 0.11$ mC, $q_3 = 0.32$ mC; (b) $V_1 = V_2 = 21$ V, $V_3 = 79$ V; (c) $U_1 = 2.2$ mJ, $U_2 = 1.1$ mJ, $U_3 = 13$ mJ **47.** (a) $q_1 = q_2 = 0.33$ mC, $q_3 = 0.40$ mC; (b) $V_1 = 33$ V, $V_2 = 67$ V, $V_3 = 100$ V; (c) $U_1 = 5.6$ mJ, $U_2 = 11$ mJ, $U_3 = 20$ mJ **53.** Pyrex **55.** (a) 6.2 cm; (b) 280 pF **57.** 0.63 m² **59.** (a) 2.85 m³; (b) 1.01×10^4 **61.** (a) $\epsilon_0 A/(d-b)$; (b) $d/(d-b)$; (c) $-q^2 b/2\epsilon_0 A$, sucked in **65.** $\dfrac{\epsilon_0 A}{4d}\left(\kappa_1 + \dfrac{2\kappa_2\kappa_3}{\kappa_2 + \kappa_3}\right)$ **67.** (a) 13.4 pF; (b) 1.15 nC; (c) 1.13×10^4 N/C; (d) 4.33×10^3 N/C **69.** (a) 7.1; (b) 0.77 μC **71.** (a) 0.606; (b) 0.394

Chapter 27

CP **1.** 8 A, rightward **2.** (a)–(c) rightward **3.** a and c tie, then b **4.** Device 2 **5.** (a) and (b) tie, then (d), then (c) **Q** **1.** a, b, and c tie, then d (zero) **3.** b, a, c **5.** tie of A, B, and C, then a tie of $A + B$ and $B + C$, then $A + B + C$ **7.** (a)–(c) 1 and 2 tie, then 3 **9.** C, A, B **11.** b, a, c **13.** (a) conductors: 1 and 4; semiconductors: 2 and 3; (b) 2 and 3; (c) all four **EP** **1.** 1.25×10^{15} **3.** 6.7 μC/m² **5.** 14-gauge **7.** (a) 2.4×10^{-5} A/m²; (b) 1.8×10^{-15} m/s **9.** 0.67 A, toward the negative terminal **11.** (a) 0.654 μA/m²; (b) 83.4 MA **13.** 13 min **15.** (a) $J_0 A/3$; (b) $2J_0 A/3$ **17.** 2.0×10^{-8} $\Omega \cdot$m **19.** 100 V **21.** (a) 1.53 kA; (b) 54.1 MA/m²; (c) 10.6×10^{-8} $\Omega \cdot$m, platinum **23.** (a) 253°C; (b) yes **25.** (a) 0.38 mV; (b) negative; (c) 3 min 58 s **27.** 54 Ω **29.** 3.0 **31.** (a) 1.3 mΩ; (b) 4.6 mm **33.** (a) 6.00 mA; (b) 1.59×10^{-8} V; (c) 21.2 nΩ **35.** 2000 K **37.** (a) copper: 5.32×10^5 A/m², aluminum: 3.27×10^5 A/m²; (b) copper: 1.01 kg/m, aluminum: 0.495 kg/m **39.** 0.40 Ω **41.** (a) $R = \rho L/\pi ab$ **43.** 14 kC **45.** 11.1 Ω **47.** (a) 1.0 kW; (b) 25¢ **49.** 0.135 W **51.** (a) 1.74 A; (b) 2.15 MA/m²; (c) 36.3 mV/m; (d) 2.09 W **53.** (a) 1.3×10^5 A/m²; (b) 94 mV **55.** (a) \$4.46 for a 31-day month; (b) 144 Ω; (c) 0.833 A **57.** 660 W **59.** (a) 3.1×10^{11}; (b) 25 μA; (c) 1300 W, 25 MW **61.** 27 cm/s **63.** (a) 120 Ω; (b) 107 Ω; (c) 5.3×10^{-3}/C°; (d) 5.9×10^{-3}/C°; (e) 276 Ω

Chapter 28

CP **1.** (a) rightward; (b) all tie; (c) b, then a and c tie; (d) b, then a and c tie **2.** (a) all tie; (b) R_1, R_2, R_3 **3.** (a) less; (b) greater; (c) equal **4.** (a) $V/2$, i; (b) V, $i/2$ **5.** (a) 1, 2, 4, 3; (b) 4, tie of 1 and 2; then 3 **Q** **1.** 3, 4, 1, 2 **3.** (a) no; (b) yes; (c) all tie (the circuits are the same) **5.** parallel, R_2, R_1, series **7.** (a) equal; (b) more **9.** (a) less; (b) less; (c) more **11.** C_1, 15 V; C_2, 35 V; C_3, 20 V; C_4, 20 V; C_5, 30 V **13.** 60 μC **15.** c, b, a **17.** (a) all tie; (b) 1, 3, 2 **19.** 1, 3, and 4 tie (8 V on each resistor), then 2 and 5 tie (4 V on each resistor) **EP** **1.** (a) \$320; (b) 4.8¢ **3.** 11 kJ

5. (a) counterclockwise; (b) battery 1; (c) B   **7.** (a) 80 J;
(b) 67 J; (c) 13 J converted to thermal energy within battery
9. (a) 14 V; (b) 100 W; (c) 600 W; (d) 10 V, 100 W
11. (a) 50 V; (b) 48 V; (c) B is connected to the negative
terminal   **13.** 2.5 V   **15.** (a) 6.9 km; (b) 20 Ω
17. 8.0 Ω **19.** 10^{-6} **21.** the cable   **23.** (a) 1000 Ω;
(b) 300 mV; (c) 2.3×10^{-3} **25.** (a) 3.41 A or 0.586 A;
(b) 0.293 V or 1.71 V   **27.** 5.56 A   **29.** 4.0 Ω and 12 Ω
31. 4.50 Ω **33.** 0.00, 2.00, 2.40, 2.86, 3.00, 3.60, 3.75,
3.94 A   **35.** $V_d - V_c = +0.25$ V, by all paths   **37.** three
39. (a) 2.50 Ω; (b) 3.13 Ω **41.** nine   **43.** (a) left branch:
0.67 A down; center branch: 0.33 A up; right branch: 0.33 A
up; (b) 3.3 V   **47.** (a) 120 Ω; (b) $i_1 = 51$ mA, $i_2 = i_3 =$
19 mA, $i_4 = 13$ mA   **49.** (a) 19.5 Ω; (b) 0; (c) ∞; (d) 82.3 W,
57.6 W   **51.** (a) Cu: 1.11 A, Al: 0.893 A; (b) 126 m
53. (a) 13.5 kΩ; (b) 1500 Ω; (c) 167 Ω; (d) 1480 Ω
55. 0.45 A   **57.** (a) 12.5 V; (b) 50 A   **59.** -0.9%
65. (a) 0.41τ; (b) 1.1τ **67.** 4.6   **69.** (a) 0.955 μC/s;
(b) 1.08 μW; (c) 2.74 μW; (d) 3.82 μW   **71.** 2.35 MΩ
73. 0.72 MΩ **75.** 24.8 Ω to 14.9 kΩ **77.** (a) at $t = 0$,
$i_1 = 1.1$ mA, $i_2 = i_3 = 0.55$ mA; at $t = \infty$, $i_1 = i_2 = 0.82$ mA,
$i_3 = 0$; (c) at $t = 0$, $V_2 = 400$ V; at $t = \infty$, $V_2 = 600$ V;
(d) after several time constants ($\tau = 7.1$ s) have elapsed
79. (a) $V_T = -ir + \mathscr{E}$; (b) 13.6 V; (c) 0.060 Ω
81. (a) 6.4 V; (b) 3.6 W; (c) 17 W; (d) -5.6 W; (e) a

Chapter 29

CP **1.** $a, +z$; $b, -x$; c, $F_B = 0$ **2.** 2, then tie of 1 and 3
(zero); (b) 4   **3.** (a) $+z$ and $-z$ tie, then $+y$ and $-y$ tie, then
$+x$ and $-x$ tie (zero); (b) $+y$ **4.** (a) electron;
(b) clockwise   **5.** $-y$ **6.** (a) all tie; (b) 1 and 4 tie, then 2
and 3 tie   **Q** **1.** (a) all tie; (b) 1 and 2 (charge is negative)
3. a, no, $\mathbf{v}$ and $\mathbf{F}_B$ must be perpendicular; b, yes; c, no, $\mathbf{B}$ and
$\mathbf{F}_B$ must be perpendicular   **5.** (a) $\mathbf{F}_E$; (b) $\mathbf{F}_B$ **7.** (a) right;
(b) right   **9.** (a) negative; (b) equal; (c) equal; (d) half a
circle   **11.** (a) $\mathbf{B}_1$; (b) $\mathbf{B}_1$ into page; $\mathbf{B}_2$ out of page; (c) less
13. all   **15.** all tie   **17.** (a) positive; (b) (1) and (2) tie, then
(3) which is zero   **EP** **1.** M/QT **3.** (a) 9.56×10^{-14} N, 0;
(b) $0.267°$ **5.** (a) east; (b) 6.28×10^{14} m/s^2; (c) 2.98 mm
7. $0.75\mathbf{k}$ T   **9.** (a) 3.4×10^{-4} T, horizontal and to the left
as viewed along $\mathbf{v}_0$; (b) yes, if velocity is the same as the
electron's velocity   **11.** $(-11.4\mathbf{i} - 6.00\mathbf{j} + 4.80\mathbf{k})$ V/m
13. 680 kV/m   **17.** (b) 2.84×10^{-3} **19.** (a) 1.11×10^7 m/s;
(b) 0.316 mm   **21.** (a) 0.34 mm; (b) 2.6 keV
23. (a) 2.05×10^7 m/s; (b) 467 μT; (c) 13.1 MHz; (d) 76.3 ns
25. (a) 2.60×10^6 m/s; (b) 0.109 μs; (c) 0.140 MeV;
(d) 70 kV   **29.** (a) 1.0 MeV; (b) 0.5 MeV   **31.** $R_d = \sqrt{2}R_p$;
$R_\alpha = R_p$.   **33.** (a) $B\sqrt{mq/2V}\,\Delta x$; (b) 8.2 mm   **37.** (a) $-q$;
(b) $\pi m/qB$ **39.** $B_{min} = \sqrt{mV/2ed^2}$ **41.** (a) 22 cm;
(b) 21 MHz   **43.** neutron moves tangent to original path,
proton moves in a circular orbit of radius 25 cm   **45.** 28.2 N,
horizontally west   **47.** 20.1 N   **49.** $Bitd/m$, away from
generator   **51.** $-0.35\mathbf{k}$ N   **53.** 0.10 T, at 31° from the
vertical   **55.** 4.3×10^{-3} N·m, negative y **59.** $qvaB/2$
61. (a) 540 Ω, in series; (b) 2.52 Ω, in parallel
63. (a) 12.7 A; (b) 0.0805 N·m   **65.** (a) 0.184 A·m^2;

(b) 1.45 N·m   **67.** (a) 20 min; (b) 5.9×10^{-2} N·m
69. (a) $(8.0 \times 10^{-4}$ N·m$)(-1.2\mathbf{i} - 0.90\mathbf{j} + 1.0\mathbf{k})$;
(b) -6.0×10^{-4} J

Chapter 30

CP **1.** a, c, b **2.** b, c, a **3.** d, tie of a and c, then b
4. d, a, tie of b and c (zero)   **Q** **1.** c, d, then a and b tie
3. 2 and 4   **5.** a, b, c **7.** b, d, c, a (zero)   **9.** (a) 1, $+x$;
2, $-y$; (b) 1, $+y$; 2, $+x$ **11.** outward   **13.** c and
d tie, then b, a **15.** d, then tie of a and e, then b, c
17. 0 (dot product is zero)   **EP** **1.** 32.1 A
3. (a) 3.3 μT; (b) yes   **5.** (a) $(0.24\mathbf{i})$ nT; (b) 0; (c) $(-43\mathbf{k})$ pT;
(d) $(0.14\mathbf{k})$ nT   **7.** (a) 16 A; (b) west to east   **9.** 0
11. (a) 0; (b) $\mu_0 i/4R$, into the page; (c) same as (b)
13. $\mu_0 i\theta (1/b - 1/a)/4\pi$, out of page   **15.** (a) 1.0 mT, out of
the figure; (b) 0.80 mT, out of the figure   **25.** 200 μT, into
page   **27.** (a) it is impossible to have other than $B = 0$
midway between them; (b) 30 A   **29.** 4.3 A, out of page
35. $0.338\mu_0 i^2/a$, toward the center of the square
37. (b) to the right   **39.** (b) 2.3 km/s   **41.** $+5\mu_0 i$
47. (a) $\mu_0 ir/2\pi c^2$; (b) $\mu_0 i/2\pi r$; (c) $\dfrac{\mu_0 i}{2\pi(a^2 - b^2)}\cdot\dfrac{a^2 - r^2}{r}$;
(d) 0   **49.** $3i/8$, into page   **53.** 0.30 mT   **55.** 108 m
61. 0.272 A   **63.** (a) 4; (b) 1/2   **65.** (a) 2.4 A·m^2;
(b) 46 cm   **67.** (a) $\mu_0 i(1/a + 1/b)/4$, into page; (b) $\frac{1}{2}i\pi(a^2 + b^2)$,
into page   **69.** (a) 79 μT; (b) 1.1×10^{-6} N·m
71. (b) $(0.060$ **j**$)$ A·m^2; (c) $(9.6 \times 10^{-11}$**j**$)$ T, $(-4.8 \times$
10^{-11}**j**$)$ T   **73.** (a) B from sum: 7.069×10^{-5} T; $\mu_0 in =$
5.027×10^{-5} T; 40% difference; (b) B from sum: $1.043 \times$
10^{-4} T; $\mu_0 in = 1.005 \times 10^{-4}$ T; 4% difference; (c) B from
sum: 2.506×10^{-4} T; $\mu_0 in = 2.513 \times 10^{-4}$ T; 0.3%
difference   **75.** (a) $\mathbf{B} = (\mu_0/2\pi)\,[i_1/(x - a) + i_2/x]\mathbf{j}$;
(b) $\mathbf{B} = (\mu_0/2\pi)\,(i_1/a)\,(1 + b/2)\mathbf{j}$

Chapter 31

CP **1.** b, then d and e tie, and then a and c tie (zero)   **2.** a
and b tie, then c (zero)   **3.** c and d tie, then a and b tie
4. b, out; c, out; d, into; e, into   **5.** d and e   **6.** (a) 2, 3, 1
(zero); (b) 2, 3, 1   **7.** a and b tie, then c **Q** **1.** (a) all tie
(zero); (b) all tie (nonzero); (c) 3, then tie of 1 and 2 (zero)
3. out   **5.** (a) into; (b) counterclockwise; (c) larger
7. (a) leftward; (b) rightward   **9.** c, a, b **11.** (a) 1, 3, 2;
(b) 1 and 3 tie, then 2   **13.** a, tie of b and c   **15.** (a) more;
(b) same; (c) same; (d) same (zero)   **17.** a, 2; b, 4; c, 1; d, 3
EP **1.** 57 μWb   **3.** 1.5 mV   **5.** (a) 31 mV; (b) right to
left   **7.** (a) 0.40 V; (b) 20 A   **9.** (b) 58 mA   **11.** 1.2 mV
13. 1.15 μWb   **15.** 51 mV, clockwise when viewed along the
direction of $\mathbf{B}$ **17.** (a) 1.26×10^{-4} T, 0, -1.26×10^{-4} T;
(b) 5.04×10^{-8} V   **19.** (b) no   **21.** 15.5 μC
23. (a) 24 μV; (b) from c to b **25.** (b) design it so that
$Nab = (5/2\pi)$ m^2 **27.** (a) 0.598 μV; (b) counterclockwise
29. (a) $\dfrac{\mu_0 ia}{2\pi}\left(\dfrac{2r + b}{2r - b}\right)$; (b) $2\mu_0 iabv/\pi R(4r^2 - b^2)$
31. $A^2 B^2/R\Delta t$ **33.** (a) 48.1 mV; (b) 2.67 mA; (c) 0.128 mW
35. $v_t = mgR/B^2 L^2$ **37.** 268 W   **39.** (a) 240 μV; (b) 0.600

mA; (c) 0.144 μW; (d) 2.88×10^{-8} N; (e) same as (c)
41. 1, -1.07 mV; 2, -2.40 mV; 3, 1.33 mV **43.** at a:
4.4×10^7 m/s^2, to the right; at b: 0; at c: 4.4×10^7 m/s^2, to the
left **45.** 0.10 μWb **47.** (a) 800; (b) 2.5×10^{-4} H/m
49. (a) $\mu_0 i/W$; (b) $\pi\mu_0 R^2/W$ **51.** (a) decreasing;
(b) 0.68 mH **53.** (a) 0.10 H; (b) 1.3 V **55.** (a) 16 kV;
(b) 3.1 kV; (c) 23 kV **57.** (b) so that the changing magnetic
field of one does not induce current in the other;
(c) $1/L_{eq} = \sum\limits_{j=1}^{N} (1/L_j)$ **59.** 6.91 **61.** 1.54 s **63.** (a) 8.45
ns; (b) 7.37 mA **65.** $(42 + 20t)$ V **67.** 12.0 A/s
69. (a) $i_1 = i_2 = 3.33$ A; (b) $i_1 = 4.55$ A, $i_2 = 2.73$ A;
(c) $i_1 = 0$, $i_2 = 1.82$ A; (d) $i_1 = i_2 = 0$ **71.** $\mathscr{E}L_1/R(L_1 + L_2)$
73. (a) $i(1 - e^{-Rt/L})$ **75.** $1.23\tau_L$ **77.** (a) 240 W;
(b) 150 W; (c) 390 W **79.** (a) 97.9 H; (b) 0.196 mJ
81. (a) 10.5 mJ; (b) 14.1 mJ **83.** (a) 34.2 J/m^3; (b) 49.4
mJ **85.** 1.5×10^8 V/m **87.** $(\mu_0/2\pi)\ln(b/a)$ **89.** (a) 1.3
mT; (b) 0.63 J/m^3 **91.** (a) 1.0 J/m^3; (b) 4.8×10^{-15} J/m^3
93. (a) 1.67 mH; (b) 6.00 mWb **95.** 13 H **99.** magnetic
field exists only within the cross section of solenoid 1

Chapter 32

CP **1.** d, b, c, a (zero) **2.** (a) 2; (b) 1 **3.** (a) away;
(b) away; (c) less **4.** (a) toward; (b) toward; (c) less
5. a, c, b, d (zero) **6.** tie of b, c, and d, then a
Q **1.** (a) a, c, f; (b) bar gh **3.** supplied **5.** (a) all down;
(b) 1 up, 2 down, 3 zero **7.** (a) 1 up, 2 up, 3 down;
(b) 1 down, 2 up, 3 zero **9.** (a) rightward; (b) leftward
11. (a) decreasing; (b) decreasing **13.** (a) tie of a and b, then
c, d; (b) none (plate lacks circular symmetry, so $\mathbf{B}$ is not
tangent to a circular loop); (c) none **15.** 1/4 **EP** **1.** (b)
sign is minus; (c) no, compensating positive flux through open
end near magnet **3.** 47 μWb, inward **5.** 55 μT **7.** (a)
600 MA; (b) yes; (c) no **9.** (a) 31.0 μT, 0°; (b) 55.9 μT,
73.9°; (c) 62.0 μT, 90° **11.** 4.6×10^{-24} J **13.** (a) 5.3 $\times$
10^{11} V/m; (b) 20 mT; (c) 660 **15.** (a) 7; (b) 7; (c) $3h/2\pi$, 0;
(d) $3eh/4\pi m$, 0; (e) $3.5h/2\pi$; (f) 8 **17.** (b) in the direction
of the angular momentum vector **19.** $\Delta\mu = e^2 r^2 B/4m$
21. 20.8 mJ/T **23.** yes **25.** (a) 4 K; (b) 1 K
29. (a) 3.0 μT; (b) 5.6×10^{-10} eV **31.** (a) 8.9 A·m^2;
(b) 13 N·m **35.** (a) 0.14 A; (b) 79 μC **37.** 2.4 $\times$
10^{13} V/m·s **39.** 1.9 pT **41.** 7.5×10^5 V/s
43. 7.2×10^{12} V/m·s **45.** (a) 2.1×10^{-8} A, downward;
(b) clockwise **47.** (a) 0.63 μT; (b) 2.3×10^{12} V/m·s
49. (a) 2.0 A; (b) 2.3×10^{11} V/m·s; (c) 0.50 A;
(d) 0.63 μT·m **51.** (a) 7.60 μA; (b) 859 kV·m/s;
(c) 3.39 mm; (d) 5.16 pT

Chapter 33

CP **1.** (a) $T/2$, (b) T, (c) $T/2$, (d) $T/4$ **2.** (a) 5 V;
(b) 150 μJ **3.** (a) 1; (b) 2 **4.** (a) C, B, A; (b) 1, A;
2, B; 3, S; 4, C; (c) A **5.** (a) increases; (b) decreases
6. (a) 1, lags; 2, leads; 3, in phase; (b) 3 ($\omega_d = \omega$ when
$X_L = X_C$) **7.** (a) increase (circuit is mainly capacitive;
increase C to decrease X_C to be closer to resonance for

maximum P_{av}); (b) closer **8.** step-up **Q** **1.** (a) $T/4$,
(b) $T/4$, (c) $T/2$ (see Fig. 33-2), (d) $T/2$ (see Eq. 31-40)
3. b, a, c **5.** (a) 3, 1, 2; (b) 2, tie of 1 and 3
7. slower **9.** (a) 1 and 4; (b) 2 and 3 **11.** (a) 3,
then 1 and 2 tie; (b) 2, 1, 3 **13.** (a) negative; (b) lead
15. (a)–(c) rightward, increase **EP** **1.** 9.14 nF
3. 45.2 mA **5.** (a) 6.00 μs; (b) 167 kHz; (c) 3.00 μs
7. (a) 89 rad/s; (b) 70 ms; (c) 25 μF **9.** 38 μH
11. 7.0×10^{-4} s **15.** (a) 3.0 nC; (b) 1.7 mA; (c) 4.5 nJ
17. (a) 3.60 mH; 1.33 kHz; (c) 0.188 ms **19.** 600, 710,
1100, 1300 Hz **21.** (a) $Q/\sqrt{3}$; (b) 0.152 **25.** (a) 1.98 μJ;
(b) 5.56 μC; (c) 12.6 mA; (d) $-46.9°$; (e) $+46.9°$ **27.** (a) 0;
(b) $2i(t)$ **29.** (a) 356 μs; (b) 2.50 mH; (c) 3.20 mJ
31. 8.66 mΩ **33.** $(L/R)\ln 2$ **35.** (a) $\pi/2$ rad; (b) $q =$
$(I/\omega') e^{-Rt/2L} \sin \omega' t$ **39.** (a) 0.0955 A; (b) 0.0119 A
41. (a) 4.60 kHz; (b) 26.6 nF; (c) $X_L = 2.60$ kΩ, $X_C =$
0.650 kΩ **43.** (a) 0.65 kHz; (b) 24 Ω **45.** (a) 39.1 mA;
(b) 0; (c) 33.9 mA **47.** (a) 6.73 ms; (b) 2.24 ms;
(c) capacitor; (d) 59.0 μF **49.** (a) $X_C = 0$, $X_L = 86.7$ Ω,
$Z = 182$ Ω, $I = 198$ mA, $\phi = 28.5°$ **51.** (a) $X_C = 37.9$ Ω,
$X_L = 86.7$ Ω, $Z = 167$ Ω, $I = 216$ mA, $\phi = 17.1°$
53. (a) 2.35 mH; (b) they move away from 1.40 kHz
55. 1000V **57.** (a) 36.0 V; (b) 27.3 V; (c) 17.0 V;
(d) -8.34 V **59.** (a) 224 rad/s; (b) 6.00 A; (c) 228 rad/s,
219 rad/s; (d) 0.040 **61.** (a) 707 Ω; (b) 32.2 mH;
(c) 21.9 nF **63.** (a) resonance at $f = 1/2\pi\sqrt{LC} = 85.7$ Hz;
(b) 15.6 μF; (c) 225 mA **65.** (a) 796 Hz; (b) no change;
(c) decreased; (d) increased **69.** 141 V **71.** (a) taking;
(b) supplying **73.** 0, 9.00 W, 3.14 W, 1.82 W
75. 177 Ω, no **77.** 7.61 A **83.** (a) 117 μF; (b) 0;
(c) 90.0 W, 0; (d) 0°, 90°; (e) 1, 0 **85.** (a) 2.59 A;
(b) 38.8 V, 159 V, 224 V, 64.2 V, 75.0 V; (c) 100 W for R,
0 for L and C. **87.** (a) 2.4 V; (b) 3.2 mA, 0.16 A
89. (a) 1.9 V, 5.9 W; (b) 19 V, 590 W; (c) 0.19 kV, 59 kW
91. (a) $X_C = [(2\pi)(45 \times 10^{-6} \text{ F})f]^{-1}$; (c) 17.7 Hz
93. (a) $X_L = (2\pi)(40 \times 10^{-3} \text{ H})f$; (c) 796 Hz
95. (b) 61 Hz; (c) 90 Ω and 61 Hz

Chapter 34

CP **1.** (a) (Use Fig. 34-5.) On right side of rectangle, $\mathbf{E}$ is in
negative y direction; on left side, $\mathbf{E} + d\mathbf{E}$ is greater and in
same direction; (b) $\mathbf{E}$ is downward. On right side, $\mathbf{B}$ is in
negative z direction; on left side, $\mathbf{B} + d\mathbf{B}$ is greater and in
same direction. **2.** positive direction of x **3.** (a) same;
(b) decrease **4.** a, d, b, c (zero) **5.** a **6.** (a) yes; (b) no
Q **1.** (a) positive direction of z; (b) x **3.** (a) same;
(b) increase; (c) decrease **5.** both 20° clockwise from the y
axis **7.** two **9.** b, 30°; c, 60°; d, 60°; e, 30°; f, 60°
11. d, b, a, c **13.** (a) b; (b) blue; (c) c **15.** 1.5
EP **1.** (a) 4.7×10^{-3} Hz; (b) 3 min 32 s **3.** (a) 4.5×10^{24}
Hz; (b) 1.0×10^4 km or 1.6 Earth radii **7.** it would steadily
increase; (b) the summed discrepancies between the apparent
time of eclipse and those observed from x; the radius of Earth's
orbit **9.** 5.0×10^{-21} H **11.** 1.07 pT **17.** 4.8×10^{-29}
W/m^2 **19.** 4.51×10^{-10} **21.** 89 cm **23.** 1.2 MW/m^2
25. 820 m **27.** (a) 1.03 kV/m; 3.43 μT **29.** (a) 1.4 $\times$

10^{-22} W; (b) 1.1×10^{15} W **31.** (a) 87 mV/m; (b) 0.30 nT; (c) 13 kW **33.** 3.3×10^{-8} Pa **35.** (a) 4.7×10^{-6} Pa; (b) 2.1×10^{10} times smaller **37.** 5.9×10^{-8} Pa **39.** (a) 3.97 GW/m²; (b) 13.2 Pa; (c) 1.67×10^{-11} N; (d) 3.14×10^3 m/s² **41.** $I(2 - frac)/c$ **43.** $p_{r\perp} \cos^2 \theta$ **45.** 1.9 mm/s **47.** (b) 580 nm **49.** (a) 1.9 V/m; (b) 1.7×10^{-11} Pa **51.** 1/8 **53.** 3.1% **55.** 20° or 70° **57.** 19 W/m² **59.** (a) 2 sheets; (b) 5 sheets **61.** 180° **63.** 1.26 **65.** 1.07 m **69.** (a) 0; (b) 20°; (c) still 0 and 20° **73.** 1.41 **75.** 1.22 **77.** 182 cm **79.** (a) no; (b) yes; (c) about 43° **81.** (a) 35.6°; (b) 53.1° **83.** (b) 23.2° **85.** (a) 53°; (b) yes **87.** (a) 55.5°; (b) 55.8°

Chapter 35

CP Kaleidoscope answer: two mirrors that form a V with an angle of 60° **1.** $0.2d$, $1.8d$, $2.2d$ **2.** (a) real; (b) inverted; (c) same **3.** (a) e; (b) virtual, same **4.** virtual, same as object, diverging **Q 1.** c **3.** (a) a; (b) c **5.** (a) no; (b) yes (fourth is off mirror ed) **7.** (a) from infinity to the focal point; (b) decrease continually **9.** d (infinite), tie of a and b, then c **11.** mirror, equal; lens, greater **13.** (a) all but variation 2; (b) for 1, 3, and 4: right, inverted; for 5 and 6: left, same **15.** (a) less; (b) less **EP 1.** (a) virtual; (b) same; (c) same; (d) $D + L$ **3.** 40 cm **7.** (a) 7; (b) 5; (c) 1 to 3; (d) depends on the position of O and your perspective **11.** new illumination is 10/9 of the old **15.** 10.5 cm **19.** (a) 2.00; (b) none **23.** 1.14 **25.** (b) separate the lenses by a distance $f_2 - |f_1|$, where f_2 is the focal length of the converging lens **27.** 45 mm, 90 mm **29.** (a) $+40$ cm; (b) at infinity **33.** (a) 40 cm, real; (b) 80 cm, real; (c) 240 cm, real; (d) -40 cm, virtual; (e) -80 cm, virtual; (f) -240 cm, virtual **35.** same orientation, virtual, 30 cm to left of second lens, $m = 1$ **37.** (a) final image coincides in location with the object; it is real, inverted, and $m = -1.0$ **39.** (a) coincides in location with the original object and is enlarged 5.0 times; (c) virtual; (d) yes **45.** $i = \dfrac{(2 - n)r}{2(n - 1)}$, to the right of the right side of the sphere **47.** 2.1 mm **49.** (b) when image is at near point **51.** (b) farsighted **53.** -125

Chapter 36

CP 1. b (least n), c, a **2.** (a) top; (b) bright intermediate illumination (phase difference is 2.1 wavelengths) **3.** (a) 3λ, 3; (b) 2.5λ, 2.5 **4.** a and d tie (amplitude of resultant wave is $4E_0$), then b and c tie (amplitude of resultant wave is $2E_0$) **5.** (a) 1 and 4; (b) 1 and 4 **Q 1.** a, c, b **3.** (a) 300 nm; (b) exactly out of phase **5.** c **7.** (a) increase; (b) 1λ **9.** down **11.** (a) maximum; (b) minimum; (c) alternates **13.** d **15.** (a) 0.5 wavelength; (b) 1 wavelength **17.** bright **19.** all **EP 1.** (a) 5.09×10^{14} Hz; (b) 388 nm; (c) 1.97×10^8 m/s **5.** 2.1×10^8 m/s **7.** the time is longer for the pipeline containing air, by about 1.55 ns **9.** 22°, refraction reduces θ **11.** (a) pulse 2; (b) $0.03L/c$ **13.** (a) 1.70 (or 0.70); (b) 1.70 (or 0.70); (c) 1.30

(or 0.30); (d) brightness is identical, close to fully destructive interference **15.** (a) 0.833; (b) intermediate, closer to fully constructive interference **17.** $(2m + 1)\pi$ **19.** 2.25 mm **21.** 648 nm **23.** 1.6 mm **25.** 16 **27.** 0.072 mm **29.** 8.75λ **31.** 0.03% **33.** 6.64 μm **35.** $y = 17 \sin(\omega t + 13°)$ **39.** (a) 1.17 m, 3.00 m, 7.50 m; (b) no **41.** $I = \frac{1}{9}I_m[1 + 8 \cos^2(\pi d \sin \theta/\lambda)]$, $I_m =$ intensity of central maximum **43.** $L = (m + \frac{1}{2})\lambda/2$, for $m = 0, 1, 2, \ldots$ **45.** 0.117 μm, 0.352 μm **47.** $\lambda/5$ **49.** 70.0 nm **51.** none **53.** (a) 552 nm; (b) 442 nm **55.** 338 nm **59.** $2n_2L \cos \theta_r = (m + \frac{1}{2})\lambda$, for $m = 0, 1, 2, \ldots$, where $\theta_r = \sin^{-1}[(\sin \theta_i)/n_2]$ **61.** intensity is diminished by 88% at 450 nm and by 94% at 650 nm **63.** (a) dark; (b) blue end **65.** 1.89 μm **67.** 1.00025 **69.** (a) 34; (b) 46 **73.** 588 nm **75.** 1.0003 **77.** $I = I_m \cos^2(2\pi x/\lambda)$

Chapter 37

CP 1. (a) expand; (b) expand **2.** (a) second side maximum; (b) 2.5 **3.** (a) red; (b) violet **4.** diminish **5.** (a) increase; (b) same **6.** (a) left; (b) less **Q 1.** (a) contract; (b) contract **3.** with megaphone (larger opening, less diffraction) **5.** four **7.** (a) larger; (b) red **9.** (a) decrease; (b) same; (c) in place **11.** (a) A; (b) left; (c) left; (d) right **EP 1.** 690 nm **3.** 60.4 μm **5.** (a) 2.5 mm; (b) 2.2×10^{-4} rad **7.** (a) 70 cm; (b) 1.0 mm **9.** 41.2 m from the central axis **11.** 160° **15.** (d) 53°, 10°, 5.1° **19.** (a) 1.3×10^{-4} rad; (b) 10 km **21.** 50 m **23.** 30.5 μm **25.** 1600 km **27.** (a) 17.1 m; (b) 1.37×10^{-10} **29.** 27 cm **31.** 4.7 cm **33.** (a) 0.347°; (b) 0.97° **35.** (a) red; (b) 130 μm **37.** five **41.** $\lambda D/d$ **43.** (a) 5.05 μm; (b) 20.2 μm **45.** (a) 3.33 μm; (b) 0, $\pm 10.2°$, $\pm 20.7°$, $\pm 32.0°$, $\pm 45.0°$, $\pm 62.2°$ **47.** all wavelengths shorter than 635 nm **49.** 13,600 **51.** 500 nm **53.** (a) three; (b) 0.051° **55.** 523 nm **61.** 470 nm to 560 nm **63.** 491 **65.** 3650 **67.** (a) 1.0×10^4 nm; (b) 3.3 mm **69.** (a) 0.032°/nm, 0.076°/nm, 0.24°/nm; (b) 40,000, 80,000, 120,000 **71.** (a) $\tan \theta$; (b) 0.89 **73.** 0.26 nm **75.** 6.8° **77.** (a) 170 pm; (b) 130 pm **81.** 0.570 nm **83.** 30.6°, 15.3° (clockwise); 3.08°, 37.8° (counterclockwise)

Chapter 38

CP 1. (a) same (speed of light postulate); (b) no (the start and end of the flight are spatially separated); (c) no (again, because of the spatial separation) **2.** (a) Sally's; (b) Sally's **3.** a, negative; b, positive; c, negative **4.** (a) right; (b) more **5.** (a) equal; (b) less **Q 1.** all tie (pulse speed is c) **3.** (a) C_1; (b) C_1 **5.** (a) 3, 2, 1; (b) 1 and 3 tie, then 2 **7.** (a) negative; (b) positive **9.** c, then b and d tie, then a **11.** (a) 3, tie of 1 and 2, then 4; (b) 4, tie of 1 and 2, then 3; (c) 1, 4, 2, 3 **13.** greater than f_1 **EP 1.** (a) 3×10^{-18}; (b) 8.2×10^{-8}; (c) 1.1×10^{-6}; (d) 3.7×10^{-5}; (e) 0.10 **3.** $0.75c$ **5.** $0.99c$ **7.** 55 m **9.** 1.32 m **11.** 0.63 m **13.** 6.4 cm **15.** (a) 26 y; (b) 52 y; (c) 3.7 y **17.** (b) 0.999 999 15c **19.** (a) $x' = 0$, $t' = 2.29$ s; (b) $x' =$

6.55×10^8 m, $t' = 3.16$ s   **21.** (a) 25.8 μs; (b) small flash
23. (a) 1.25; (b) 0.800 μs   **25.** 2.40 μs   **27.** (a) 0.84c, in
the direction of increasing x; (b) 0.21c, in the direction of
increasing x; the classical predictions are 1.1c and 0.15c
29. (a) 0.35c; (b) 0.62c **31.** 1.2 μs   **33.** seven
35. 22.9 MHz   **37.** +2.97 nm   **39.** (a) $\tau_0/\sqrt{1 - v^2/c^2}$
41. (a) 0.134c; (b) 4.65 keV; (c) 1.1%   **43.** (a) 0.9988, 20.6;
(b) 0.145, 1.01; (c) 0.073, 1.0027   **45.** (a) 5.71 GeV,
6.65 GeV, 6.58 GeV/c; (b) 3.11 MeV, 3.62 MeV,
3.59 MeV/c **47.** 18 smu/y   **49.** (a) 0.943c; (b) 0.866c
51. (a) 256 kV; (b) 0.746c **53.** $\sqrt{8}mc$ **55.** 6.65 $\times$ 10^6 mi,
or 270 earth circumferences   **57.** 110 km   **59.** (a) 2.7 $\times$

10^{14} J; (b) 1.8 $\times$ 10^7 kg; (c) 6.0 $\times$ 10^6 **61.** 4.00 u, probably
a helium nucleus   **63.** 330 mT
65.

SIGNAL	TIME SENT (h)	TIME REPLY RECEIVED (h)	TIME REPORTED	DISTANCE (m)
1	6.0	400	11.8	2.10×10^{14}
2	12.0	800	23.6	4.19×10^{14}
3	18.0	1200	35.5	6.29×10^{14}
4	24.0	1600	47.3	8.38×10^{14}
5	30.0	2000	59.1	1.05×10^{15}

67. (a) $vt \sin \theta$; (b) $t[1 - (v/c) \cos \theta]$; (c) 3.24c

Photo Credits

Chapter 16
Page 372: Tom van Dyke/Sygma. Page 373: Kent Knudson/FPG International. Page 387: Bettmann Archive. Page 392: Courtesy NASA.

Chapter 17
Page 399: John Visser/Bruce Coleman, Inc. Page 415: Richard Megna/Fundamental Photographs. Page 416: Courtesy T.D. Rossing, Northern Illinois University.

Chapter 18
Page 425: Stephen Dalton/Animals Animals. Page 426: Howard Sochurak/The Stock Market. Page 433: Ben Rose/The Image Bank. Page 434: Bob Gruen/Star File. Page 435: John Eastcott/Yva Momativk/DRK Photo. Page 441: Philippe Plailly/Science Photo Library/Photo Researchers. Page 444: Courtesy NASA.

Chapter 19
Page 453: Tom Owen Edmunds/The Image Bank. Page 459: AP/Wide World Photos. Page 471: Peter Arnold/Peter Arnold, Inc. Page 472: Courtesy Daedalus Enterprises, Inc.

Chapter 20
Page 484: Bryan and Cherry Alexander Photography.

Chapter 21
Page 509: Steven Dalton/Photo Researchers. Page 517: Richard Ustinich/The Image Bank. Page 523 (left): Cary Wolinski/Stock, Boston. Page 523 (right): Courtesy of Professor Bernard Hallet, Quaternary Research Center, University of Washington, Seattle. Page 534: Jeff Werner.

Chapter 22
Page 537: Michael Watson. Page 538: Fundamental Photographs. Page 539: Courtesy Xerox. Page 540: Johann Gabriel Doppelmayr, Neuentdeckte Phaenomena von Bewünderswurdigen Würckungen der Natur, Nuremberg, 1744. Page 547: Courtesy Lawrence Berkeley Laboratory.

Chapter 23
Page 554: Quesada/Burke Studios. Page 565: Russ Kinne/Comstock, Inc. Page 566: Courtesy Environmental Elements Corporation.

Chapter 24
Page 579: E.R. Degginger/Bruce Coleman, Inc. Page 589 (top): Courtesy E. Philip Krider, Institute for Atmospheric Physics, University of Arizona, Tucson. Page 589 (bottom): C. Johnny Autery.

Chapter 25
Pages 601 and 605: Courtesy NOAA. Page 617: Courtesy Westinghouse Corporation.

Chapter 26
Page 628: Goivaux Communication/Phototake. Page 629: Paul Silvermann/Fundamental Photographs. Page 638: ©The Harold E. Edgerton 1992 Trust/Courtesy Palm Press, Inc. Page 639: Courtesy The Royal Institute, England.

Chapter 27
Page 651: UPI/Corbis-Bettmann. Page 657: The Image Works. Page 663: Laurie Rubie/Tony Stone Images/New York, Inc. Page 666 (left): Courtesy AT&T Bell Laboratories. Page 666 (right): Courtesy Shoji Tonaka/International Superconductivity Technology Center, Tokyo, Japan.

Chapter 28
Page 673: Norbert Wu. Page 674: Courtesy Southern California Edison Company.

Chapter 29
Page 700: Johnny Johnson/Earth Scenes/Animals Animals. Page 701: Schneps/The Image Bank. Page 703: Courtesy Lawrence Berkeley Laboratory, University of California. Page 704: Courtesy Dr. Richard Cannon, Southeast Missouri State University, Cape Girardeau. Page 708: Courtesy John Le P. Webb, Sussex University, England. Page 710: Courtesy Dr. L.A. Frank, University of Iowa. Page 712: Courtesy Fermi National Accelerator Laboratory.

Chapter 30
Page 728: Michael Brown/Florida Today/Gamma Liaison. Page 730: Courtesy Education Development Center.

Chapter 31
Page 752: Dan McCoy/Black Star. Page 756: Courtesy Fender Musical Instruments Corporation. Page 760: Courtesy Jenn-Air Co. Page 765: Courtesy The Royal Institute, England.

Chapter 32
Page 786: Bob Zehring. Page 787: Runk/Schoenberger/Grant Heilman Photography. Page 794: Peter Lerman. Page 797 (top): Courtesy Ralph W. DeBlois. Page 797 (bottom): R.E. Rosenweig, Research and Science Laboratory, courtesy Exxon Co. USA.

Chapter 33
Page 811: Courtesy Haverfield Helicopter Co. Page 814: Courtesy Hewlett Packard. Page 827 (left): Steve Kagan/Gamma Liaison. Page 827 (right): Ted Cowell/Black Star.

Chapter 34
Page 841: Courtesy Hansen Publications. Page 850: ©1992 Ben and Miriam Rose, from the collection of the Center for Creative Photography, Tucson. Page 854: Diane Schiumo/Fundamental Photographs. Page 855: PSSC Physics, 2nd edition; ©1975 D.C. Heath and Co. with Education Development Center, Newton, MA. Page 856: Courtesy Lockheed Advanced Development Company. Page 858 (top right): Tony Stone Images/New York, Inc. Page 858 (bottom left): Courtesy Bausch & Lomb. Page 861: Will and Deni McIntyre/Photo Researchers. Page 867: Cornell University.

Chapter 35
Page 872: Courtesy Courtauld Institute Galleries, London. Page 874 (left): Frans Lanting/Minden Pictures, Inc. Page 874 (right): Wayne Sorce. Page 882: Dr. Paul A. Zahl/Photo Researchers.

Page 884: Courtesy Matthew J. Wheeler. Page 894: Piergiorgio Scharandis/Black Star.

Chapter 36

Page 901: E.R. Degginger. Page 905: Runk Schoenberger/ Grant Heilman Photography. Page 906: From *Atlas of Optical Phenomena* by M. Cagnet et al., Springer-Verlag, Prentice Hall, 1962. Page 914: Richard Megna/Fundamental Photographs. Page 917: Courtesy Dr. Helen Ghiradella, Department of Biological Sciences, SUNY, Albany. Page 926: Courtesy Bausch & Lomb.

Chapter 37

Page 929: Georges Seurat, French, 1859–1891, *A Sunday on La Grande Jatte,* 1884. Oil on canvas; 1884–86, 207.5 × 308 cm; Helen Birch Bartlett Memorial Collection, 1926. Photograph ©1996, The Art Institute of Chicago. All Rights Reserved. Page 930: Ken Kay/Fundamental Photographs. Pages 931, 937 (bottom left) and 940: From *Atlas of Optical Phenomena* by Cagnet, Francon, Thierr, Springer-Verlag, Berlin, 1962. Page 937 (bottom right): AP/Wide World Photos, Inc. Page 938: Cath Ellis/ Science Photo Library/Photo Researchers. Page 945 (left): Department of Physics, Imperial College/Science Photo Library/ Photo Researchers. Page 945 (top right): Peter L. Chapman/Stock, Boston. Page 953: Kjell B. Sandved/Bruce Coleman, Inc. Page 954: Courtesy Professor Robert Greenler, Physics Department, University of Wisconsin.

Chapter 38

Page 958: T. Tracy/FPG International. Page 959: Courtesy Hebrew University of Jerusalem, Israel.